2010 ANNUAL STATISTIC REPORT ON ENVIRONMENT IN CHINA

MINISTRY OF ENVIRONMENTAL PROTECTION OF THE PEOPLE'S REPUBLIC OF CHINA

中国环境统计年报

中华人民共和国环境保护部 编

中国环境科学出版社 · 北京
CHINA ENVIRONMENTAL SCIENCE PRESS · BEIJING

图书在版编目（CIP）数据

中国环境统计年报．2010/中华人民共和国环境保护部编．—北京：中国环境科学出版社，2011.12

ISBN 978-7-5111-0802-9

Ⅰ．①中…　Ⅱ．①中…　Ⅲ．①环境统计—统计资料—中国—2010—年报　Ⅳ．①X508.2-54

中国版本图书馆 CIP 数据核字（2011）第 246576 号

责任编辑　贾卫列
责任校对　扣志红
封面设计　玄石至上

出版发行　中国环境科学出版社
（100062　北京东城区广渠门内大街 16 号）
网　　址：http://www.cesp.com.cn
联系电话：010-67112765（编辑管理部）
发行热线：010-67125803，010-67113405（传真）
印　　刷　北京市联华印刷厂
经　　销　各地新华书店
版　　次　2011 年 12 月第 1 版
印　　次　2011 年 12 月第 1 次印刷
开　　本　889×1194　1/16
印　　张　17.25
字　　数　400 千字
定　　价　100.00 元

《中国环境统计年报·2010》编委会

组织编写　环境保护部污染物排放总量控制司

中国环境监测总站

资料提供　各省、自治区、直辖市环境保护厅（局）

批　　准　中华人民共和国环境保护部

目　　录

1

全国环境统计概要

QUANGUO HUANJING TONGJI GAIYAO

ANNUAL STATISTIC REPORT ON ENVIRONMENT IN CHINA

2010

综　述

2010年，在党中央、国务院的领导下，全国环境保护系统深入贯彻落实科学发展观，大力推进生态文明建设，积极探索中国环保新道路，把环境保护与推动发展方式转变、污染减排与促进经济结构战略性调整、环境治理与保障和改善民生更加有机地结合起来，以解决影响环境质量和损害群众健康的突出环境问题为重点，扎实推进环保各项工作，较好地完成了2010年各项工作任务。与2005年相比，2010年全国化学需氧量排放量和二氧化硫排放量分别下降12.5%和14.3%，两项主要污染物均超额完成了“十一五”的总量减排目标。

2010年全国废水排放总量617.3亿吨，比上年增加4.7%。其中，工业废水排放量237.5亿吨，比上年增加1.3%。城镇生活污水排放量379.8亿吨，比上年增加6.9%。废水中化学需氧量（COD）排放量1 238.1万吨，比上年减少3.1%。废水中氨氮排放量120.3万吨，比上年减少1.9%。工业用水重复利用率85.7%，比上年提高0.7个百分点。

全国废气中二氧化硫（SO_2）排放量2 185.1万吨，比上年减少1.3%。烟尘排放量829.1万吨，比上年减少2.2%。工业粉尘排放量448.7万吨，比上年减少14.3%。氮氧化物排放量1 852.4万吨，比上年增加9.4%。

全国工业固体废物产生量24.1亿吨，比上年增加18.1%；工业固体废物综合利用量16.2亿吨，比上年增加16.9%；工业固体废物贮存量2.4亿吨，比上年增加14.5%。工业固体废物处置量5.7亿吨，比上年增加20.5%；工业固体废物排放量498万吨，比上年减少29.9%。工业固体废物综合利用率为66.7%，比上年减少0.3个百分点。

全国共统计了城市污水处理厂2 881座，比上年增加689座。城市生活污水处理率达到72.9%，比上年提高9.6个百分点。

1.1 统计企业基本情况

2010 年，发表调查了 112 799 家工业企业，对其他工业企业的污染排放量按比率进行估算。

1.1.1 工业企业污染物治理情况

发表调查的企业中，共有 15 635 套废水污染物在线监测仪器，80 332 套废水治理设施，共投入运行费 545.3 亿元。共去除化学需氧量 1 415.4 万吨，氨氮 82.6 万吨，石油类 35.4 万吨，挥发酚 7.3 万吨，氰化物 0.7 万吨。

企业在用的工业锅炉和炉窑数分别为 8.5 万台和 8.3 万台，共安装 10 227 套废气污染物在线监测仪器，18.7 万套废气治理设施（其中，脱硫设施 26 611 套），投入运行费 1 054.5 亿元。共去除二氧化硫 3 304.0 万吨（其中，燃料燃烧过程中去除 2 231.3 万吨），烟尘 3 894.1 万吨，粉尘 9 501.7 万吨。

1.1.2 工业企业污染治理投资情况

2010 年，调查工业企业中，共实施工业污染治理项目 6 597 个，其中废水治理项目 2 942 个，废气治理项目 2 519 个，工业固体废物治理项目 372 个。当年竣工的项目数 5 866 个，其中废水治理项目 2 575 个，废气治理项目 2 294 个，工业固体废物治理项目 330 个。

工业污染治理项目本年完成投资总额为 397.0 亿元，其中，工业废水治理项目完成投资 130.1 亿元，新增设计处理能力 907.2 万吨/日。工业废气治理项目完成投资 188.8 亿元，新增设计处理能力（标态）29 028 万米^3/时。工业固体废物治理项目完成投资 14.3 亿元，新增设计处理能力 24.7 万吨/日。

1.2 废水

1.2.1 废水及主要污染物排放情况

（1）废水排放情况

2010 年，全国废水排放总量 617.3 亿吨，比上年增加 4.7%。

工业废水排放量 237.5 亿吨，比上年增加 1.3%；工业废水排放量占废水排放总量的 38.5%，比上年有所降低。

生活污水排放量 379.8 亿吨，比上年增加 7.0%；生活污水排放量占废水排放总量的 61.5%，高于上年。

自 2001 年以来，废水排放总量呈持续上升趋势。其中，生活污水排放量始终呈增长趋势，而工业废水排放量近年来总体上稳中有降。

表 1　全国废水及其主要污染物排放量年际对比

项目 年份	废水排放量（亿吨）			化学需氧量排放量（万吨）			氨氮排放量（万吨）		
	合计	工业	生活	合计	工业	生活	合计	工业	生活
2001	433.0	202.7	230.3	1 404.8	607.5	797.3	125.2	41.3	83.9
2002	439.5	207.2	232.3	1 366.9	584.0	782.9	128.8	42.1	86.7
2003	460.0	212.4	247.6	1 333.6	511.9	821.7	129.7	40.4	89.3
2004	482.4	221.1	261.3	1 339.2	509.7	829.5	133.0	42.2	90.8
2005	524.5	243.1	281.4	1 414.2	554.7	859.4	149.8	52.5	97.3
2006	536.8	240.2	296.6	1 428.2	542.3	885.9	141.3	42.5	98.8
2007	556.8	246.6	310.2	1 381.8	511.0	870.8	132.4	34.1	98.3
2008	571.7	241.7	330.0	1 320.7	457.6	863.1	127.0	29.7	97.3
2009	589.7	234.5	355.2	1 277.5	439.7	837.8	122.6	27.3	95.3
2010	617.3	237.5	379.8	1 238.1	434.8	803.3	120.3	27.3	93.0
增长率（%）	4.7	1.3	6.9	−3.1	−1.1	−4.1	−1.9	0	−2.4

注：增长率指 2010 年与 2009 年相比，下同。

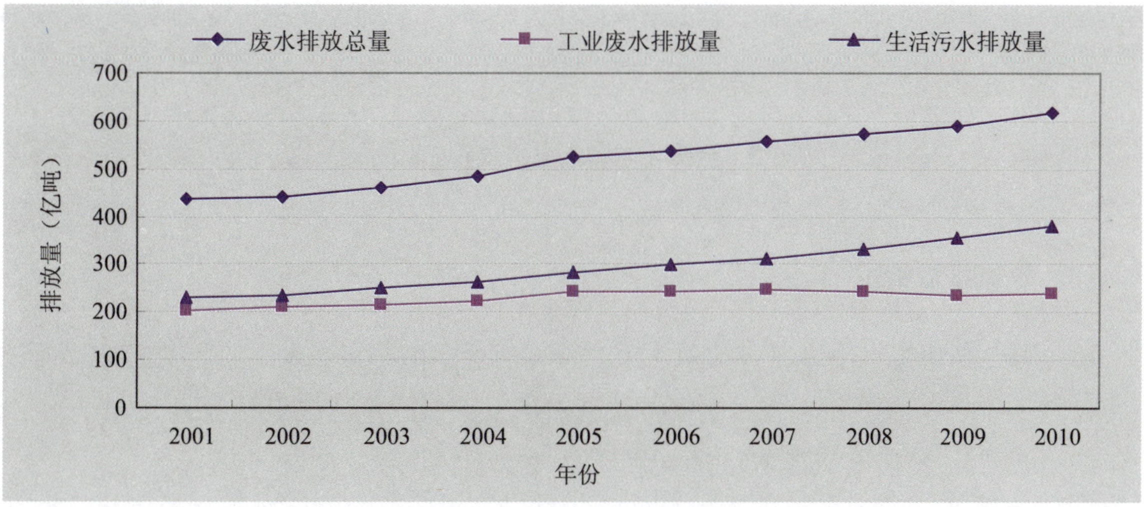

图 1　全国废水排放量年际对比

（2）化学需氧量排放情况

2010 年，全国废水中化学需氧量排放量 1 238.1 万吨，比上年减少 3.1%。

工业废水中化学需氧量排放量 434.8 万吨，比上年减少 1.1%；工业化学需氧量排放量占废水化学需氧量排放总量的 35.1%。

生活污水中化学需氧量排放量 803.3 万吨，比上年减少 4.1%；生活化学需氧量排放量占废水化学需氧量排放总量的 64.9%。

“十一五”期间，全国废水中化学需氧量排放总量、工业废水中化学需氧量排放量和生活污水中化学需氧量排放量均呈现逐年下降趋势。2010 年全国化学需氧量排放总量较 2005 年下降了 12.5%，超额完成了“十一五”总量减排任务。

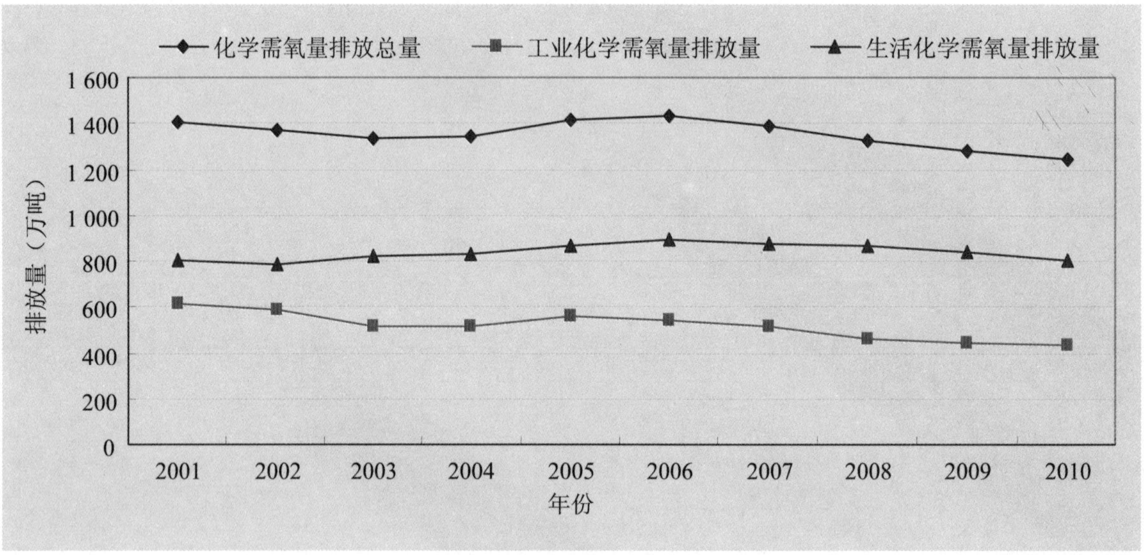

图 2　全国化学需氧量排放量年际对比

（3）氨氮排放情况

2010 年，全国废水中氨氮排放量 120.3 万吨，比上年减少 1.9%。工业氨氮排放量 27.3 万吨，与上年持平；工业氨氮占氨氮排放总量的 22.7%。生活氨氮排放量 93.0 万吨，比上年减少 2.4%；生活氨氮占氨氮排放总量的 77.3%。

“十一五”期间，全国废水中氨氮排放总量、工业废水中氨氮排放量和生活污水中氨氮排放量均呈现下降趋势，且工业氨氮排放量下降较快。

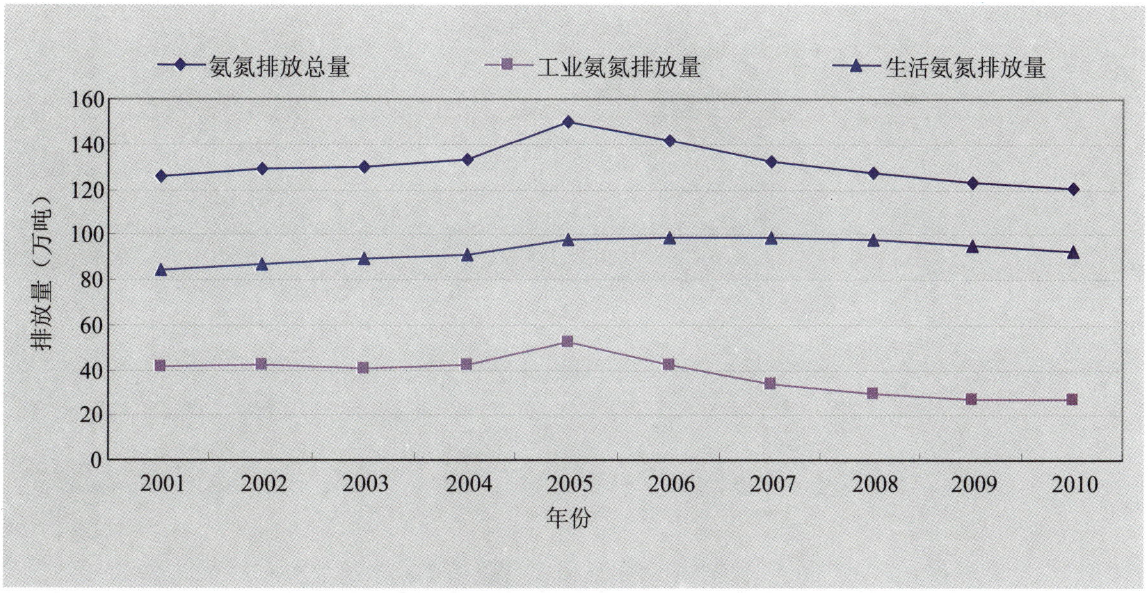

图 3　全国废水中氨氮排放量年际对比

（4）废水中其他主要污染物排放情况

2010 年，全国工业废水中石油类排放量 1.01 万吨，比上年减少 6.6%；挥发酚排放量 966.0 吨，比上年减少 7.5%；氰化物排放量 241.8 吨，比上年减少 3.4%。工业废水中五项重金属（汞、镉、六价铬、铅、砷）排放量自“十一五”以来总体呈下降趋势。

表 2　全国废水中其他有毒有害污染物排放量年际对比　　单位：吨

年度	汞	镉	六价铬	铅	砷
2001	5.6	110.5	121.4	489.9	408.4
2002	4.8	105.6	111.1	484.8	346.2
2003	5.5	84.5	103.1	568.5	373.7
2004	3.0	56.3	150.8	366.2	306.1
2005	2.7	62.1	105.6	378.3	453.2
2006	2.6	49.4	96.4	339.1	245.2
2007	1.2	39.3	69.0	319.7	187.4
2008	1.36	39.5	75.3	240.9	215.0
2009	1.39	32.3	55.4	182.2	197.3
2010	1.05	30.1	54.8	140.8	118.1
增长率（%）	−24.5	−6.8	−1.1	−22.7	−40.1

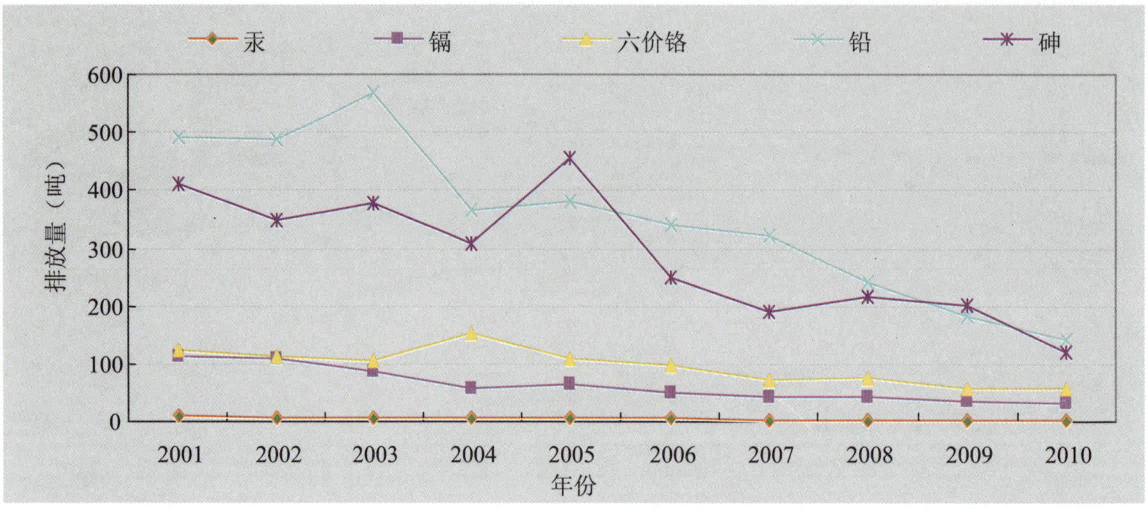

图 4　工业废水中五项重金属历年排放趋势

1.2.2　各地区废水及主要污染物排放情况

（1）各地区废水排放情况

2010 年，废水排放量大于 30 亿吨的省份依次为广东、江苏、山东、浙江、河南、广西，6 个省份废水排放总量为 278.1 亿吨，占全国废水排放量的 45.1%。工业废水排放量最大的是江苏，生活污水排放量最大的是广东，与上年相同。

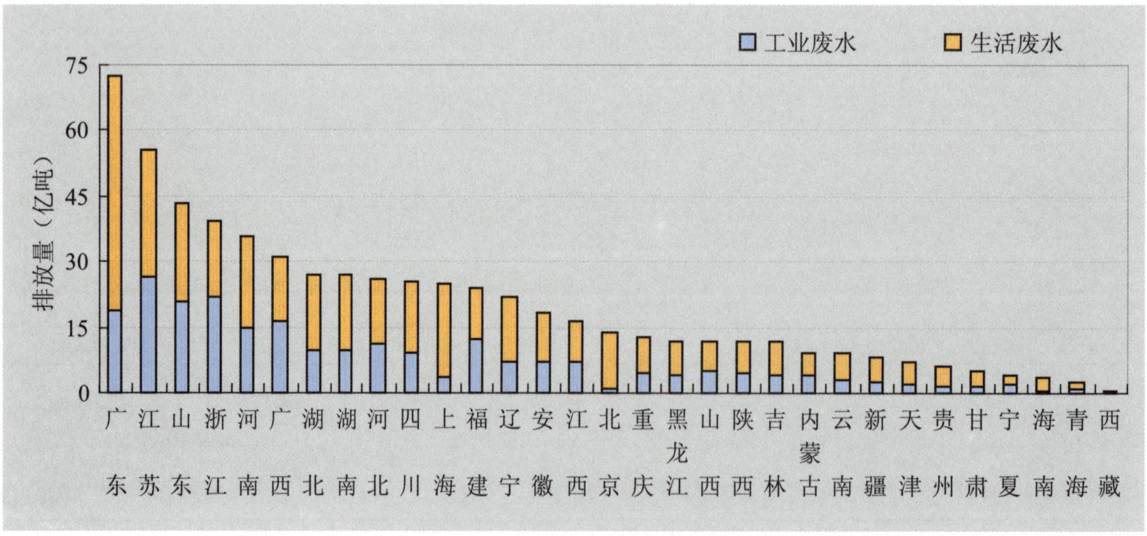

图 5　各地区废水排放情况

（2）各地区化学需氧量排放情况

化学需氧量排放量前 10 位的省份依次为广西、广东、湖南、江苏、四川、山东、河南、湖北、河北和辽宁，10 个省份的化学需氧量排放量为 702.3 万吨，占全国化学需氧量排放量的 56.7%。工业化学需氧量排放量最大的是广西，生活化学需氧量排放量最大的是广东。

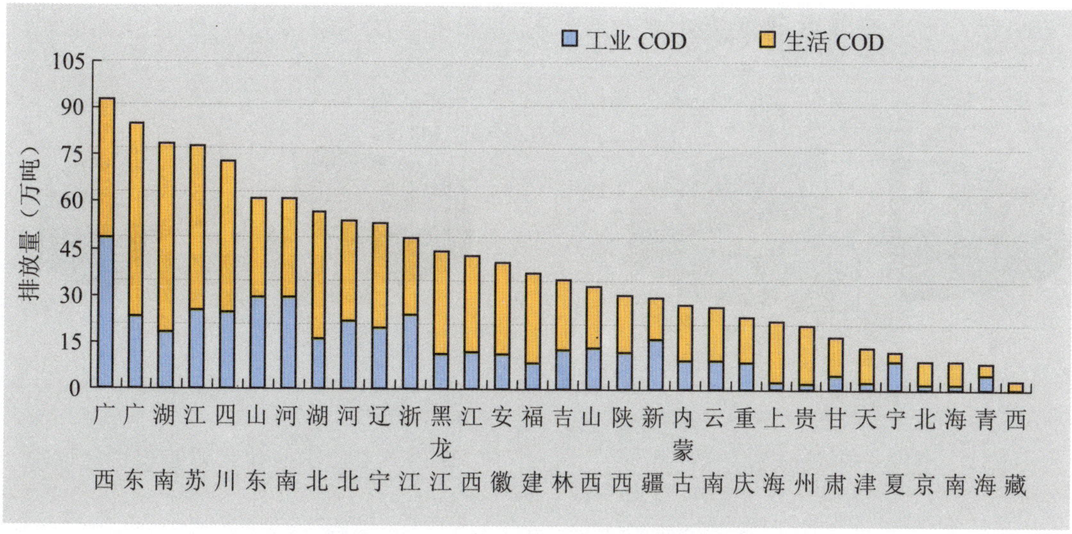

图 6　各地区化学需氧量排放情况

（3）各地区氨氮排放情况

氨氮排放量前 10 位的省份依次为广东、湖南、河南、山东、江苏、四川、湖北、辽宁、河北和广西，10 个省份的氨氮排放量为 66.3 万吨，占全国氨氮排放量的 55.1%。工业氨氮排放量最大的是河南，生活氨氮排放量最大的是广东。

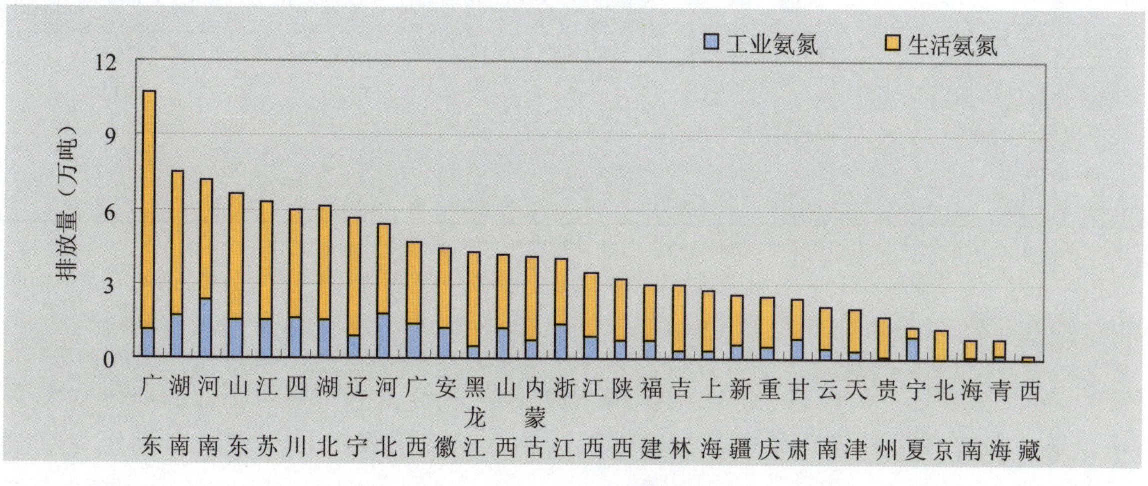

图 7　各地区氨氮排放情况

1.2.3 工业行业废水及主要污染物排放情况

（1）工业行业废水排放情况

2010 年，在统计的 39 个工业行业中，废水排放量位于前 4 位的行业依次为：造纸与纸制品业；化学原料及化学制品制造业；纺织业、农副食品加工业。4 个行业的废水排放量为 109.1 亿吨，占重点调查统计企业废水排放量的 51.5%。

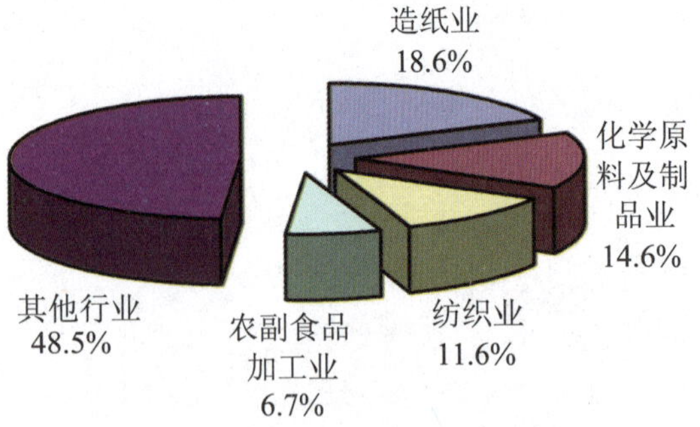

图 8　工业行业废水排放情况

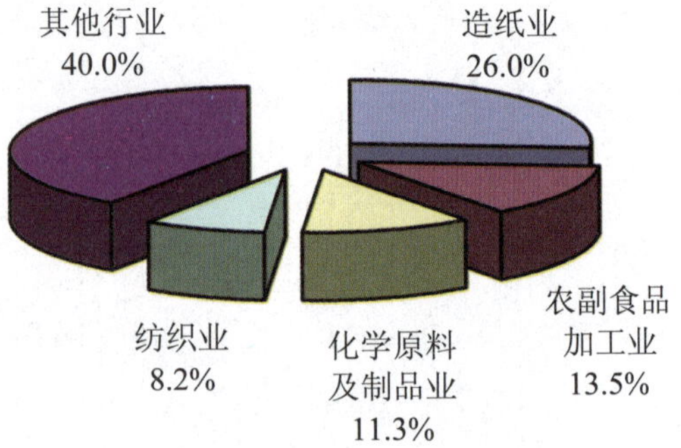

图 9　工业行业化学需氧量排放情况

（2）工业行业化学需氧量排放情况

2010 年，化学需氧量排放量位于前 4 位的行业依次为：造纸与纸制品业、农副食品加工业、化学原料及化学制品制造业、纺织业。4 个行业的化学需氧量排放量为 219.5 万吨，污染贡献率占 60.0%。

表 3　重点行业化学需氧量污染贡献率变化趋势　　单位：%

行　业	2003 年	2004 年	2005 年	2006 年	2007 年	2008 年	2009 年	2010 年
造纸业	34.5	33.0	32.4	33.6	34.7	32.8	28.9	26.0
农副食品加工业	14.4	13.3	13.7	12.8	12.8	14.9	13.9	13.6
化学原料及制品业	10.8	11.2	11.5	11.7	10.3	10.6	11.3	12.2
纺织业	5.6	6.7	6.1	6.8	7.6	8.0	8.3	8.2
累　计	65.3	64.2	63.7	64.9	65.4	66.3	62.3	60.0

注：① 污染贡献率指该行业某种污染物排放量与所有统计行业中此污染物排放总量之比，下同。
② 因 2002 年后《国民经济行业分类》标准执行 GB/T 4754—2002，行业分类有所变化，故本表起始年份为 2003 年。

与 2009 年相比，2010 年造纸与纸制品业、农副食品加工业及纺织业的污染贡献率均有所下降，但化学原料及化学制品制造业的污染贡献率不降反升。2010 年造纸与纸制品业、农副食品加工业、化学原料及化学制品制造业三个行业的经济贡献率均变化不大，但纺织业经济贡献率下降较快。

（3）工业行业氨氮排放情况

2010 年，氨氮排放量位于前 4 位的行业依次为：化学原料及化学制品制造业、造纸与纸制品业；农副食品加工业、纺织业。4 个行业氨氮排放量为 14.0 万吨，占重点调查统计企业氨氮排放量的 56.9%。

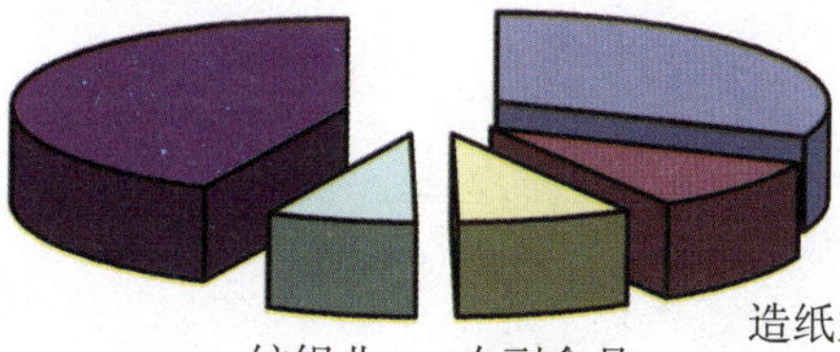

图 10　工业行业氨氮排放情况

（4）工业行业重金属等污染物排放情况

2010 年，重金属（汞、镉、六价铬、铅、砷）排放量位于前 4 位的行业依次为：有色金属矿采选业；有色金属冶炼及压延加工业；化学原料及化学制品制造业；金属制品业。4 个行业重金属排放量为 291.9 吨，占重点调查统计企业排放量的 84.6%。

图 11　工业行业重金属排放情况

（5）工业行业石油类污染物排放情况

2010 年，石油类排放量位于前 4 位的行业依次为：黑色金属冶炼及压延加工业；化学原料及化学制品制造业；石油加工炼焦及核燃料加工业；煤炭开采和洗选业。4 个行业石油类排放量为 5 725.9 吨，占重点调查统计企业石油类排放量的 56.5%。

图 12　工业行业石油类污染物排放情况

1.2.4　七大流域接纳废水及污染治理情况

1.2.4.1　接纳废水及主要污染物情况

2010 年，辽河、海河、淮河、长江、黄河、松花江和珠江七大流域共有 88 608 家工业企业纳入重点调查统计范围，占全部重点调查统计企业数的 78.6%。

（1）废水

七大流域共接纳废水 514.6 亿吨，比上年增加 6.7%，占全国废水排放总量的 83.4%；接纳工业废水 190.3 亿吨，比上年增加 3.2%，占全国工业废水排放量的 80.1%；接纳生活污水 324.3 亿吨，比上年增加 8.8%，占全国生活污水排放量的 85.4%。

表 4　七大流域废水及污染物接纳情况

年度	废水（亿吨）			化学需氧量（万吨）			氨氮（万吨）		
	总计	工业	生活	总计	工业	生活	总计	工业	生活
2005	379.3	172.5	206.8	1 059.8	406.3	653.6	110.9	40.4	70.5
2006	412.0	180.8	231.2	1 136.4	422.5	713.9	113.0	36.2	76.8
2007	433.6	191.0	242.6	1 114.1	410.5	703.6	107.1	28.9	78.2
2008	447.3	185.3	262.0	1 070.0	365.6	704.4	103.1	24.9	78.2
2009	482.5	184.4	298.1	1 053.4	356.5	696.9	102.1	23.4	78.7
2010	514.6	190.3	324.3	1 025.3	350.8	674.5	100.6	23.0	77.6
增长率（%）	6.7	3.2	8.8	−2.7	−1.6	−3.2	−1.5	−1.7	−1.4

注：①从 2004 年起，本年报中松花江流域和珠江流域统计范围较往年有所扩大。其中，松花江流域包括松花江流域和黑龙江流域，珠江流域包括珠江流域和粤桂琼沿海诸河流域。

②从 2006 年起，本年报中流域数据的汇总方法有所变化，流域规划所含区县的全部数据，不再沿用以前的按“排水去向”汇总数据的方法，汇总的区县数有所减少。以下湖泊数据同。

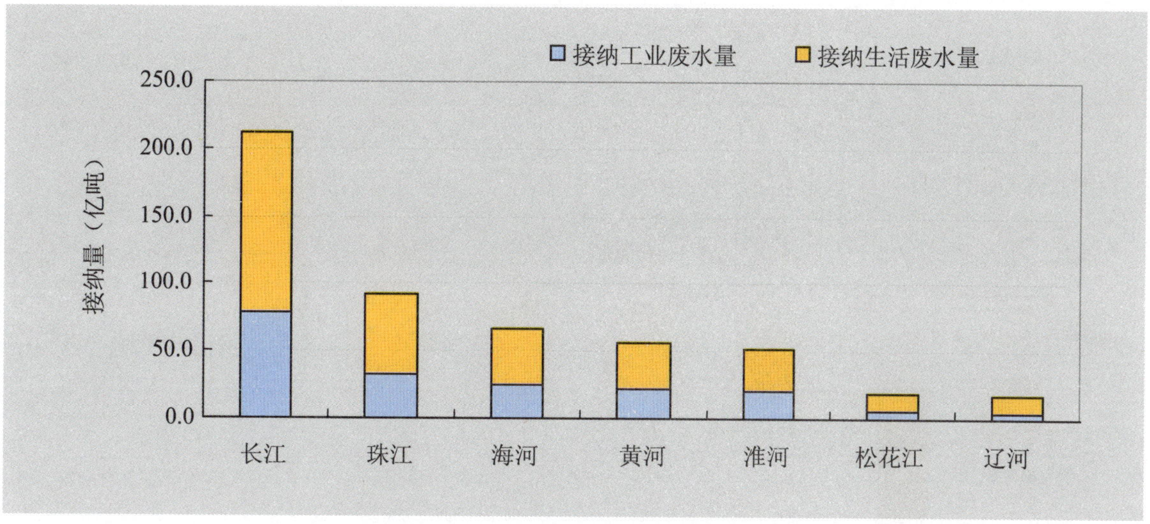

图 13　七大流域废水接纳情况

长江接纳的废水量占七大流域接纳总量的 41.0%，列第一位；其次是珠江，占 17.9%；第三位是海河，占 12.8%。

（2）化学需氧量

七大流域接纳化学需氧量 1 025.3 万吨，比上年降低 2.7%，占全国化学需氧量排放量的 82.8%；接纳工业化学需氧量 350.8 万吨，比上年降低 1.6%，占全国工业化学需氧量排放量的 80.7%；接纳生活化学需氧量 674.5 万吨，比上年降低 3.2%，占全国生活化学需氧量排放量的 84.0%。

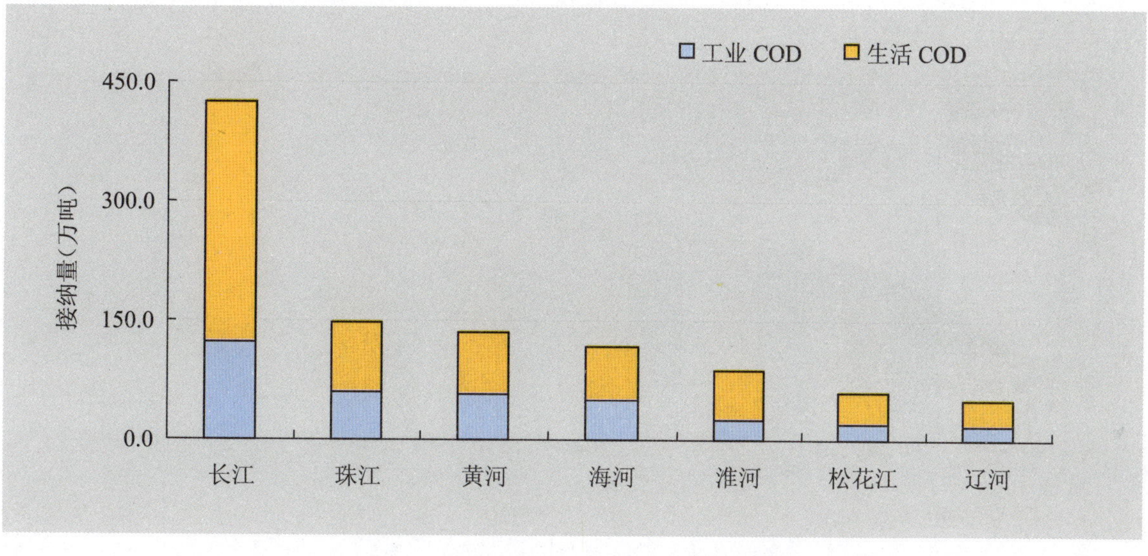

图 14　七大流域化学需氧量接纳情况

（3）氨氮

七大流域接纳氨氮 100.6 万吨，比上年减少 1.5%，占全国氨氮排放量的 83.6%；接纳工业氨氮 23.0 万吨，比上年减少 1.7%，占全国工业氨氮排放量的 84.2%；接纳生活氨氮 77.6 万吨，与上年减少 1.4%，占全国生活氨氮排放量的 83.4%。

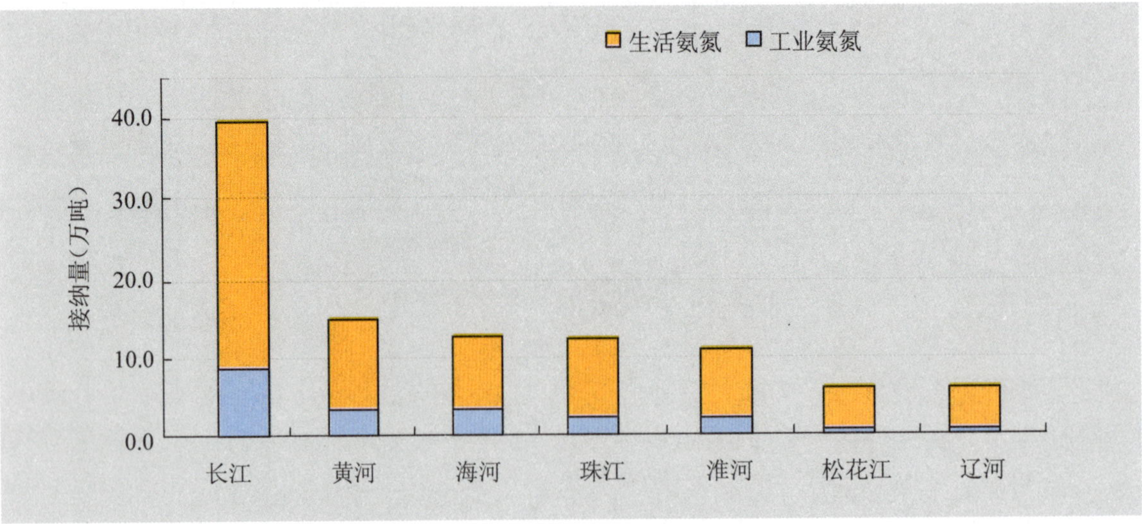

图 15　七大流域氨氮接纳情况

1.2.4.2　废水及主要污染物治理与投资情况

2010 年，七大流域共有废水治理设施 65 093 套，年运行费用 446.2 亿元，共去除化学需氧量 1 108.1 万吨、氨氮 63.5 万吨、石油类 31.0 万吨、挥发酚 6.7 万吨、氰化物 0.5 万吨。

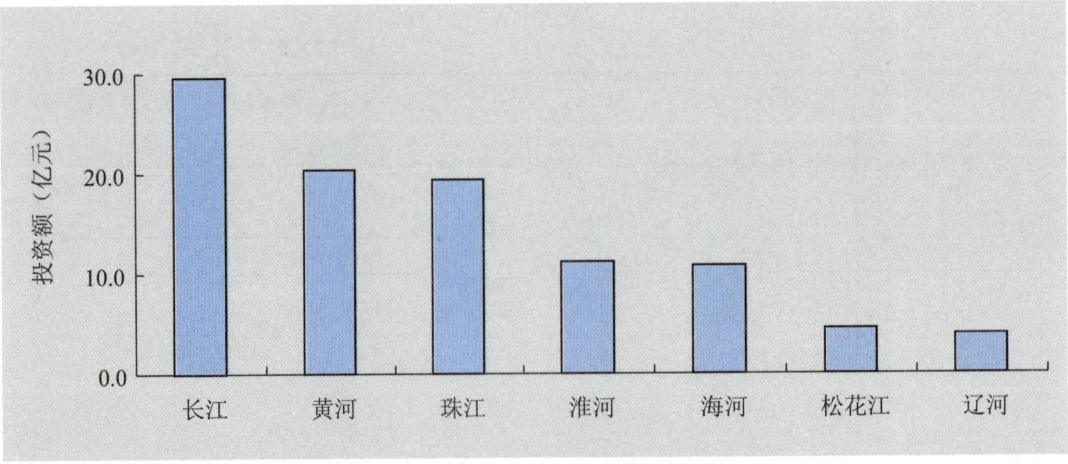

图 16　七大流域工业废水治理投资情况

七大流域实施工业废水治理项目 1 935 个，竣工 1 666 个，工业废水治理项目完成投资 100.2 亿元，占全国工业废水治理项目完成投资额的 77.0%。

七大流域经处理的工业废水为 510.6 亿吨，工业废水排放达标率为 96.1%。纳入统计的污水处理厂 2 543 座，比上年增加 649 座，共形成 10 633 万吨/日的处理能力，处理生活污水 242.8 亿吨/年。城市生活污水处理率为 74.9%，高于全国平均水平 2.0 个百分点。

1.2.5 五大湖泊接纳废水及污染治理情况

(1) 接纳废水及主要污染物情况

2010 年，滇池、巢湖、太湖、洞庭湖和鄱阳湖流域重点调查统计企业 7 131 家，接纳废水排放量 45.8 亿吨，比上年增加 9.3%；其中工业废水 21.8 亿吨，比上年增加 7.9%；生活污水 24.0 亿吨，比上年增加 10.6%。接纳化学需氧量 51.2 万吨，比上年减少 6.6%；

表 5　五大湖泊废水及污染物接纳情况

年度	废水（亿吨）			化学需氧量（万吨）			氨氮（万吨）		
	总计	工业	生活	总计	工业	生活	总计	工业	生活
2006	41.0	22.7	18.3	66.4	29.6	36.8	5.7	2.2	3.5
2007	41.1	21.5	19.6	60.5	25.6	34.9	5.7	2.1	3.6
2008	40.6	20.2	20.4	57.1	23.7	33.4	5.1	1.6	3.5
2009	41.9	20.2	21.7	54.8	23.0	31.8	4.5	1.4	3.1
2010	45.8	21.8	24.0	51.2	23.2	28.0	4.4	1.2	3.2
增长率（%）	9.3	7.9	10.6	−6.6	0.9	−11.9	−2.2	−14.3	3.2

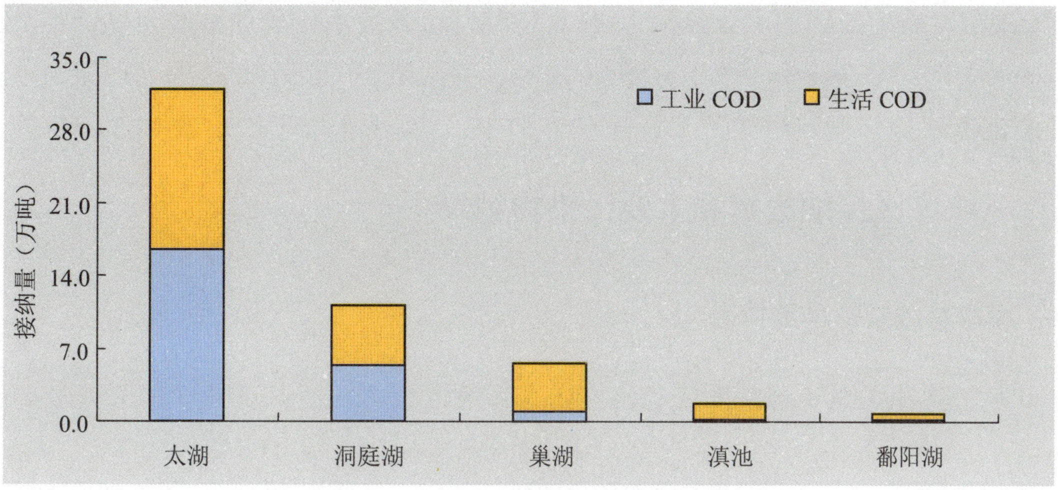

图 17　五大湖泊化学需氧量接纳情况

其中工业化学需氧量 23.2 万吨，比上年增加 0.9%；生活化学需氧量 28.0 万吨，比上年减少 11.9%。接纳氨氮 4.4 万吨，比上年减少 2.2%；其中工业氨氮 1.2 万吨，比上年减少 14.3%；生活氨氮 3.2 万吨，比上年增加 3.2%。

（2）废水及主要污染物治理与投资情况

2010 年，五大湖泊流域共有废水治理设施 5 735 套，年运行费用 37.5 亿元，共去除化学需氧量 107.7 万吨、氨氮 3.2 万吨、石油类 0.8 万吨、挥发酚 815.3 吨、氰化物 52.1 吨。

五大湖泊流域工业废水施工治理项目数 157 个，竣工项目数 140 个，工业废水治理项目完成投资 6.1 亿元，占全国工业废水治理项目完成投资额的 4.7%。工业废水治理竣工项目新增设计处理能力 24.4 万吨/日。

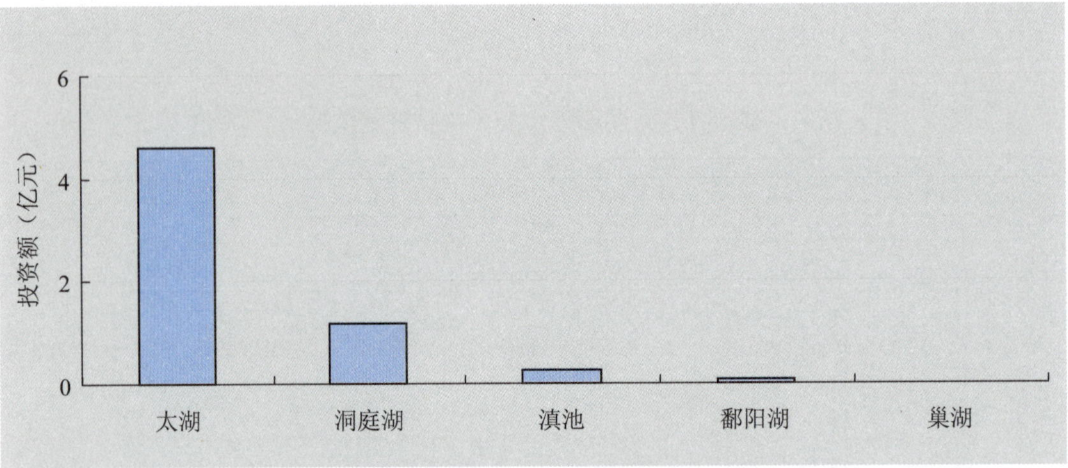

图 18　五大湖泊工业废水治理投资情况

五大湖泊流域经处理的工业废水为 26.6 亿吨，工业废水排放达标率为 98.5%。纳入统计的污水处理厂共 281 座，比上年增加 13 座，共形成 1 074 万吨/日的处理能力，处理生活污水 20.9 亿吨。城市生活污水处理率为 87.1%，高于全国平均水平 14.2 个百分点。

1.2.6　三峡库区接纳废水和主要污染物情况

（1）废水及污染物排放情况

2010 年，重点调查了三峡库区（含库区、影响区及上游区共 314 个区县）11 764 家企业。

三峡库区共排放废水 48.2 亿吨，比上年减少 3.8%。其中，工业废水 16.1 亿吨，比上年减少 16.3%；生活污水 32.1 亿吨，比上年增加 4.0%。

三峡库区排放化学需氧量为 118.9 万吨，比上年减少 1.6%。其中，工业化学需氧量为 36.2 万吨，比上年减少 1.6%；生活化学需氧量为 82.7 万吨，比上年减少 1.6%。

三峡库区排放氨氮为 10.6 万吨，比上年增加 0.7%。其中，工业氨氮为 2.4 万吨，比上年增加 5.3%；生活氨氮为 8.2 万吨，比上年减少 0.6%。

表 6　三峡库区及其上游主要污染物排放情况

年度	废水排放量（亿吨）			化学需氧量排放量（万吨）			氨氮排放量（万吨）		
	合计	工业	生活	合计	工业	生活	合计	工业	生活
2003	35.7	20.2	15.6	123.8	55.4	68.3	8.5	2.6	5.9
2004	34.3	18.5	15.7	103.3	34.2	69.1	8.6	2.7	6.0
2005	38.0	19.2	18.8	104.1	33.8	70.4	8.8	2.4	6.5
2006	46.5	21.4	25.1	127.1	43.8	83.4	11.1	3.6	7.5
2007	47.0	20.4	26.6	125.5	41.5	84.0	10.5	3.1	7.4
2008	49.1	19.4	29.7	121.8	37.3	84.5	10.5	2.7	7.8
2009	50.1	19.2	30.9	120.8	36.8	84.0	10.5	2.3	8.3
2010	48.2	16.1	32.1	118.9	36.2	82.7	10.6	2.4	8.2
增长率（%）	−3.8	−16.3	4.0	−1.6	−1.6	−1.6	0.7	5.3	−0.6

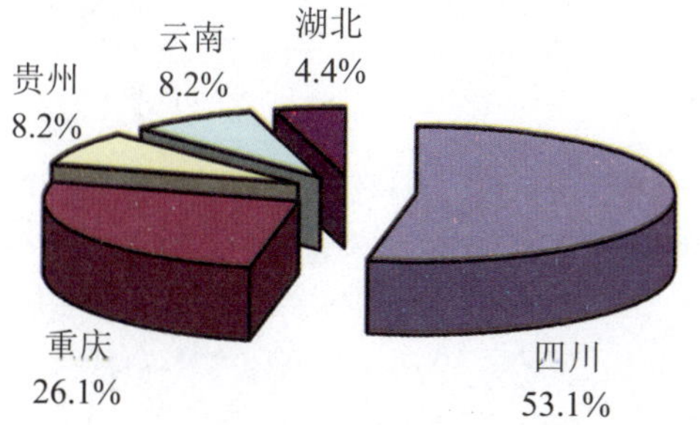

图 19　三峡库区省市废水排放构成情况

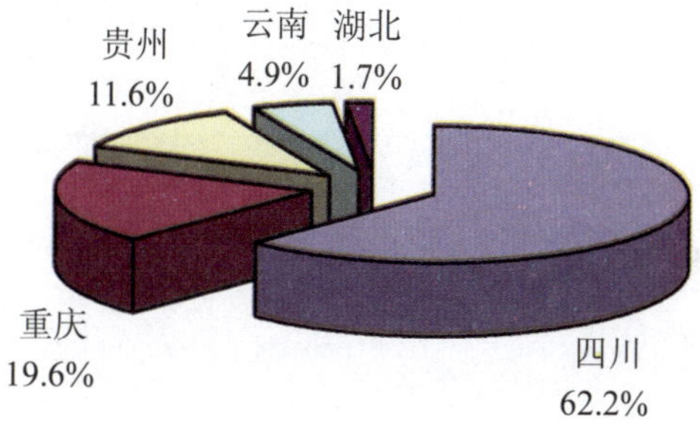

图 20　三峡库区省市化学需氧量排放构成情况

三峡库区排放废水量最大的是四川，其次为重庆、贵州、云南、湖北；化学需氧量排放最大的是四川，其次为重庆、贵州、云南、湖北。

（2）废水及污染物治理与投资情况

2010 年，三峡地区工业废水治理施工项目数 264 个，竣工项目数 231 个，工业废水治理项目完成投资 11.2 亿元，占全国工业废水治理项目完成投资额的 8.6%。工业废水治理竣工项目新增设计处理能力 52.8 万吨/日。

三峡地区共处理工业废水 33.3 亿吨，工业废水排放达标率为 95.4%。纳入统计的污水处理厂 338 座，比上年增加 98 座，共形成 897 万吨/日的处理能力，处理生活污水 22.8

亿吨/年，城镇生活污水处理率 71.0%。

1.2.7 “南水北调”东线工程沿线地区废水及主要污染物排放情况

“南水北调”东线工程途经 6 个地区的 23 个市（地级市）、105 个县（县级市、县城和区）。

2010 年，沿线重点调查工业企业数 8 570 家，排放工业废水 21.5 亿吨、工业化学需氧量 35.1 万吨、工业氨氮 2.7 万吨。沿线各地区工业废水平均排放达标率为 97.4%。

沿线工业废水治理施工项目数 191 个，竣工项目数 166 个，工业废水治理项目完成投资额 10.4 亿元，新增工业废水治理能力 135.1 万吨/日。

沿线生活污水排放 30.3 亿吨，生活化学需氧量为 55.1 万吨，生活氨氮为 7.8 万吨。沿线地区统计了污水处理厂 222 座，污水处理能力 1 095 万吨/日，处理生活污水量 24.7 亿吨。生活污水平均处理率为 81.5%，高于全国平均水平 8.6 个百分点。

1.2.8 入海陆源废水及主要污染物排放情况

2010 年，入海陆源的统计范围为我国沿海 11 个地区的 163 个县（区、市）。四大海域的重点调查工业企业数为 21 539 家，占全国重点调查统计工业企业数的 19.1%。

表 7　近岸海域主要污染物接纳情况

年度	废水（亿吨）			化学需氧量（万吨）			氨氮（万吨）		
	总计	工业	生活	总计	工业	生活	总计	工业	生活
2003	74.6	34.4	40.2	168.4	56.4	112.0	16.4	3.9	12.5
2004	85.3	37.5	47.8	166.8	53.4	113.4	16.6	3.6	13.0
2005	91.7	40.3	51.4	179.9	63.4	116.5	18.6	4.5	14.1
2006	100.4	43.1	57.3	196.1	61.5	134.6	19.5	4.2	15.3
2007	97.2	42.6	54.6	180.5	62.7	117.8	17.0	3.6	13.4
2008	106.4	44.7	61.7	175.4	54.2	121.2	17.4	3.2	14.2
2009	109.3	41.9	67.4	171.2	52.1	119.1	16.6	2.7	13.9
2010	121.7	43.6	78.1	173.0	55.6	117.4	16.8	3.0	13.8
增长率（%）	11.3	4.1	15.9	1.1	6.7	−1.4	1.2	11.1	−0.7

四大海域入海陆源的废水排放总量为 121.7 亿吨，比上年增加 11.3%。其中，工业废水排放量为 43.6 亿吨，比上年增加 4.1%，占入海陆源废水排放总量的 35.8%；直排入海的工业废水量为 11.8 亿吨。生活污水排放量为 78.1 亿吨，比上年增加 15.9%，占入海陆源废水排放总量的 64.2%。工业废水接纳量最大的海域是东海，生活污水接纳量最大的海域是南海。

四大海域入海陆源排放的化学需氧量 173.0 万吨，比上年增加 1.1%。其中，工业化学需氧量 55.6 万吨，比上年增加 6.7%，占化学需氧量排放量的 32.1%；生活化学需氧量 117.4 万吨，比上年减少 1.4%，占化学需氧量排放量的 67.9%。工业化学需氧量接纳量最大的海域是东海，生活化学需氧量接纳量最大的海域是南海。

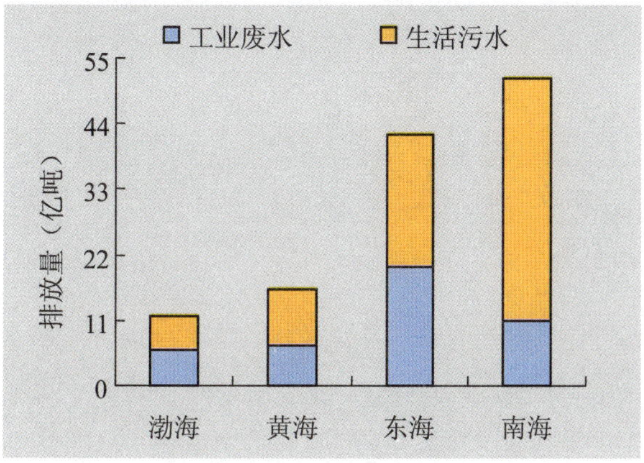

图 21　四大海域入海陆源废水排放情况

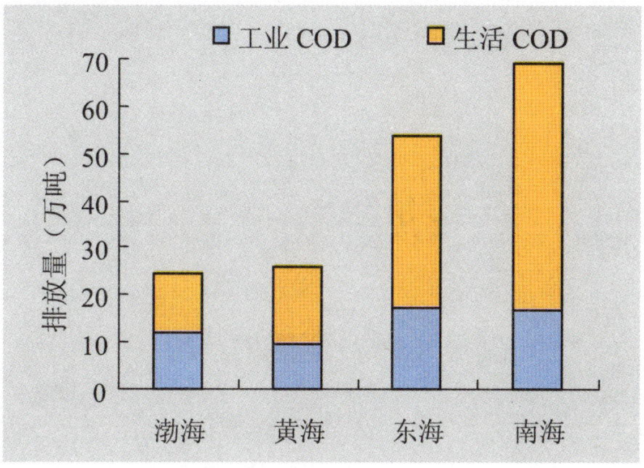

图 22　四大海域入海陆源化学需氧量排放情况

四大海域入海陆源排放的氨氮 16.8 万吨，比上年增加 1.2%。其中，工业氨氮为 3.0 万吨，比上年增加 11.1%；生活氨氮为 13.8 万吨，比上年减少 0.7%。工业氨氮接纳量最大的海域是东海，生活氨氮接纳量最大的海域是南海。

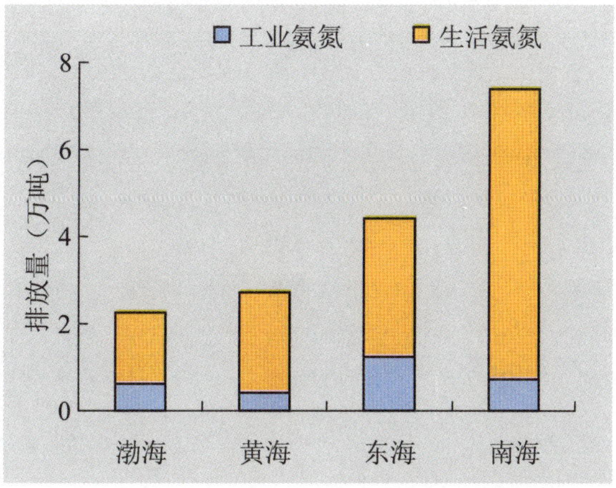

图 23　四大海域入海陆源氨氮排放情况

四大海域入海陆源共有废水治理设施 15 727 套，年运行费用 98.6 亿元，共去除化学需氧量 296.5 万吨、氨氮 16.6 万吨、石油类 7.2 万吨、挥发酚 2 929 吨、氰化物 1 559 吨。四大海域入海陆源工业废水治理施工项目数 678 个，竣工项目数 613 个，工业废水治理项目完成投资额 22.7 亿元，占全国工业废水治理项目完成投资额的 17.5%。工业废水治理竣工项目新增设计处理能力 130.9 万吨/日。工业废水排放达标率为 96.2%。

四大海域入海陆源纳入统计的污水处理厂 424 座，新增 113 座，共形成 2 833.3 万吨/日的处理能力，处理生活污水 59.4 亿吨/年。生活污水处理率为 76.1%，高于全国平均水平 3.2 个百分点。

1.3 废气

1.3.1 废气及废气中主要污染物排放情况

（1）煤炭及燃料油使用情况

2010 年，全国环境统计的煤炭消费总量 35.6 亿吨，比上年增加 13.3%。

工业煤炭消费量 33.8 亿吨，比上年增加 14.5%。其中，工业煤耗中燃料煤消费量为 24.1 亿吨，原料煤消费量为 9.7 亿吨。生活煤炭消费量 1.8 亿吨，比上年减少 5.7%。

工业（不含车船用）燃料油消费量 2 087 万吨，比上年增加 3.5%。其中，重油 1 159 万吨，柴油 885 万吨。

表 8　全国环境统计煤炭、燃料油消耗量　　单位：万吨

年度＼项目	煤炭消耗量				燃料油消费量（不含车船用）		
	合计	工业		生活	合计		
		燃料煤	原料煤			重油	柴油
2001	142 217	91 234	30 571	20 412	2 646	2 034	387
2002	152 812	97 264	36 524	19 024	2 773	2 043	495
2003	172 430	110 728	42 624	19 078	2 624	2 141	343
2004	195 611	125 972	50 026	19 613	2 734	2 295	365
2005	226 164	143 627	60 796	21 741	3 447	2 412	383
2006	250 452	162 089	67 987	20 376	2 666	2 049	571
2007	285 377	187 815	78 642	18 920	3 207	2 613	557
2008	297 663	199 862	80 307	19 328	2 335	1 850	475
2009	314 687	210 140	85 081	19 466	2 016	1 585	406
2010	356 368	240 534	97 465	18 369	2 087	1 159	885
增长率（%）	13.3	14.5	14.6	−5.7	3.5	−26.9	118.0

（2）二氧化硫排放情况

2010 年，全国工业废气排放量 519 168 亿米^3（标态），比上年增加 19.1%。

全国二氧化硫排放量为 2 185.1 万吨，比上年减少 1.3%。其中，工业二氧化硫排放量为 1 864.4 万吨，基本与上年持平，占全国二氧化硫排放量的 85.3%；生活二氧化硫排放量 320.7 万吨，比上年减少 7.9%，占全国二氧化硫排放量的 14.7%。

“十一五”期间，全国废气中二氧化硫排放总量、工业废气中二氧化硫排放量和生活废气中二氧化硫排放量均呈现逐年下降趋势。2010 年全国二氧化硫排放总量较 2005 年下降了 14.3%，超额完成了“十一五”总量减排任务。

表 9　全国近年废气中主要污染物排放量　　单位：万吨

年度	二氧化硫			烟尘			工业粉尘	氮氧化物		
	合计	工业	生活	合计	工业	生活		合计	工业	生活
2001	1 947.8	1 566.6	381.2	1 069.8	851.9	217.9	990.6	—	—	—
2002	1 926.6	1 562.0	364.6	1 012.7	804.2	208.5	941.0	—	—	—
2003	2 158.7	1 791.4	367.3	1 048.7	846.2	202.5	1 021.0	—	—	—
2004	2 254.9	1 891.4	363.5	1 094.9	886.5	208.4	904.8	—	—	—
2005	2 549.3	2 168.4	380.9	1 182.5	948.9	233.6	911.2	—	—	—
2006	2 588.8	2 237.6	351.2	1 088.8	864.5	224.3	808.4	1 523.8	1 136.0	387.8
2007	2 468.1	2 140.0	328.1	986.6	771.1	215.5	698.7	1 643.4	1 261.3	382.0
2008	2 321.2	1 991.3	329.9	901.6	670.7	230.9	584.9	1 624.5	1 250.5	374.0
2009	2 214.4	1 865.9	348.5	847.7	604.4	243.3	523.6	1 692.7	1 284.8	407.9
2010	2 185.1	1 864.4	320.7	829.1	603.2	225.9	448.7	1 852.4	1 465.6	386.8
增长率（%）	−1.3	−0.1	−7.9	−2.2	−0.1	−7.2	−14.3	9.4	14.1	−5.2

注：我国从 2006 年开始统计氮氧化物排放量，生活排放量中含交通源排放的氮氧化物。

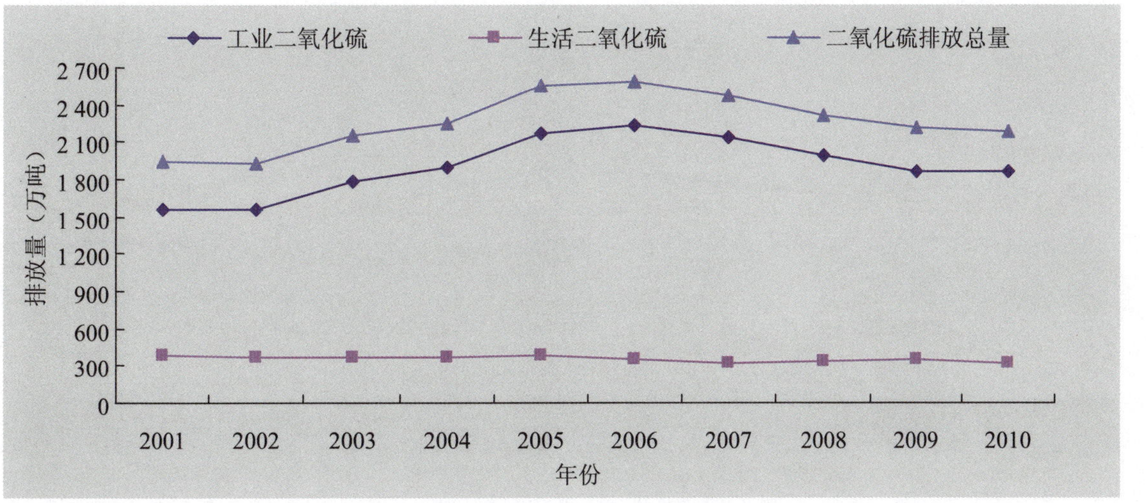

图 24　全国二氧化硫排放量年际变化

（3）氮氧化物排放情况

2010 年，氮氧化物排放量为 1 852.4 万吨，比上年增加 9.4%。其中，工业氮氧化物排放量为 1 465.6 万吨，比上年增加 14.1%，占全国氮氧化物排放量的 79.1%；生活氮氧化物排放量为 386.8 万吨，比上年减少 5.2%，占全国氮氧化物排放量的 20.9%；其中交通源氮氧化物排放量为 290.6 万吨，占全国氮氧化物排放量的 15.7%。

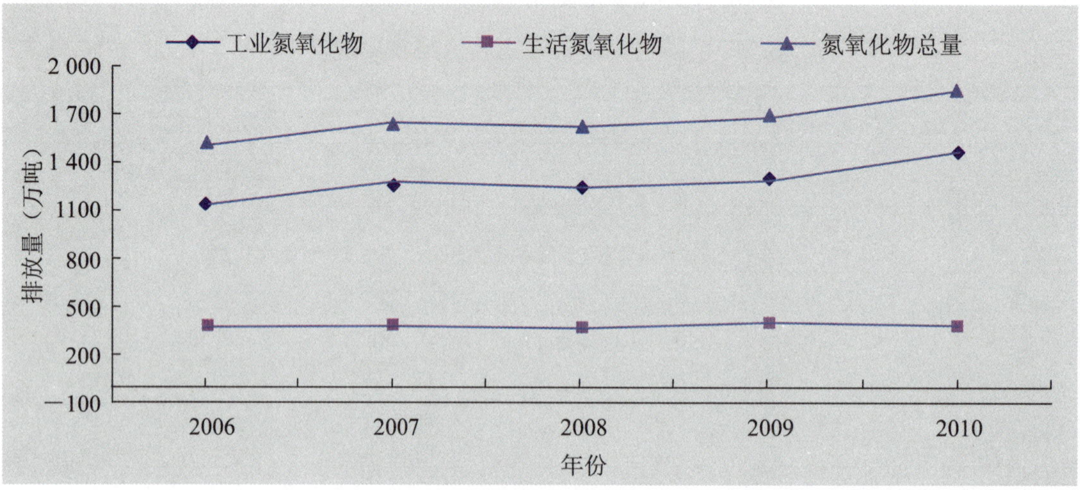

图 25　全国氮氧化物排放量年际变化

（4）烟尘及工业粉尘排放情况

2010 年，烟尘排放量为 829.1 万吨，比上年减少 2.1%。其中，工业烟尘排放量为 603.2

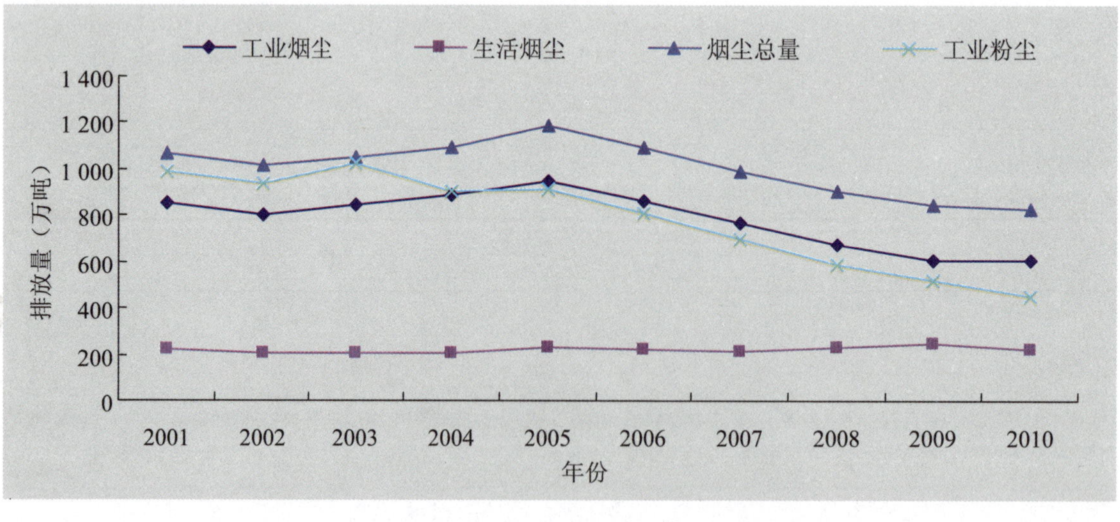

图 26　全国烟尘和工业粉尘排放量年际变化

万吨，基本与上年持平，占全国烟尘排放量的 72.8%；生活烟尘排放量为 225.9 万吨，比上年减少 7.2%，占全国烟尘排放量的 27.2%。

2010 年，工业粉尘排放量为 448.7 万吨，比上年减少 14.3%。

1.3.2 各地区废气中主要污染物排放情况

（1）二氧化硫排放情况

2010 年，二氧化硫排放量超过 100 万吨的省份依次为山东、内蒙古、河南、山西、河北、贵州、四川、广东、江苏和辽宁，10 个省份的二氧化硫排放量占全国排放量的 55.6%。工业二氧化硫排放量最大的是山东，占全国工业二氧化硫排放量的 7.4%；生活二氧化硫排放量最大的是贵州，占全国生活二氧化硫排放量的 15.9%。

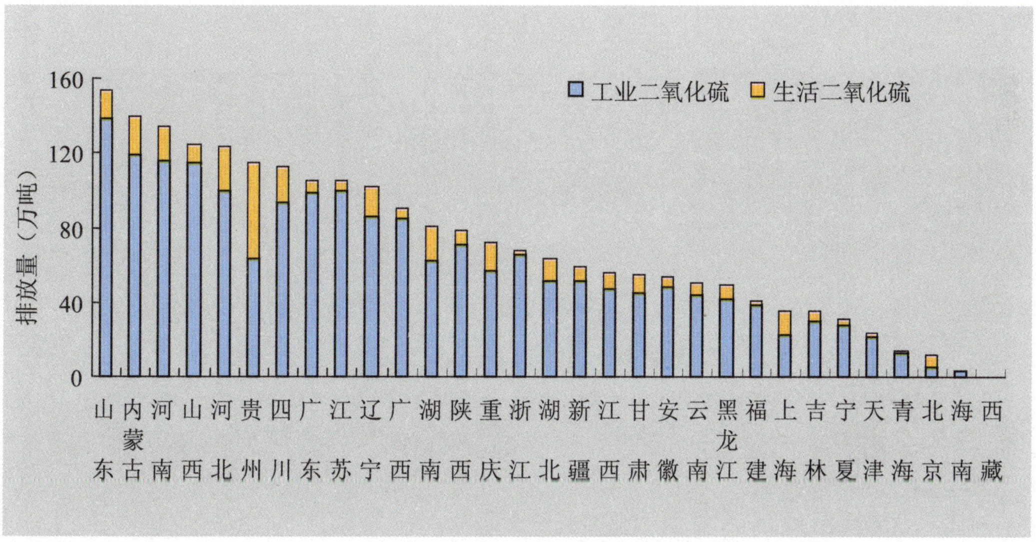

图 27　各地区二氧化硫排放情况

（2）氮氧化物排放情况

氮氧化物排放量超过 100 万吨的省份依次为山东、广东、内蒙古、江苏、河南和河北，6 个省份氮氧化物排放量占全国氮氧化物排放量的 41.0%。工业和生活氮氧化物排放量最大的分别是山东和广东，分别占全国工业和生活氮氧化物排放量的 8.0%和 12.0%。

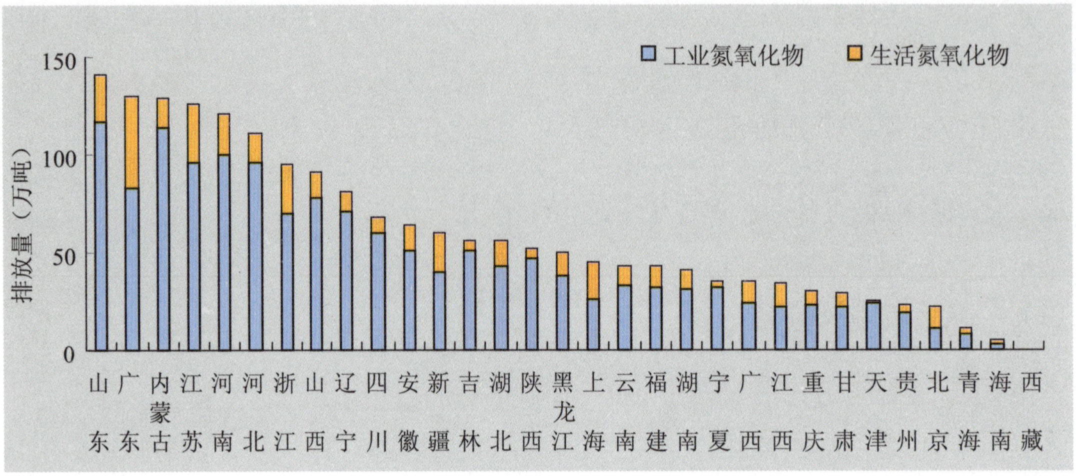

图 28　各地区氮氧化物排放情况

（3）烟尘排放情况

烟尘排放量超过 50 万吨的省份依次为内蒙古、辽宁、山西、河南和河北，5 个省份烟尘排放量占全国烟尘排放量的 35.6%。工业和生活烟尘排放量最大的分别是内蒙古和辽宁，分别占全国工业和生活烟尘排放量的 7.9%和 10.4%。

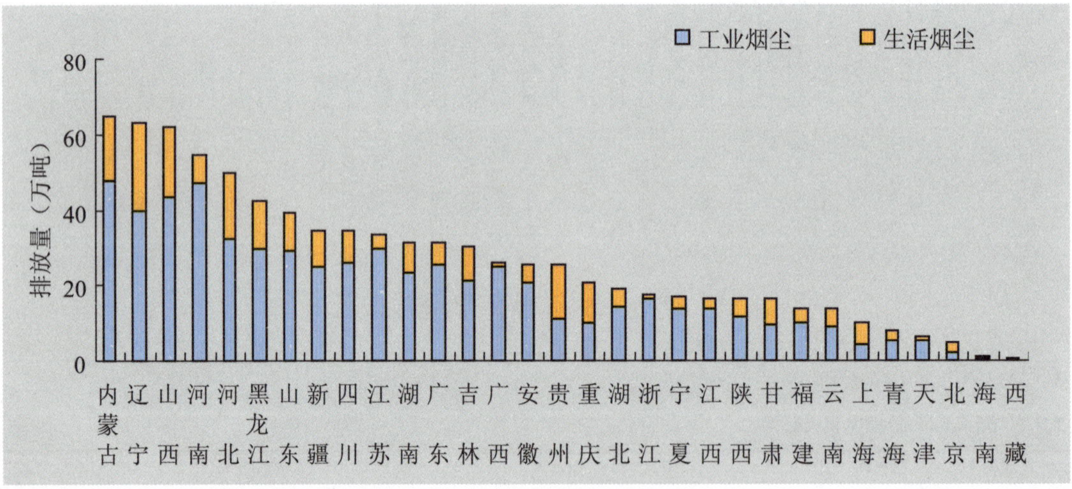

图 29　各地区烟尘排放情况

（4）工业粉尘排放情况

工业粉尘排放量超过 30 万吨的省份依次为湖南、山西、河北和广西，4 个省份工业粉尘排放量占全国工业粉尘排放量的 31.2%。

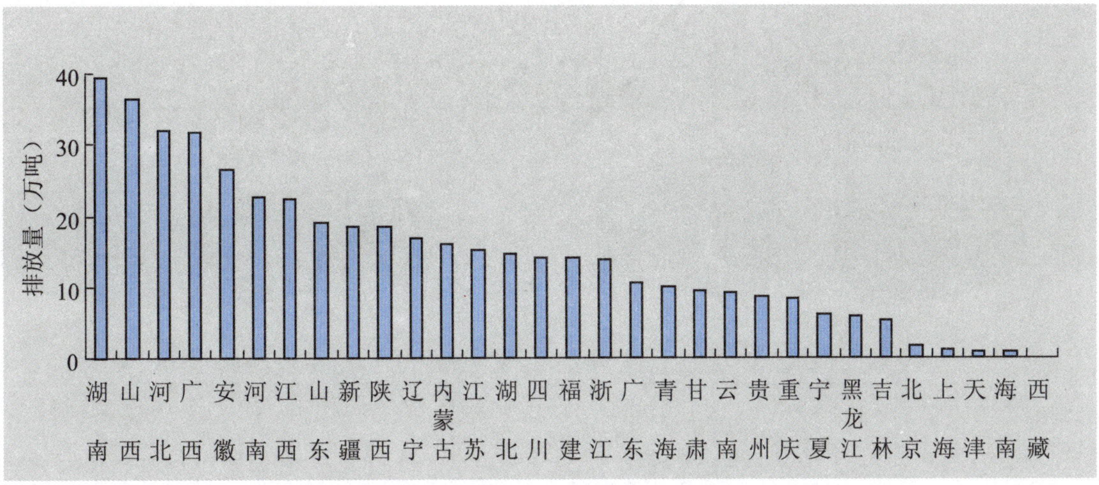

图 30　各地区工业粉尘排放情况

1.3.3　工业行业废气中主要污染物排放情况

（1）二氧化硫排放情况

2010 年，二氧化硫排放量排行前三名的行业依次为：电力、热力的生产及供应业；黑色金属冶炼及压延加工业；非金属矿物制品业。3 类行业共排放二氧化硫 1 245.1 万吨，占统计工业行业二氧化硫排放量的 73.0%。

与上年相比，3 类行业中，黑色金属冶炼及压延加工业、非金属矿物制品业的二氧化硫污染贡献率有所升高，电力、热力的生产及供应业略有下降。近几年黑色金属冶炼及压延加工业的污染贡献率持续升高。

表 10　重点行业二氧化硫污染贡献率年际变化　　单位：%

行　业	2001 年	2002 年	2003 年	2004 年	2005 年	2006 年	2007 年	2008 年	2009 年	2010 年
电力、热力业	53.5	54.9	61.7	57.1	58.9	59.0	58.2	57.8	55.1	52.8
黑色金属冶炼业	5.4	5.9	5.1	6.5	7.2	7.3	8.2	8.8	10.0	10.4
非金属矿物制品业	11.6	11.4	9.5	9.8	9.0	9.1	9.3	9.2	9.5	9.9
总计	70.5	72.2	76.3	73.4	75.1	75.4	75.7	75.8	74.6	73.0

（2）氮氧化物排放情况

2010 年，氮氧化物排放量位于排名前 3 位的行业依次为：电力、热力的生产及供应业；非金属矿物制品业；黑色金属冶炼及压延加工业。3 类行业占统计行业氮氧化物排放量的 83.5%，其中电力、热力的生产及供应业占 65.1%。

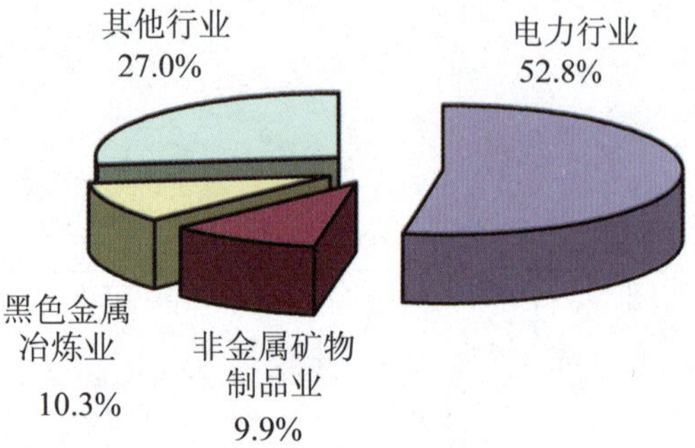

图 31 工业行业二氧化硫排放情况

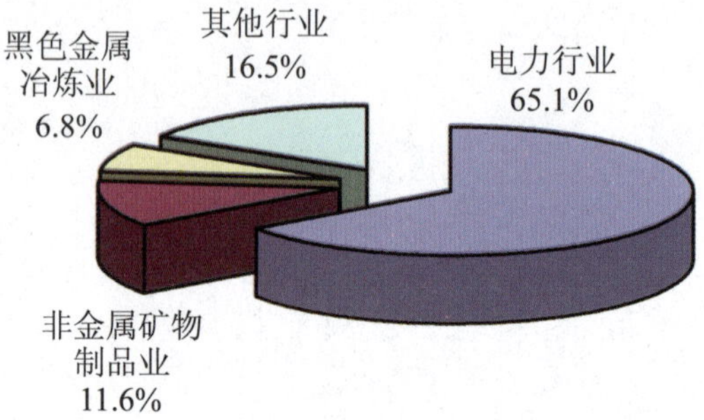

图 32 工业行业氮氧化物排放情况

（3）烟尘排放情况

2010 年，烟尘排放量位于排名前 3 位的行业依次为：电力、热力的生产及供应业；非金属矿物制品业；黑色金属冶炼及压延加工业。与上年相同。3 类行业占统计行业烟尘排放量的 65.9%，其中电力、热力的生产及供应业占 36.2%。

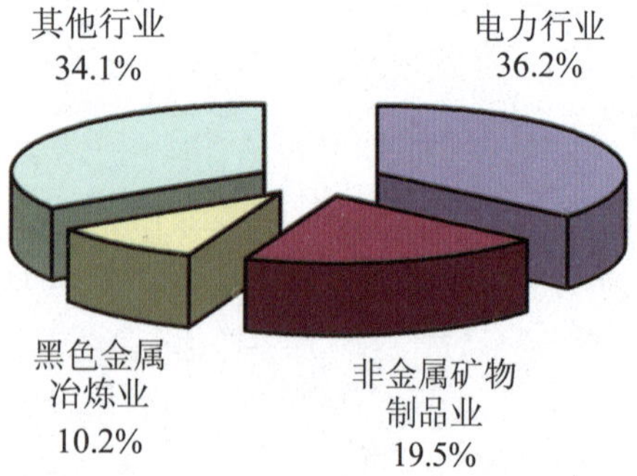

图 33 工业行业烟尘排放情况

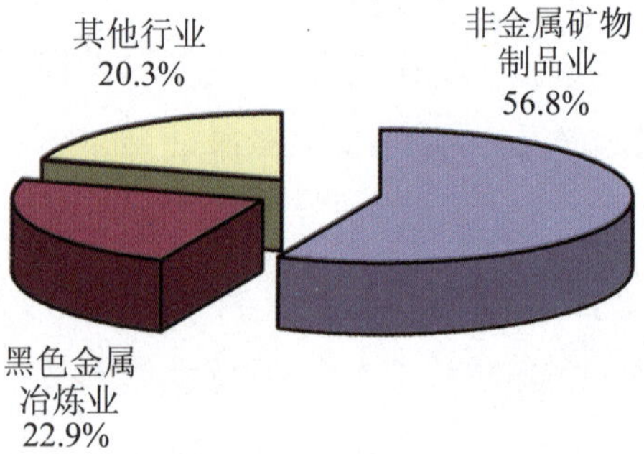

图 34 工业行业粉尘排放情况

（4）工业粉尘排放情况

2010 年，工业粉尘排放量位于排名前 2 位的行业依次为非金属矿物制品业、黑色金属冶炼及压延加工业。两个行业占统计行业工业粉尘排放量的 79.7%。其中，非金属矿物制品业占 56.8%，黑色金属冶炼及压延加工业占 22.9%。

1.3.4 火电厂二氧化硫排放情况

2010 年，纳入重点调查统计范围的电力企业 2 386 家。其中，独立火电厂 1 642 家，自备电厂 744 家。

独立火电厂共消耗燃料煤 16.6 亿吨，占全国工业煤炭消耗量的 49.2%。二氧化硫排放量为 835 万吨，比上年减少 4.8%，占全国工业二氧化硫排放量的 44.8%。独立火电厂二氧化硫排放量大于 50 万吨的省份依次为山东、山西、内蒙古、河南和江苏，占全国独立火电厂二氧化硫排放量的 37.3%。

独立火电厂共安装了 3 266 套脱硫设施，比上年增加 134 套。去除二氧化硫 1 900 万吨，比上年增加 35.0%；二氧化硫去除率达到 69.5%，比上年升高 7.9 个百分点。

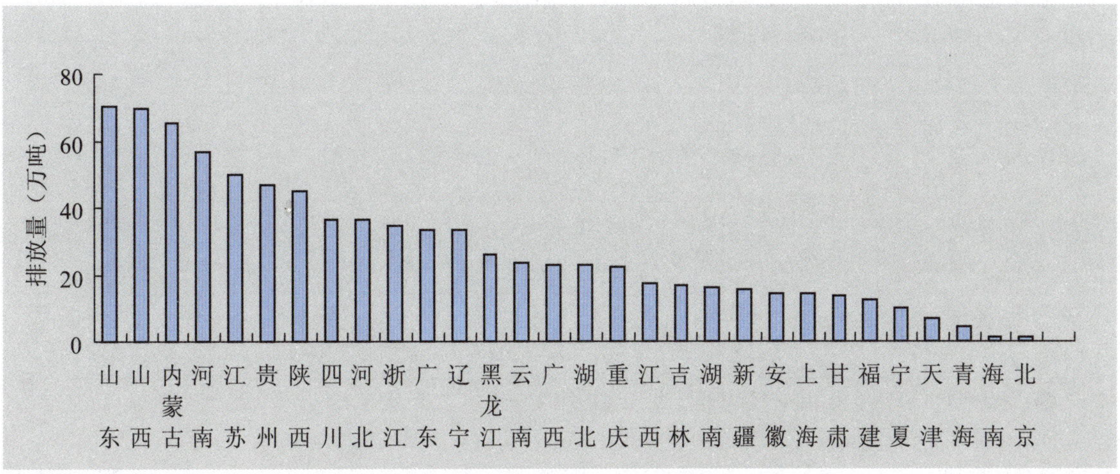

图 35　各地区独立火电厂二氧化硫排放情况

1.4　工业固体废物

1.4.1　工业固体废物产生、排放及利用情况

2010 年，全国工业固体废物产生量 240 944 万吨，比上年增加 18.1%；工业固体废物排放量 498 万吨，比上年减少 30.0%。全国危险废物产生量 1 587 万吨，比上年增加 11.0%。

工业固体废物综合利用量 161 772 万吨，比上年增加 16.9%；工业固体废物贮存量 23 918 万吨，比上年增加 14.5%。其中危险废物贮存量 166 万吨，比上年减少 24.2%；工业固体废物处置量 57 264 万吨，比上年增加 20.5%，其中危险废物处置量 513 万吨，比上年增加 19.9%。

“十一五”期间，全国工业固体废物产生量、综合利用量和处置量均呈现逐年上升的趋势，贮存量基本持平，而工业固体废物排放量呈现逐年下降的趋势。

表 11　全国工业固体废物产生及处理情况　　单位：万吨

年度	产生量		排放量		综合利用量		贮存量		处置量	
	合计	危险废物	合计	危险废物	合计	危险废物	合计	危险废物	合计	危险废物
2001	88 746	952	2 894	2.1	47 290	442	30 183	307	14 491	229
2002	94 509	1 000	2 635	1.7	50 061	392	30 040	383	16 618	242
2003	100 428	1 170	1 941	0.3	56 040	427	27 667	423	17 751	375
2004	120 030	995	1 762	1.1	67 796	403	26 012	343	26 635	275
2005	134 449	1 162	1 655	0.6	76 993	496	27 876	337	31 259	339
2006	151 541	1 084	1 302	20.0	92 601	566	22 398	267	42 883	289
2007	175 632	1 079	1 197	0.1	110 311	650	24 119	154	41 350	346
2008	190 127	1 357	782	0.07	123 482	819	21 883	196	48 291	389
2009	203 943	1 430	710	…	138 186	831	20 929	219	47 488	428
2010	240 944	1 587	498	…	161 772	977	23 918	166	57 264	513
增长率（%）	18.1	11.0	−30.0	0	16.9	17.6	14.5	−24.2	20.5	19.9

注：①“综合利用量”和“处置量”指标中含有综合利用和处置往年贮存量。

②“…”表示数字小于规定单位。

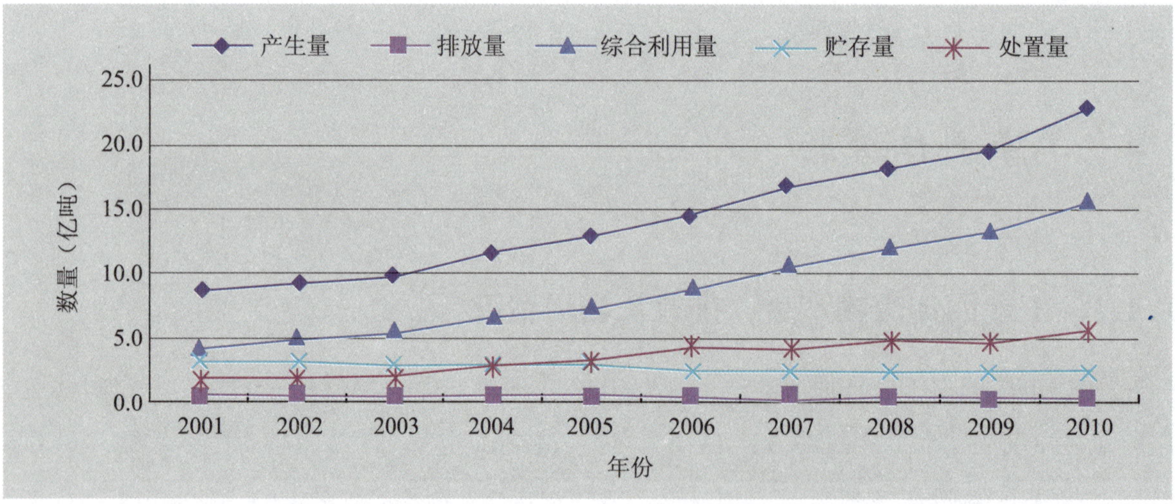

图 36　全国工业固体废物产生、处理及排放量年际变化

1.4.2　各地区工业固体废物排放及处理情况

2010 年，工业固体废物排放量超过 50 万吨的省份有重庆、山西、新疆和贵州。这 4 个省份的工业固体废物排放量占全国工业固体废物排放量的 70.6%。

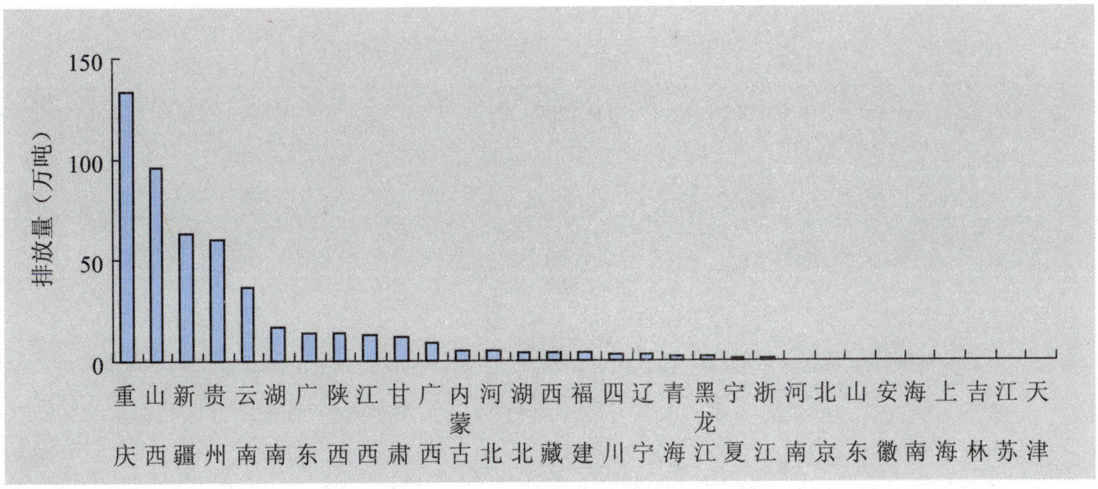

图 37　各地区工业固体废物排放情况

1.4.3　工业行业固体废物排放情况

2010 年，工业固体废物排放量超过 60 万吨的行业依次为煤炭开采和洗选业、有色金属矿采选业。2 个行业工业固体废物排放量占统计工业行业固体废物排放总量的 60.7%。

1.4.4　各地区危险废物集中处置情况

2010 年，纳入统计的危险废物集中处置厂共 546 座。除西藏外，其余省份均建成了危险废物集中处置厂。

危险废物集中处置厂运行费用为 38.9 亿元，比上年增加 21.0%；危险废物日处置能力为 17 795 吨。其中，焚烧处置能力为 13 147 吨，填埋处置能力为 3 768 吨；危险废物实际处置量为 181.2 万吨，比上年减少 8.2%。其中，焚烧量 134.4 万吨，比上年减少 13.6%，填埋量 42.1 万吨，比上年增加 24.6%；危险废物综合利用量为 150.1 万吨，比上年减少 3.2%。

1.5　环境污染治理投资情况

2010 年，环境污染治理投资为 6 654.2 亿元，比上年增加 47.0%，占当年 GDP 的 1.67%。其中，城市环境基础设施建设投资 4 224.2 亿元，比上年增加 68.2%；工业污染源治理投资 397.0 亿元，比上年减少 10.3%；建设项目“三同时”环保投资 2 033.0 亿元，比上年增加 47.0%。

表 12　全国近年环境污染治理投资情况　　单位：亿元

年度	城市环境基础设施建设投资	工业污染源治理投资	建设项目“三同时”环保投资	投资总额
2001	595.7	174.5	336.4	1 106.6
2002	785.3	188.4	389.7	1 363.4
2003	1 072.4	221.8	333.5	1 627.3
2004	1 141.2	308.1	460.5	1 909.8
2005	1 289.7	458.2	640.1	2 388.0
2006	1 314.9	483.9	767.2	2 566.0
2007	1 467.8	552.4	1 367.4	3 387.6
2008	1 801.0	542.6	2 146.7	4 490.3
2009	2 512.0	442.5	1 570.7	4 525.2
2010	4 224.2	397.0	2 033.0	6 654.2
增长率（%）	68.2	−10.3	29.4	47.0

1.5.1　城市环境基础设施建设

2010 年，在城市环境基础设施建设投资中，燃气工程建设投资 290.8 亿元，比上年增加 59.6%；集中供热工程建设投资 433.2 亿元，比上年增加 17.5%；排水工程建设投资 901.6 亿元，比上年增加 23.5%；园林绿化工程建设投资 2 297.0 亿元，比上年增加 151.1%；市容环境卫生工程建设投资 301.6 亿元，比上年减少 4.7%。

燃气、集中供热、排水、园林绿化和市容环境卫生投资分别占城市环境基础设施建设总投资的 6.9%、10.3%、21.3%、54.4%和 7.1%，排水设施和园林绿化投资为城市环境基础设施建设投资的重点。

表 13　全国近年城市环境基础设施建设投资构成　　单位：亿元

年度	投资总额					
		燃气	集中供热	排水	园林绿化	市容环境卫生
2001	595.7	75.5	82.0	224.5	163.2	50.6
2002	789.1	88.4	121.4	275.0	239.5	64.8
2003	1 072.4	133.5	145.8	375.2	321.9	96.0
2004	1 141.2	148.3	173.4	352.3	359.5	107.8
2005	1 289.7	142.4	220.2	368.0	411.3	147.8
2006	1 314.9	155.1	223.6	331.5	429.0	175.8
2007	1 467.8	160.4	230.0	410.0	525.6	141.8
2008	1 801.2	163.5	269.7	496.0	649.9	222.0
2009	2 512.0	182.2	368.7	729.8	914.9	316.5
2010	4 224.2	290.8	433.2	901.6	2 297.0	301.6

1.5.2 工业污染源污染治理投资

2010年，在工业污染源污染治理投资中，废水治理资金130.1亿元，比上年减少12.9%；废气治理资金188.8亿元，比上年减少18.8%；工业固体废物治理资金14.3亿元，比上年减少34.7%；噪声治理资金1.5亿元，比上年增加7.8%。

废水、废气、固废、噪声以及其他污染要素治理投资，分别占工业源治理总投资的32.8%、47.6%、3.6%、0.4%和15.7%，废水和废气仍是工业污染治理的重点。

表14 全国近年工业源污染治理投资构成 单位：万元

年度	废水	废气	固废	噪声	其他
2001	729 214.3	657 940.4	186 967.2	6 424.4	164 733.7
2002	714 935.1	697 864.3	161 287.3	10 463.5	299 112.6
2003	873 747.7	921 222.4	161 763.4	10 139.2	251 408.3
2004	1 055 868.1	1 427 974.9	226 464.8	13 416.1	357 335.6
2005	1 337 146.9	2 129 571.3	274 181.3	30 613.3	810 395.9
2006	1 511 164.5	2 332 697.1	182 630.5	30 145.1	782 847.9
2007	1 960 721.8	2 752 642.2	182 531.9	18 278.6	606 837.9
2008	1 945 977.4	2 656 986.8	196 850.6	28 382.9	597 746.7
2009	1 494 606.0	2 324 616.0	218 535.7	14 100.0	372 664.2
2010	1 301 148.7	1 888 456.5	142 692.2	15 193.2	621 777.6

1.5.3 建设项目“三同时”环保投资

2010年，建设项目“三同时”环保投资2 033.0亿元，比上年增加29.4%。建设项目“三同时”环保投资占环境污染治理投资总额的比例为30.6%，占建设项目投资总额的4.1%。

表15 建设项目“三同时”投资情况

年 度	环保投资额（亿元）	占建设项目投资总额（%）	占全社会固定资产投资总额（%）	占环境治理投资总额（%）
2001	336.4	3.6	0.9	30.4
2002	389.7	5.2	0.9	28.6
2003	333.5	3.9	0.6	20.5
2004	460.5	3.9	0.7	24.1
2005	640.1	4.0	0.7	26.8
2006	767.2	1.0	0.7	29.9
2007	1 367.4	5.0	1.0	40.4
2008	2 146.7	6.4	1.3	47.8
2009	1 570.7	3.2	0.7	34.7
2010	2 033.0	4.1	0.7	30.6

1.6 工业污染物排放达标情况

1.6.1 工业废水排放达标率

2010 年，全国工业废水排放达标率为 95.3%，比上年提高 1.1 个百分点。工业废水排放达标率高于 95%的省份依次为天津、北京、福建、河北、山东、江苏、上海、安徽、海南、陕西、河南、广西、湖北、四川和浙江。

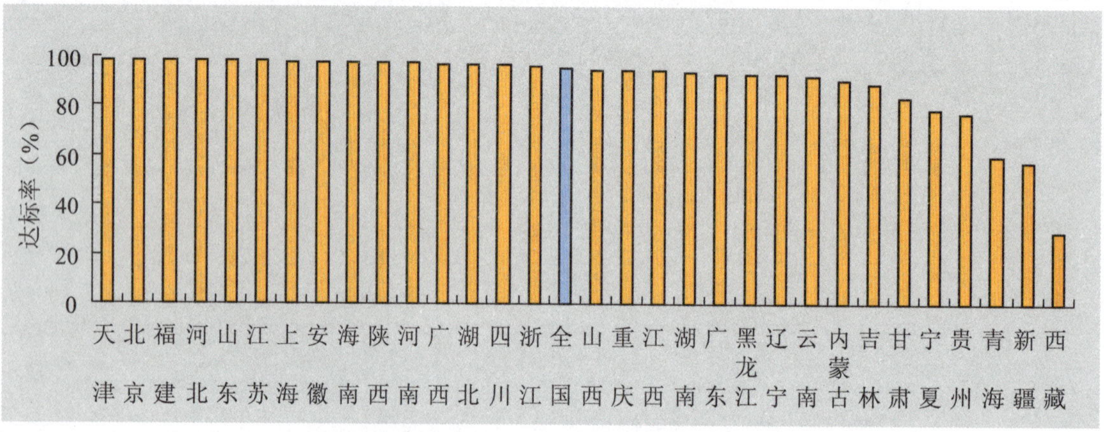

图 38 各地区工业废水排放达标情况

1.6.2 工业二氧化硫排放达标率

2010 年，全国工业二氧化硫排放达标率为 92.1%，比上年提高 1.1 个百分点。工业二氧化硫排放达标率高于 95%的省份依次为天津、北京、河北、山西、上海、海南、福建、山东、浙江、河南、云南、黑龙江、陕西和江苏。

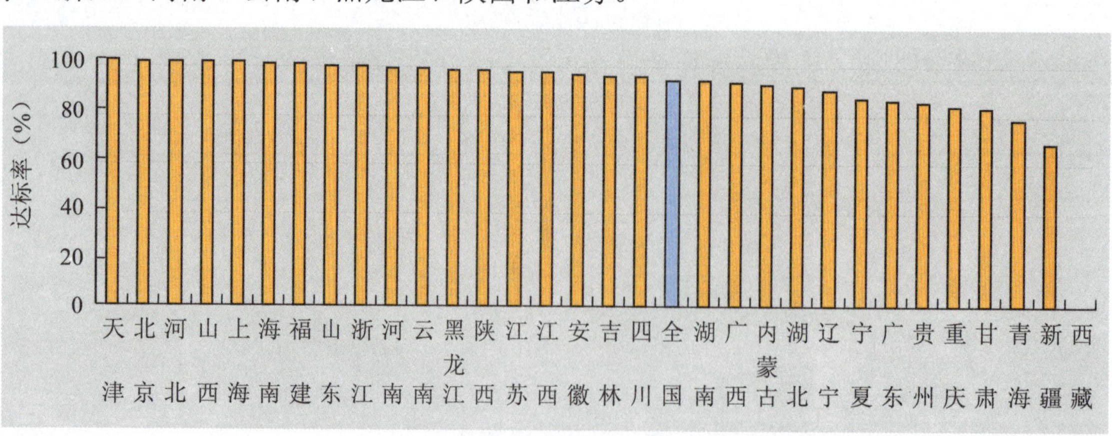

图 39 各地区工业二氧化硫排放达标情况

1.6.3　工业烟尘排放达标率

2010 年，全国工业烟尘排放达标率为 90.6%，与上年基本持平。达标率高于 95%的省份依次为天津、河北、北京、山东、江苏、陕西、浙江、河南、上海、山西、广西、海南、吉林、湖北、安徽、江西和福建。

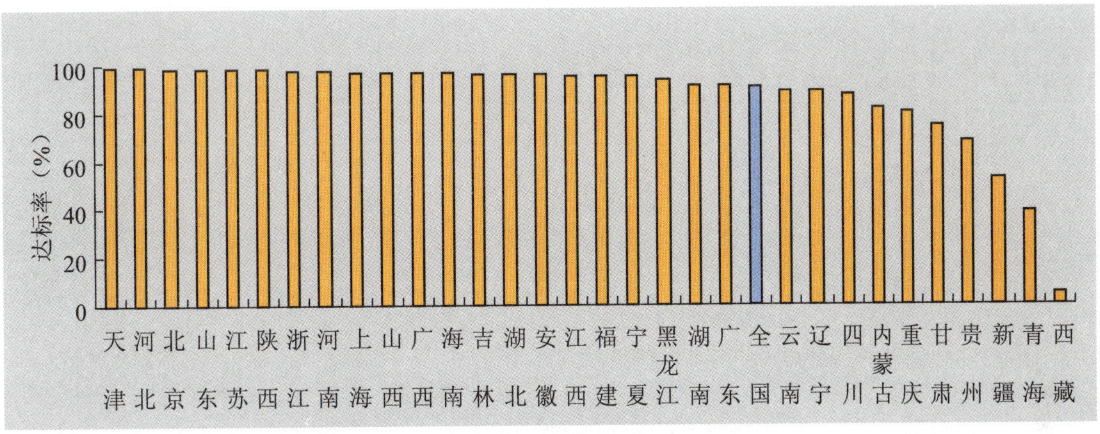

图 40　各地区工业烟尘排放达标情况

1.6.4　工业粉尘排放达标率

2010 年，全国工业粉尘排放达标率为 91.4%，比上年提高 1.5 个百分点。达标率高于 95%的省份依次为北京、天津、海南、河北、上海、陕西、福建、湖北、江苏、广西、安徽、浙江、山东、河南、江西和四川。

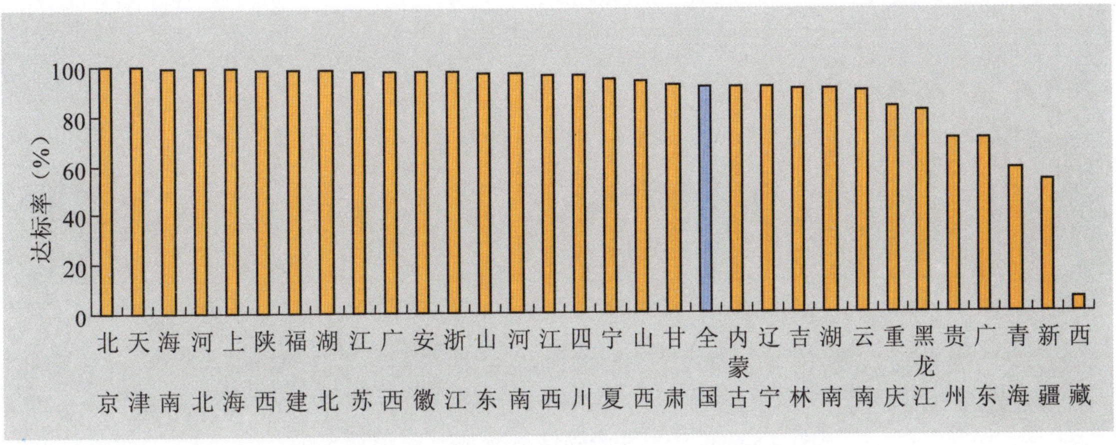

图 41　各地区工业粉尘排放达标情况

1.6.5 工业氮氧化物排放达标率

2010 年，全国工业氮氧化物排放达标率为 87.9%，与上年基本持平。达标率高于 95%的省份依次为天津、上海、福建、北京、浙江、广东、江苏和湖北。

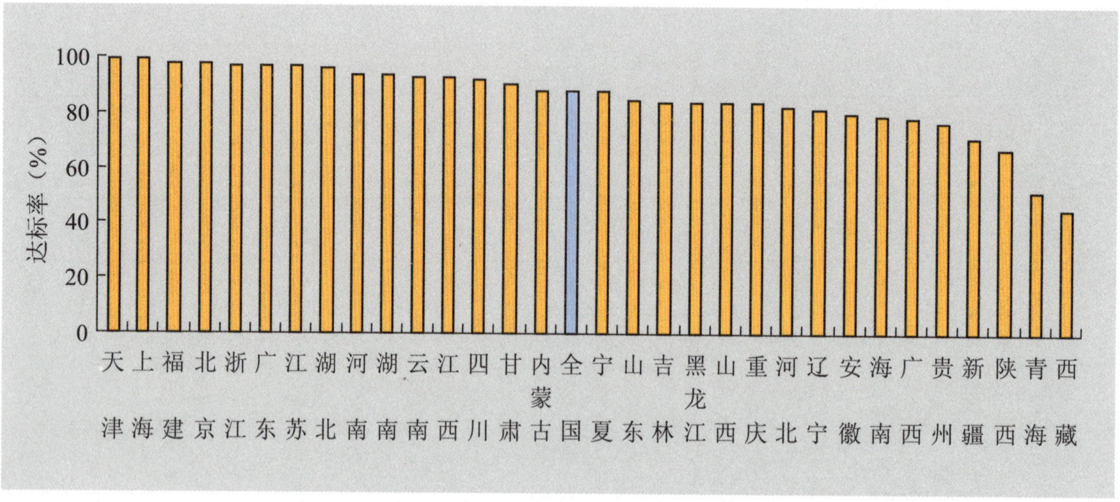

图 42 各地区工业氮氧化物排放达标情况

1.6.6 工业固体废物综合利用率

2010 年，全国工业固体废物综合利用率为 66.7%，与上年基本持平。综合利用率高于 90%的省份依次为天津、上海、江苏、山东、浙江和广东。

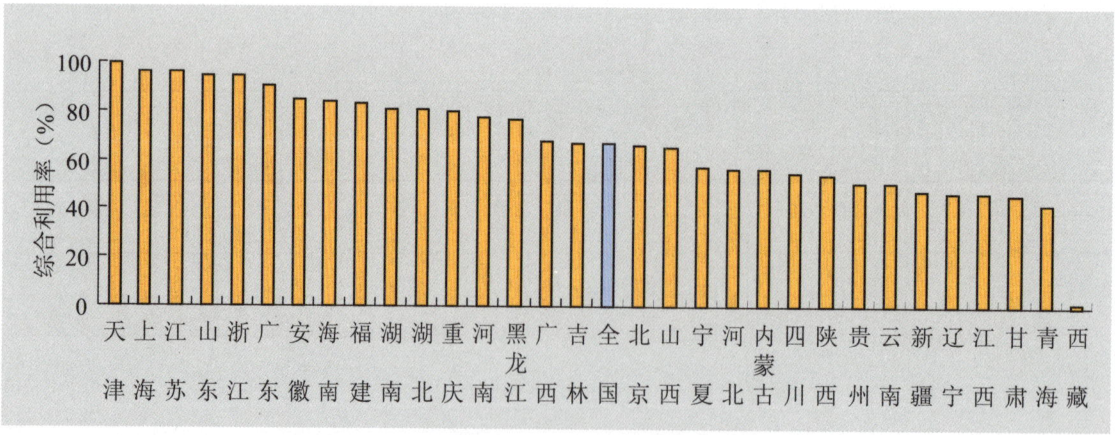

图 43 各地区工业固体废物综合利用情况

1.7 城镇生活污水处理情况

2010 年，全国共统计 2 881 座城市污水处理厂，比上年增加 689 座；设计处理能力为 12 331 万吨/日，比上年新增 1 854 万吨/日。全年共处理废水 337.2 亿吨，其中，生活污水 276.9 亿吨，占总处理水量的 82.1%。共去除化学需氧量 882.6 万吨，氨氮 64.1 万吨；城镇生活污水处理率达到 72.9%，比上年提高 9.6 个百分点。

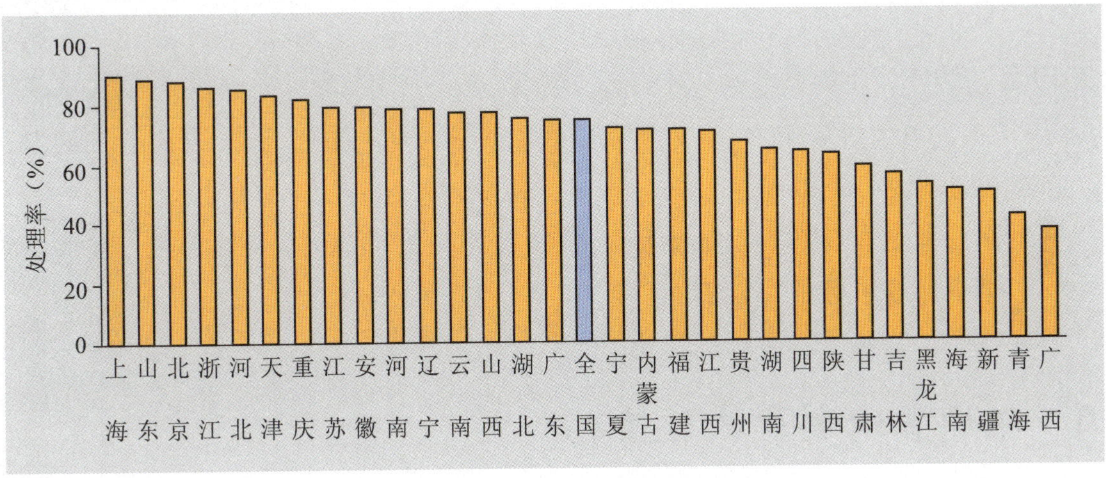

图 44 各地区城镇生活污水处理率

1.8 重点城市主要污染物排放情况

2010 年，113 个重点城市废水排放量为 370 亿吨，占全国废水排放量的 60.0%。其中，工业废水排放量 128 亿吨，生活污水排放量 242 亿吨。重点城市工业废水排放达标率为 95.7%，高于全国平均水平 0.4 个百分点。

重点城市化学需氧量排放量为 579 万吨，占全国化学需氧量排放量的 46.8%。其中，工业化学需氧量排放量 191 万吨，生活化学需氧量排放量 388 万吨。氨氮排放量为 59 万吨，占全国氨氮排放总量的 49.0%。其中，工业氨氮排放量 13 万吨，生活氨氮排放量 46 万吨。

重点城市二氧化硫排放量为 1 082 万吨，占全国二氧化硫排放量的 49.5%。其中，工业二氧化硫排放量 941 万吨，生活二氧化硫排放量 141 万吨。氮氧化物排放量 999 万吨，占全国氮氧化物排放量的 53.9%。其中，工业氮氧化物排放量 781 万吨，生活氮氧化物排放量 218 万吨。烟尘排放量 363 万吨，占全国烟尘排放量的 43.8%。其中，工业烟尘排放量 267 万吨，生活烟尘排放量 96 万吨。工业粉尘排放量 189 万吨，占全国工业粉尘排放量的 42.1%。工业固体废物排放量 233 万吨，占全国工业固体废物排放量的 46.8%。

重点城市共有污水处理厂 1 644 座，城市生活污水处理率为 76.4%，高出全国城镇生活污水处理率 3.5 个百分点。

1.9 医院主要污染物排放情况

2010 年，纳入调查的县及县以上医院 10 332 家，共有 255 万张床位。废水排放量为 4.5 亿吨，化学需氧量排放量为 6.2 万吨，氨氮排放量为 0.8 万吨，医疗废物产生量为 33.6 万吨，放射源总数为 1.1 万枚。

调查的医院共设有 9 703 套废水处理设施，废水日处理能力为 411 万吨，废水处理率为 95.9%，废水排放达标率为 89.0%。

1.10 环境管理制度执行情况

1.10.1 环保建议、提案

2010 年，各级人大建议数 4 971 件，其中已办理 4 827 件，处理率 97.1%。各级政协提案数 6 918 件，其中已办理 6 773 件，处理率 97.9%。

1.10.2 环境信访

2010 年，全国环保系统共收到群众来信 70.1 万封，涉及环境污染与生态破坏有关问题的有 67.7 万件。其中，反映水污染的有 9.2 万件，大气污染的 26.3 万件，固体废物污染的有 1.3 万件，噪声污染的有 26.2 万件，反映“三产”等其他污染的 4.6 万件。来信处理率 93.4%。

群众来访 3.5 万批次，6.6 万人次。其中，涉及环境污染与生态破坏有关问题的有：水污染 0.8 万批次，大气污染 1.4 万批次，固体废物污染 0.1 万批次，噪声污染 0.8 万批次，反映其他污染 0.4 万批次。来访处理率 94.8%。

表 16 环境信访工作情况

年度	来信总数（封）	水污染（件）	大气污染（件）	固体废物污染（件）	噪声与震动（件）	来访批次（批）	来访人次（次）
2001	369 712	47 536	144 880	6 762	154 780	80 575	95 033
2002	435 420	47 438	160 332	7 567	171 770	90 746	109 353
2003	525 988	60 815	194 148	11 698	201 143	85 028	120 246
2004	595 852	68 012	234 569	10 674	254 089	86 892	130 340
2005	608 245	66 660	234 908	10 890	255 638	88 237	142 360
2006	616 122	73 133	242 298	8 538	263 146	71 287	110 592
2007	123 357	23 788	45 986	3 762	40 638	43 909	77 399
2008	705 127	106 521	286 699	14 135	239 737	43 862	84 971
2009	696 134	100 497	260 168	15 010	242 521	42 170	73 798
2010	701 073	91 967	262 953	12 908	262 389	34 683	65 948

1.10.3 环境法制

2010 年，全国办理环境行政处罚案件 11.7 万起，环境行政复议案件 694 起，环境行政诉讼案件 228 起，环境犯罪案件 11 起。当年做出环境行政处罚决定的案件 11.2 万起，做出环境行政复议决定的案件 598 起，做出判决的环境行政诉讼案件 168 起，做出判决的环境犯罪案件 8 起，分别占当年办理案件的 95.9%、86.2%、73.7%、72.7%。

1.10.4 机构建设

2010 年，全国环保系统机构总数 12 849 个。其中，国家级机构 43 个，省级机构 371 个，地市级环保机构 1 937 个，县级环保机构 8 606 个，乡镇环保机构 1 892 个。各级环保行政机构 3 175 个，各级环境监察机构 3 060 个，各级环境监测机构 2 587 个。

全国环保系统共有 19.4 万人。其中，环保机关人员 4.6 万人，占环保系统总人数的 23.7%；环境监察人员 6.2 万人，占环保系统总人数的 32.2%；环境监测人员 5.5 万人，占环保系统总人数的 28.2%。

1.10.5 环境科技

2010 年，全国各地共完成课题研究 1 133 项，课题研究总经费达 4.5 亿元，比上年减少 48.0%；56 项课题研究荣获省部级以上科学技术奖励。其中，有 1 项获得国家级奖励。

我国从事环境科技活动人数达 1.02 万人，其中科技人员 0.72 万人，占 70.6%。全年环境科研业务费支出 4.0 亿元。

表 17　环保机关、监察机构、监测站年末实有人员情况

年份	年末实有人数（人）	环保机关		监察机构		监测站	
		实有人数（人）	占本级环保人员总数的比例（%）	实有人数（人）	占本级环保人员总数的比例（%）	实有人数（人）	占本级环保人员总数的比例（%）
2001	142 766	39 175	27.4	37 934	26.6	43 629	30.6
2002	154 233	40 709	26.4	41 878	27.2	46 515	30.2
2003	156 542	40 598	25.9	44 250	28.3	45 813	29.3
2004	160 246	42 134	26.3	47 189	29.4	45 849	28.6
2005	166 774	44 024	26.4	50 040	30.0	46 984	28.2
2006	170 290	44 141	25.9	52 845	31.2	47 689	28.2
2007	176 988	43 626	24.6	57 427	32.4	49 335	27.9
2008	183 555	44 847	24.4	59 477	32.1	51 753	28.3
2009	188 991	45 626	24.1	60 896	32.2	52 944	28.0
2010	193 911	45 938	23.7	62 468	32.2	54 698	28.2
国家级	2 584	309	12.0	415	16.1	154	6.0
省　级	12 427	2 461	19.8	1 078	8.7	2 990	24.1
地市级	42 462	8 633	20.3	9 180	21.6	15 697	37.0
县　级	129 284	34 535	26.7	51 795	40.1	35 857	27.7

1.10.6　环境监测

2010 年，全国监测用房总面积为 189.1 万米 ^{2}，环境监测仪器 169.8 万台/套。

2010 年，全国环境空气监测点位 4 468 个，酸雨监测点位 1 612 个，沙尘暴监测点位数 117 个，地表水水质监测断面 11 349 个，近岸海域监测点位 836 个，开展污染源监督性监测的重点企业 48 024 个。

1.10.7　自然生态保护

2010 年，全国各类自然保护区共计 2 588 个，比上年增加 47 个。自然保护区面积 14 944.1 万公顷，约占国土面积的 15.6%。国家级、省级、地市级、县级自然保护区个数分别占全国自然保护区总数的 12.3%、33.2%、16.2%、38.3%，其面积分别占自然保护区总面积的 62.0%、28.0%、3.1%、6.9%。

表 18　全国自然保护区数量

单位：个

年度	自然保护区数	国家级	省级	地市级	县级
2001	1 551	171	526	269	585
2002	1 757	188	609	304	656
2003	1 999	226	654	340	779
2004	2 194	226	733	396	839
2005	2 349	243	773	421	912
2006	2 395	265	793	422	915
2007	2 531	303	780	462	986
2008	2 538	303	806	432	997
2009	2 541	319	827	416	979
2010	2 588	319	859	418	992

表 19　全国自然保护区面积

单位：万公顷

年度	自然保护区面积	国家级	省级	地市级	县级
2001	12 989.0	5 903.8	5 725.9	423.2	936.0
2002	13 294.5	6 042.1	5 907.3	463.1	882.0
2003	14 398.0	8 871.3	3 995.6	429.0	1 102.1
2004	14 822.6	8 871.3	4 290.2	488.3	1 172.8
2005	14 994.9	8 898.9	4 487.0	501.5	1 107.5
2006	15 153.5	9 169.7	4 441.8	522.4	1 019.6
2007	15 188.2	9 365.6	4 260.1	537.6	1 024.9
2008	14 894.3	9 120.3	4 240.2	497.1	1 036.8
2009	14 774.7	9 267.1	4 004.5	471.2	1 031.9
2010	14 944.1	9 267.6	4 174.8	468.2	1 033.4

1.10.8　环境影响评价

2010 年，建设项目环境影响评价执行率 99.9%。审批环境影响评价的项目 39.02 万个，其中，编制环境影响报告书、填报环境影响报告表和填报环境影响登记表的分别占 6.1%、43.5%和 50.3%。申报环境影响评价项目的环保投资 7 779.4 亿元，占申请环境影响评价项目投资总额的 3.0%。

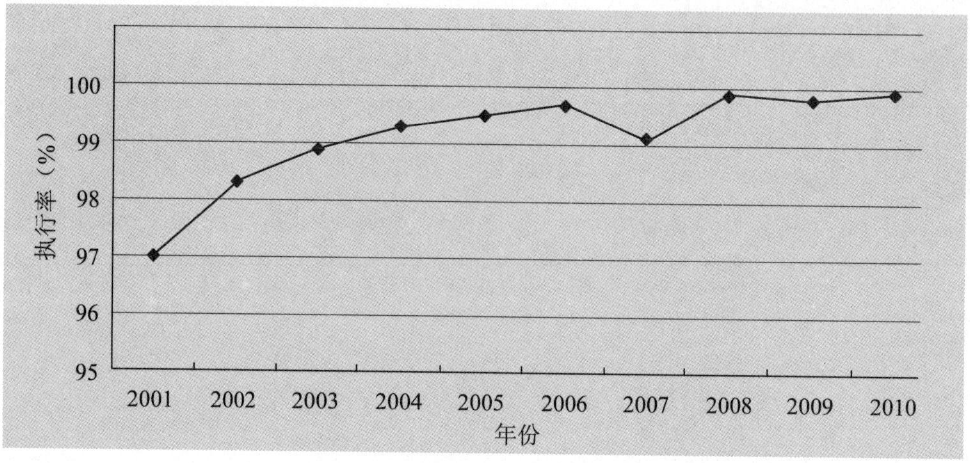

图 45 全国建设项目环境影响评价制度执行率

1.10.9 “三同时”管理

2010 年，全国完成环保验收项目 10.7 万个，“三同时”执行率 98.0%。执行“三同时”项目用于环保工程的实际投资为 2 033.0 亿元，占项目总投资的 4.1%，比上年上升了 0.9 个百分点。

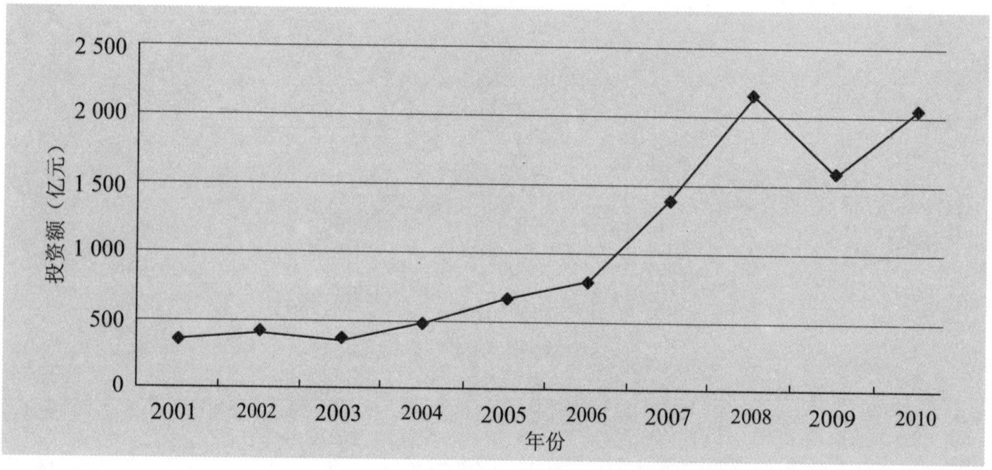

图 46 全国实际执行“三同时”建设项目环保投资情况

1.10.10 排污收费

2010 年，全国排污费开征单位 413 575 户，征收总额 188 亿元。排污费解缴入库单位 401 172 户，入库金额 178 亿元。其中，污水类解缴入库 182 048 户，入库金额 22 亿元；废气类解缴入库 193 734 户，入库金额 139 亿元；噪声类解缴入库 93 347 户，入库金额 14 亿元；危险废物解缴入库 809 户，入库金额 0.05 亿元。

11.2 核电厂周围环境电离辐射水平

浙江秦山核电基地各核电厂、广东大亚湾/岭澳核电厂和江苏田湾核电站安全运行。围辐射环境连续监测系统所测的 γ 辐射剂量率（未扣除宇宙射线响应值）年均值分别102.6 纳戈/时、122.6 纳戈/时和 101.0 纳戈/时，处于所在地区的天然本底范围内。浙秦山核电基地周围关键居民点空气、降水、地表水及部分生物样品中氚活度浓度，广大亚湾/岭澳核电厂和江苏田湾核电站排放口附近海域海水氚活度浓度与核电站运行本底值相比有所升高，但对公众造成的附加剂量远低于国家规定的限值。各核电厂周各种环境介质中除氚外其余放射性核素活度浓度与历年监测结果相比未见明显变化。

11.3 其他反应堆及核燃料循环设施周围环境电离辐射水平

中国原子能科学研究院、清华大学核能与新能源技术研究院、中国核动力研究设计、陕西省西北核技术研究所等研究设施外围环境 γ 辐射剂量率，气溶胶、沉降物、地水、土壤和生物样品中放射性核素活度浓度与历年监测结果相比未见明显变化；饮用下水总α和总β活度浓度低于《生活饮用水卫生标准》（GB 5749—2006）规定的限值。

兰州铀浓缩有限公司、陕西铀浓缩有限公司、包头核燃料元件厂、中核建中核燃料件公司、中核四〇四有限公司等核燃料循环设施及西北低中放废物处置场、北龙低中废物处置场外围环境 γ 辐射剂量率为正常环境水平，环境介质中也未监测到由企业生、加工、贮存、处理和运输等活动引起的放射性核素活度浓度升高。

11.4 铀矿冶及伴生放射性矿周围环境电离辐射

铀矿冶设施周围环境空气中氡活度浓度，气溶胶、沉降物总 α 活度浓度，地下水和物样品中放射性核素铀和镭-226 活度浓度未见异常。白云鄂博矿等部分伴生放射性矿开采、冶炼、加工活动对企业周围局部环境产生了一定程度影响。

11.5 电磁辐射设施周围环境电磁辐射

环境电磁辐射水平总体情况较好。开展监测的移动通信基站天线周围环境敏感点磁辐射水平低于《电磁辐射防护规定》（GB 8702—1988）中的公众照射导出限值；展监测的输变电设施周围环境敏感点工频电场强度和磁感应强度均低于《500 kV 超压送变电工程环境影响评价技术规范》（HJ/T 24—1998）规定的居民区工频电场、场标准。

表 20　排污费收入情况表

项目	户数（户）
排污费解缴入库合计	401 172
污水类	182 048
废气类	193 734
噪声类	93 347
危险废物	809

1.10.11　环境宣教

2010 年，环境保护新闻宣传和环保教育工作继续加强。通过开展名
教育活动，动员社会各界积极参与环境保护。

全国召开环境类新闻发布会 249 次。发布环境类新闻通稿 5.6 万篇
11 296 次。

1.11　全国辐射环境水平

2010 年，全国辐射环境质量总体保持良好。环境电离辐射水平保持
核技术利用项目周围环境电离辐射水平总体未见明显变化；环境电磁辐
较好，电磁辐射设施周围环境电磁辐射水平总体未见明显变化。

1.11.1　环境电离辐射水平

全国地级以上城市环境 γ 辐射剂量率，省会城市气溶胶、沉降物总
度，空气中氚化水活度浓度为正常环境水平。长江、黄河、珠江、松花
和辽河七大水系，西南和西北诸河、南水北调工程沿线、浙闽区河流、
放射性核素活度浓度与历年监测结果相比未见明显变化，其中天然放射
与 1983—1990 年全国环境天然放射性水平调查时的测量值处于同一水平
中式饮用水水源地总α和总β活度浓度均低于《生活饮用水卫生标准》（G
规定的限值。近岸海域海水人工放射性核素锶-90 和铯-137 活度浓度均在
（GB 3097—1997）规定限值内。开展监测的省会和部分地级城市土壤放
度与历年监测结果相比未见明显变化，其中天然放射性核素活度浓度与 1
国环境天然放射性水平调查时的测量值处于同一水平。

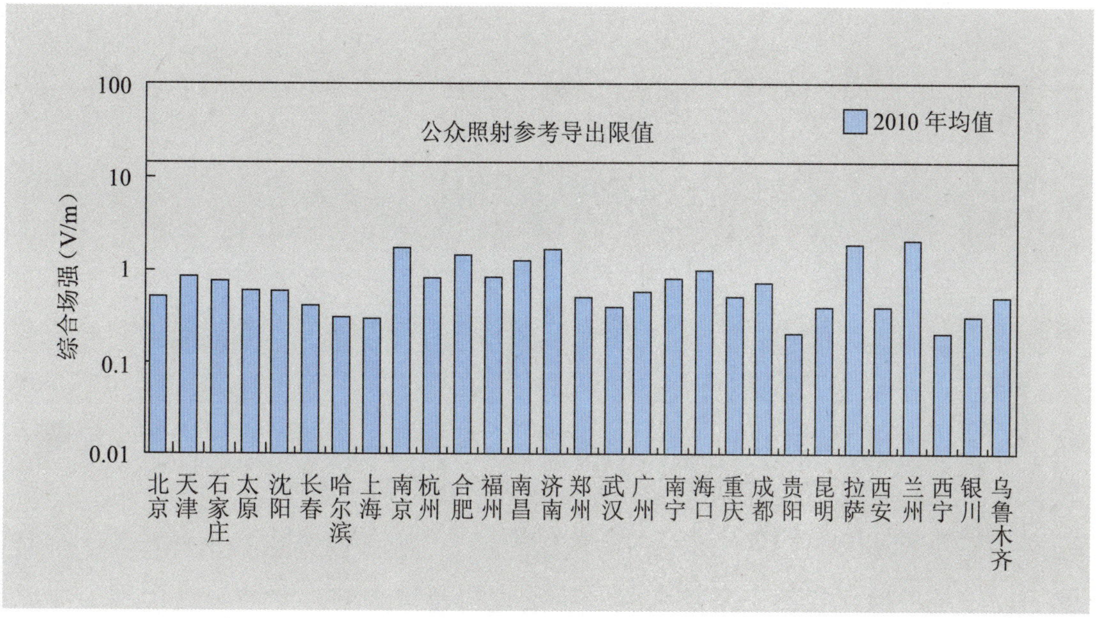

图 47　部分城市环境电磁辐射水平

简要说明

1．本年报资料根据全国 31 个省、自治区、直辖市及新疆生产建设兵团环境统计资料汇总整理而成，未包括香港、澳门特别行政区以及台湾省数据。

2．本年报主要反映中国环境保护事业发展情况。主要内容包括水环境、大气环境、固体废物、生态环境、自然灾害和环境污染治理投资等内容。主要反映中国工业废水和生活污水的排放及治理情况，废气排放及处理情况，工业固体废物的产生、处理及综合利用情况及环境污染与破坏事故情况，环境污染治理投资等情况。

3．调查方法

1）工业企业污染排放及处理利用情况年报的调查方法是对重点调查工业企业单位逐个发表填报汇总，对非重点调查工业企业的排污情况实行整体估算。

重点调查单位是指筛选出的排污量占各地区排污总量 85%以上的工业企业单位。筛选重点调查单位的原则为：①筛选指标为国家实行总量控制的各项主要污染物排放量：废水、化学需氧量、氨氮、二氧化硫、烟尘、工业粉尘及工业固体废物产生量；②排放工业废水中有重金属类有害物质的工业企业以及有危险废物产生的工业企业全部为重点调查单位。

非重点调查单位数据的估算方法是将重点调查单位的排污总量作为估算的对比基数，采取“比率估算”方法，估算出非重点调查单位的排污量。重点调查数据与非重点估算数据相加，为工业污染总排放数据。

2）生产及生活中产生的污染物实施集中处理处置情况年报的调查方法是对各集中处理处置单位逐个发表填报汇总，包括危险废物集中处置厂和城市污水处理厂。

3）生活及其他污染情况年报的调查方法是依据相关基础数据和技术参数进行估算。

4）工业企业污染治理项目建设投资情况年报的调查方法是对有在建工业污染治理项目的工业企业逐个发表填报汇总。

5）医院污染排放及处理利用情况年报的调查方法是对二级及以上的医院逐个发表填报汇总。

4．环境统计范围

1）工业企业污染排放及处理利用情况的年报综合范围为有污染物排放的工业企业。

2）工业企业污染治理项目投资情况的年报综合范围为在建的老工业污染源污染治理投资项目，不包括已纳入建设项目环境保护“三同时”管理的项目。

3）生产及生活中产生的污染物实施集中处理处置情况的年报综合范围为危险废物集中处置厂和城市污水处理厂。

4）生活及其他污染情况的年报综合范围为城镇的生活污水排放以及除工业生产以外的生活及其他活动所排放的废气中的污染物。

5）医院污染排放及处理利用情况的年报调查范围为辖区内二级及以上的医院。

5．流域汇总范围

从2004年起，本年报中流域汇总范围较往年有所扩大。其中，松花江流域包括松花江、黑龙江、乌苏里江流域及东北地区其他国际河流，珠江流域包括珠江和粤桂琼沿海诸河流域，海河流域包括海河、滦河和华北地区沿海诸河流域，辽河流域包括辽河、大凌河及辽东沿海诸河流域。

从2006年起，本年报中流域数据的汇总方法有所变化，按流域规划所含区县的数据汇总，不再沿用以前的按“排水去向”汇总数据的方法，汇总的区县数有所减少，湖泊汇总方式与流域相同。

主要环境统计指标解释附后。

2

各地区环境统计

GEDIQU HUANJING TONGJI

ANNUAL STATISTIC REPORT ON ENVIRONMENT IN CHINA

2010

各地区主要污染物排放情况（一）

Discharge of Key Pollutants by Region（1）

（2010）

年份 地区	Year Region	废水排放总量（亿吨） Total Volume of Waste Water Discharged（100 million t）	工业 Industrial	生活 Household	化学需氧量排放总量（万吨） Total Volume of COD Discharged（10 000 t）	工业 Industrial	生活 Household
	2005	524.5	243.1	281.4	1 414.2	554.7	859.5
	2006	536.8	240.2	296.6	1 428.2	542.3	885.9
	2007	556.8	246.6	310.2	1 381.8	511.1	870.8
	2008	571.7	241.7	330.0	1 320.7	457.6	863.1
	2009	589.7	234.5	355.2	1 277.5	439.7	837.8
	2010	**617.3**	**237.5**	**379.8**	**1 238.1**	**434.8**	**803.3**
北　京	Beijing	13.6	0.8	12.8	9.2	0.5	8.7
天　津	Tianjin	6.8	2.0	4.9	13.2	2.2	11.0
河　北	Hebei	26.3	11.4	14.8	54.6	21.8	32.8
山　西	Shanxi	11.8	5.0	6.8	33.3	13.8	19.5
内蒙古	Inner Mongolia	9.3	4.0	5.3	27.5	9.1	18.4
辽　宁	Liaoning	21.8	7.2	14.7	54.2	20.2	34.0
吉　林	Jilin	11.4	3.9	7.6	35.2	13.1	22.1
黑龙江	Heilongjiang	11.9	3.9	8.0	44.4	11.2	33.2
上　海	Shanghai	24.8	3.7	21.2	22.0	2.2	19.8
江　苏	Jiangsu	55.5	26.4	29.2	78.8	25.6	53.2
浙　江	Zhejiang	39.5	21.7	17.7	48.7	24.4	24.3
安　徽	Anhui	18.5	7.1	11.4	41.1	11.5	29.6
福　建	Fujian	23.9	12.4	11.4	37.3	8.3	29.0
江　西	Jiangxi	16.1	7.3	8.8	43.1	11.8	31.3
山　东	Shandong	43.6	20.8	22.8	62.1	29.5	32.5
河　南	Henan	35.9	15.0	20.8	62.0	29.6	32.4
湖　北	Hubei	27.1	9.5	17.6	57.2	16.5	40.7
湖　南	Hunan	26.8	9.6	17.3	79.8	18.8	61.0
广　东	Guangdong	72.3	18.7	53.6	85.8	23.4	62.4
广　西	Guangxi	31.3	16.5	14.7	93.7	49.3	44.4
海　南	Hainan	3.7	0.6	3.1	9.2	0.9	8.3
重　庆	Chongqing	12.8	4.5	8.3	23.5	8.7	14.8
四　川	Sichuan	25.6	9.3	16.3	74.1	25.1	49.0
贵　州	Guizhou	6.1	1.4	4.7	20.8	1.6	19.2
云　南	Yunnan	9.2	3.1	6.1	26.8	8.9	17.9
西　藏	Tibet	0.4	0.1	0.3	2.9	0.1	2.8
陕　西	Shaanxi	11.6	4.5	7.0	30.8	12.1	18.6
甘　肃	Gansu	5.1	1.5	3.6	16.8	4.5	12.3
青　海	Qinghai	2.3	0.9	1.4	8.3	4.5	3.8
宁　夏	Ningxia	4.1	2.2	1.9	12.2	9.3	2.9
新　疆	Xinjiang	8.4	2.5	5.8	29.6	16.3	13.3

各地区主要污染物排放情况（二）

Discharge of Key Pollutants by Region（2）

单位：万吨 （2010） （10 000 t）

年份 地区	Year Region	氨氮排放总量 Total Volume of Ammonia Nitrogen Discharged	工业 Industrial	生活 Household	二氧化硫排放总量 Total Volume of Sulphur Dioxide Discharged	工业 Industrial	生活 Household
	2005	149.8	52.5	97.3	2 549.4	2 168.4	381.0
	2006	141.3	42.5	98.8	2 588.8	2 237.6	351.2
	2007	132.3	34.1	98.3	2 468.1	2 140.0	328.1
	2008	127.0	29.7	97.3	2 321.2	1 991.3	329.9
	2009	122.6	27.3	95.3	2 214.4	1 865.9	348.5
	2010	**120.3**	**27.3**	**93.0**	**2 185.1**	**1 864.4**	**320.7**
北京	Beijing	1.2	…	1.2	11.5	5.7	5.8
天津	Tianjin	2.0	0.3	1.7	23.5	21.8	1.8
河北	Hebei	5.5	1.8	3.6	123.4	99.4	24.0
山西	Shanxi	4.2	1.2	3.0	124.9	114.7	10.2
内蒙古	Inner Mongolia	4.1	0.7	3.4	139.4	119.3	20.1
辽宁	Liaoning	5.6	0.9	4.7	102.2	85.9	16.3
吉林	Jilin	3.0	0.3	2.7	35.6	30.1	5.6
黑龙江	Heilongjiang	4.3	0.5	3.8	49.0	41.7	7.3
上海	Shanghai	2.8	0.3	2.4	35.8	22.1	13.7
江苏	Jiangsu	6.3	1.5	4.8	105.0	100.2	4.8
浙江	Zhejiang	4.0	1.4	2.6	67.8	65.4	2.4
安徽	Anhui	4.4	1.2	3.2	53.2	48.4	4.8
福建	Fujian	3.0	0.7	2.3	40.9	39.1	1.8
江西	Jiangxi	3.5	0.9	2.6	55.7	47.1	8.6
山东	Shandong	6.6	1.5	5.1	153.8	138.3	15.5
河南	Henan	7.2	2.3	4.9	133.9	116.3	17.6
湖北	Hubei	6.1	1.5	4.6	63.3	51.6	11.7
湖南	Hunan	7.5	1.7	5.8	80.1	62.7	17.4
广东	Guangdong	10.7	1.1	9.6	105.1	98.9	6.1
广西	Guangxi	4.7	1.4	3.3	90.4	84.8	5.6
海南	Hainan	0.8	0.1	0.7	2.9	2.8	0.1
重庆	Chongqing	2.5	0.5	2.0	71.9	57.3	14.7
四川	Sichuan	6.1	1.6	4.4	113.1	93.8	19.3
贵州	Guizhou	1.6	0.1	1.6	114.9	63.8	51.1
云南	Yunnan	2.1	0.4	1.7	50.1	44.0	6.1
西藏	Tibet	0.2	…	0.2	0.4	0.1	0.3
陕西	Shaanxi	3.2	0.7	2.5	77.9	70.7	7.2
甘肃	Gansu	2.4	0.8	1.6	55.2	45.2	9.9
青海	Qinghai	0.8	0.2	0.6	14.3	13.3	1.0
宁夏	Ningxia	1.3	0.9	0.4	31.1	28.0	3.0
新疆	Xinjiang	2.7	0.6	2.0	58.8	51.8	7.0

各地区主要污染物排放情况（三）

Discharge of Key Pollutants by Region（3）

单位：万吨　　　　（2010）　　　　（10 000 t）

年份 地区	Year Region	烟尘排放总量 Total Volume of Soot Discharged	工业 Industrial	生活 Household	工业粉尘排放量 Total Volume of Industrial Dust Discharged
	2005	1 182.5	948.9	233.6	911.2
	2006	1 088.8	864.5	224.3	808.4
	2007	986.6	771.1	215.5	698.7
	2008	901.6	670.7	230.9	584.9
	2009	847.7	604.4	243.3	523.6
	2010	**829.1**	**603.2**	**225.9**	**448.7**
北　京	Beijing	4.9	2.1	2.7	1.7
天　津	Tianjin	6.5	5.4	1.1	0.8
河　北	Hebei	50.0	32.3	17.7	32.1
山　西	Shanxi	62.1	43.2	18.9	36.5
内蒙古	Inner Mongolia	64.7	47.6	17.1	16.0
辽　宁	Liaoning	63.3	39.8	23.5	16.7
吉　林	Jilin	30.1	21.0	9.1	5.3
黑龙江	Heilongjiang	42.2	29.7	12.5	5.7
上　海	Shanghai	10.2	4.2	6.0	1.0
江　苏	Jiangsu	33.5	29.9	3.6	15.1
浙　江	Zhejiang	17.4	16.5	0.9	13.9
安　徽	Anhui	25.5	20.7	4.8	26.4
福　建	Fujian	13.9	10.0	3.9	14.0
江　西	Jiangxi	16.4	13.9	2.5	22.4
山　东	Shandong	39.2	29.1	10.0	18.9
河　南	Henan	54.7	47.4	7.3	22.7
湖　北	Hubei	19.3	14.5	4.8	14.6
湖　南	Hunan	31.2	23.5	7.7	39.4
广　东	Guangdong	31.1	25.3	5.8	10.4
广　西	Guangxi	26.2	25.0	1.1	31.8
海　南	Hainan	0.8	0.7	0.2	0.7
重　庆	Chongqing	20.8	10.2	10.6	8.4
四　川	Sichuan	34.1	26.0	8.2	14.1
贵　州	Guizhou	25.2	11.3	13.9	8.6
云　南	Yunnan	13.8	8.9	4.9	9.2
西　藏	Tibet	0.2	0.1	0.2	0.1
陕　西	Shaanxi	16.3	11.5	4.8	18.5
甘　肃	Gansu	16.3	9.8	6.5	9.3
青　海	Qinghai	7.7	5.2	2.5	9.8
宁　夏	Ningxia	17.2	13.6	3.6	6.1
新　疆	Xinjiang	34.4	24.8	9.6	18.5

各地区主要污染物排放情况（四）
Discharge of Key Pollutants by Region（4）

单位：万吨 （2010） （10 000 t）

年份 地区	Year Region	氮氧化物排放总量 Total Volume of Nitrogen Oxide Discharged	工业 Industrial	生活 Household	工业固体废物排放量 Total Volume of Industrial Solid Waste Discharged
	2005	—	—	—	1 654.7
	2006	1 523.8	1 136.0	387.8	1 302.1
	2007	1 643.4	1 261.3	382.0	1 196.7
	2008	1 624.3	1 250.3	374.0	781.8
	2009	1 692.7	1 284.8	407.9	710.5
	2010	**1 852.4**	**1 465.6**	**386.8**	**498.2**
北　京	Beijing	22.3	10.9	11.4	0.1
天　津	Tianjin	24.6	23.6	1.0	…
河　北	Hebei	111.6	95.8	15.7	4.5
山　西	Shanxi	91.5	78.1	13.4	95.4
内蒙古	Inner Mongolia	129.4	113.7	15.7	4.5
辽　宁	Liaoning	81.1	70.6	10.5	2.8
吉　林	Jilin	56.4	51.5	4.9	0.0
黑龙江	Heilongjiang	50.1	38.5	11.6	1.9
上　海	Shanghai	44.7	26.1	18.5	…
江　苏	Jiangsu	125.9	96.2	29.7	0.0
浙　江	Zhejiang	94.6	70.2	24.4	0.6
安　徽	Anhui	63.6	51.3	12.3	0.0
福　建	Fujian	43.4	32.0	11.3	3.5
江　西	Jiangxi	33.9	22.1	11.8	13.2
山　东	Shandong	140.8	116.7	24.1	…
河　南	Henan	121.2	100.0	21.2	0.2
湖　北	Hubei	55.7	42.7	13.0	4.2
湖　南	Hunan	40.8	31.3	9.4	16.4
广　东	Guangdong	129.7	83.0	46.6	14.2
广　西	Guangxi	34.6	23.6	11.0	9.1
海　南	Hainan	5.6	3.1	2.4	0.0
重　庆	Chongqing	30.0	23.1	6.9	133.6
四　川	Sichuan	67.6	59.6	8.0	3.2
贵　州	Guizhou	22.6	19.2	3.4	59.8
云　南	Yunnan	43.5	32.9	10.5	36.3
西　藏	Tibet	0.0	0.0	0.0	4.1
陕　西	Shaanxi	51.8	46.9	4.9	13.4
甘　肃	Gansu	29.2	22.4	6.8	11.4
青　海	Qinghai	11.0	7.8	3.3	2.2
宁　夏	Ningxia	34.9	32.3	2.7	0.9
新　疆	Xinjiang	60.4	40.1	20.3	62.8

各地区工业废水排放及处理情况（一）

Discharge and Treatment of Industrial Waste Water by Region（1）

单位：万吨 （2010） （10 000 t）

年份 地区	Year Region	工业废水排放量 Total Volume of Industrial Waste Water Discharged	工业废水排放达标量 Industrial Waste Water Meeting Discharge Standards	工业废水处理量 Industrial Waste Water Treated
	2005	2 431 121	2 217 093	—
	2006	2 401 946	2 178 461	4 427 385
	2007	2 466 493	2 260 719	4 894 413
	2008	2 416 511	2 233 986	5 334 970
	2009	2 344 834	2 208 743	5 459 281
	2010	**2 374 732**	**2 263 587**	**5 878 734**
北 京	Beijing	8 198	8 096	31 304
天 津	Tianjin	19 680	19 671	87 629
河 北	Hebei	114 232	112 627	938 114
山 西	Shanxi	49 881	47 221	237 229
内蒙古	Inner Mongolia	39 536	35 675	113 033
辽 宁	Liaoning	71 521	66 206	341 135
吉 林	Jilin	38 656	34 393	74 621
黑龙江	Heilongjiang	38 921	36 073	137 344
上 海	Shanghai	36 696	35 973	147 245
江 苏	Jiangsu	263 760	258 622	405 176
浙 江	Zhejiang	217 426	209 195	260 483
安 徽	Anhui	70 971	69 518	265 351
福 建	Fujian	124 168	122 525	168 250
江 西	Jiangxi	72 526	68 307	140 749
山 东	Shandong	208 257	205 010	383 197
河 南	Henan	150 406	146 449	208 384
湖 北	Hubei	94 593	91 538	329 722
湖 南	Hunan	95 605	89 583	305 098
广 东	Guangdong	187 031	174 153	315 298
广 西	Guangxi	165 211	160 139	287 315
海 南	Hainan	5 782	5 656	7 929
重 庆	Chongqing	45 180	42 798	61 331
四 川	Sichuan	93 444	90 194	195 333
贵 州	Guizhou	14 130	10 919	112 912
云 南	Yunnan	30 926	28 403	157 345
西 藏	Tibet	736	217	217
陕 西	Shaanxi	45 487	44 353	72 536
甘 肃	Gansu	15 352	12 791	32 746
青 海	Qinghai	9 031	5 412	13 199
宁 夏	Ningxia	21 977	17 304	24 845
新 疆	Xinjiang	25 413	14 569	23 667

各地区工业废水排放及处理情况（二）
Discharge and Treatment of Industrial Waste Water by Region（2）

（2010）

年 份 地 区	Year Region	废水治理设施数（套） Number of Facilities for Treatment of Waste Water（set）	废水治理设施治理能力（万吨/日） Capacity of Facilities for Treatment of Waste Water（10 000 t/d）	废水治理设施运行费用（万元） Annual Expenditure for Operation（10 000 yuan）	废水污染物在线监测仪器套数（套） On-line Apparatus for Waste Water Mornitoring（set）
	2005	69 231	16 349	2 766 907.3	—
	2006	75 830	19 553	3 885 001.7	7 749
	2007	78 210	22 076	4 280 384.5	11 029
	2008	78 725	22 897	4 529 005.8	13 159
	2009	77 018	22 703	4 784 925.0	14 738
	2010	**80 332**	**24 762**	**5 453 464.4**	**15 635**
北 京	Beijing	481	170	68 058.5	62
天 津	Tianjin	912	273	69 321.1	267
河 北	Hebei	4 008	2 750	388 592.2	784
山 西	Shanxi	2 633	798	200 389.7	352
内蒙古	Inner Mongolia	956	465	69 554.6	138
辽 宁	Liaoning	2 793	1 331	271 325.4	215
吉 林	Jilin	659	234	51 561.9	123
黑龙江	Heilongjiang	1 192	940	140 587.9	158
上 海	Shanghai	1 749	510	157 318.3	331
江 苏	Jiangsu	6 973	1 801	549 108.2	2 103
浙 江	Zhejiang	8 214	1 265	437 505.9	2 589
安 徽	Anhui	2 084	1 064	184 234.5	502
福 建	Fujian	3 153	1 135	126 817.2	357
江 西	Jiangxi	2 014	597	129 041.5	209
山 东	Shandong	5 142	1 864	487 353.0	1 298
河 南	Henan	3 105	933	208 839.7	702
湖 北	Hubei	2 093	1 037	137 920.0	582
湖 南	Hunan	3 155	1 198	129 125.3	336
广 东	Guangdong	9 651	1 394	516 684.0	2 436
广 西	Guangxi	2 405	1 455	123 367.6	418
海 南	Hainan	278	40	43 135.3	66
重 庆	Chongqing	1 498	221	56 491.5	187
四 川	Sichuan	4 437	990	328 447.2	650
贵 州	Guizhou	1 755	539	137 057.5	105
云 南	Yunnan	2 044	738	94 573.0	139
西 藏	Tibet	16	1	233.8	
陕 西	Shaanxi	4 827	373	89 128.6	206
甘 肃	Gansu	672	148	32 770.3	124
青 海	Qinghai	103	40	7 806.0	18
宁 夏	Ningxia	359	145	47 492.7	59
新 疆	Xinjiang	971	313	169 622.0	119

各地区工业废水排放及处理情况（三）

Discharge and Treatment of Industrial Waste Water by Region（3-1）

单位：吨　　（2010）　　（t）

年份 地区	Year Region	工业废水中污染物排放量 Amount of Pollutants Discharged in the Industrial Waste Water					
		汞 Mercury	镉 Cadmium	六价铬 Hexavalent Chrome	铅 Lead	砷 Arsenic	挥发酚 Volatile Hydroxy-benzene
	2005	2.687	62.058	105.601	378.284	453.201	4 166.060
	2006	2.638	49.352	96.447	339.121	245.214	3 453.131
	2007	1.210	39.320	68.996	319.748	187.427	2 926.285
	2008	1.356	39.493	75.323	240.924	214.958	1 916.138
	2009	1.387	32.317	55.424	182.214	197.299	1 044.639
	2010	**1.047**	**30.129**	**54.804**	**140.805**	**118.092**	**966.037**
北　京	Beijing			0.022	0.016	…	0.110
天　津	Tianjin	…	…	0.078	0.002	0.022	1.519
河　北	Hebei	…	…	1.242	2.567	…	7.189
山　西	Shanxi	0.013	0.016	0.495	0.188	0.025	14.044
内蒙古	Inner Mongolia	…	0.153	0.355	0.680	0.378	5.340
辽　宁	Liaoning	0.001	0.032	1.941	0.543	0.310	13.883
吉　林	Jilin	…	0.004	0.629	0.048	…	239.258
黑龙江	Heilongjiang	0.001	0.055	0.263	0.131	0.227	442.612
上　海	Shanghai	0.002	0.010	0.722	0.078	0.013	4.930
江　苏	Jiangsu	0.001	0.039	7.105	2.429	0.851	26.865
浙　江	Zhejiang	0.008	0.119	6.825	1.231	0.218	1.347
安　徽	Anhui	0.002	0.164	0.091	1.235	1.887	9.124
福　建	Fujian	0.063	0.448	8.185	2.050	1.317	10.063
江　西	Jiangxi	0.011	4.302	2.269	10.862	13.308	18.263
山　东	Shandong	0.001	0.001	0.634	1.395	0.191	8.739
河　南	Henan	0.003	0.165	3.508	2.322	0.234	16.347
湖　北	Hubei	0.008	0.347	3.230	1.906	2.915	24.574
湖　南	Hunan	0.523	13.042	3.880	32.243	58.891	38.036
广　东	Guangdong	0.024	1.661	5.083	6.661	1.855	6.631
广　西	Guangxi	0.064	3.728	1.391	21.419	23.734	20.862
海　南	Hainan	…	…	…	0.025	…	…
重　庆	Chongqing	…	0.069	3.582	0.094	…	1.350
四　川	Sichuan	0.008	0.354	0.877	1.769	0.525	9.635
贵　州	Guizhou	0.002	0.007	0.134	0.124	0.139	0.216
云　南	Yunnan	0.098	0.364	0.045	4.616	2.737	2.673
西　藏	Tibet						
陕　西	Shaanxi	0.197	0.807	1.258	8.926	1.816	2.483
甘　肃	Gansu	0.006	3.688	0.452	34.420	5.627	4.100
青　海	Qinghai	…	0.465	0.020	0.573	0.541	0.394
宁　夏	Ningxia		0.030	0.267	0.016	0.134	19.316
新　疆	Xinjiang	0.014	0.060	0.221	2.238	0.198	16.134

各地区工业废水排放及处理情况（三）（续表）

Discharge and Treatment of Industrial Waste Water by Region（3-2）

单位：吨 （2010） （t）

年 份 地 区	Year Region	工业废水中污染物排放量 Amount of Pollutants Discharged in the Industrial Waste Water 氰化物 Cyanide	化学需氧量 COD	石油类 Petroleum	氨氮 Ammonia Nitrogen
	2005	573.8	5 547 333.1	23 471.6	525 064.9
	2006	457.1	5 422 616.2	19 152.7	424 616.8
	2007	381.5	5 110 631.3	16 899.8	340 824.8
	2008	256.1	4 575 833.0	13 315.1	296 902.9
	2009	250.3	4 396 780.5	9 513.5	273 544.4
	2010	**241.8**	**4 347 668.3**	**10 139.3**	**272 752.7**
北 京	Beijing	0.1	4 882.3	57.6	392.1
天 津	Tianjin	0.1	22 217.7	97.2	3 196.8
河 北	Hebei	4.6	217 965.3	523.2	18 224.9
山 西	Shanxi	13.0	137 668.7	353.4	11 885.6
内蒙古	Inner Mongolia	1.4	91 290.1	155.2	6 929.3
辽 宁	Liaoning	4.6	201 755.3	910.7	9 121.8
吉 林	Jilin	7.7	131 084.8	658.4	2 895.5
黑龙江	Heilongjiang	4.0	112 487.4	426.7	5 274.8
上 海	Shanghai	4.7	21 574.9	415.5	3 192.4
江 苏	Jiangsu	10.2	256 291.2	859.3	14 987.5
浙 江	Zhejiang	11.5	244 084.5	196.8	14 027.0
安 徽	Anhui	6.5	114 827.0	188.2	12 062.6
福 建	Fujian	59.0	82 946.1	565.2	6 613.6
江 西	Jiangxi	6.9	117 785.8	430.1	8 674.6
山 东	Shandong	2.0	295 127.6	499.1	15 441.3
河 南	Henan	8.4	295 573.6	464.5	23 130.3
湖 北	Hubei	19.1	165 199.3	539.3	14 932.8
湖 南	Hunan	25.1	187 885.9	430.6	17 448.7
广 东	Guangdong	10.6	234 412.2	427.4	10 614.4
广 西	Guangxi	17.9	492 677.7	269.5	14 491.5
海 南	Hainan	0.0	9 233.0	1.0	512.6
重 庆	Chongqing	0.4	86 549.0	221.2	5 440.1
四 川	Sichuan	0.8	250 817.5	372.8	16 439.9
贵 州	Guizhou	0.6	16 039.5	102.5	705.7
云 南	Yunnan	4.6	89 323.0	65.9	3 774.0
西 藏	Tibet		1 134.5	…	3.3
陕 西	Shaanxi	9.1	121 395.4	321.6	7 331.3
甘 肃	Gansu	1.2	45 060.6	154.8	7 569.9
青 海	Qinghai	0.0	44 713.6	38.0	2 271.0
宁 夏	Ningxia	1.3	92 800.5	145.2	8 910.1
新 疆	Xinjiang	6.5	162 863.9	248.1	6 257.3

各地区工业废水排放及处理情况（四）

Discharge and Treatment of Industrial Waste Water by Region（4）

单位：吨　　（2010）　　(t)

年份 地区	Year Region	工业废水中污染物去除量 Amount of Pollutants Removed from Industrial Waste Water					
		氰化物 Cyanide	化学需氧量 COD	新增设施去除的 Removed by New-added Facilities of Treatment	石油类 Petroleum	氨氮 Ammonia Nitrogen	挥发酚 Volatile Hydroxy-benzene
	2005	17 564	10 882 642	412 286	271 878	483 370	73 548
	2006	16 060	10 992 651	564 778	302 388	552 649	137 743
	2007	14 789	12 653 704	582 746	315 447	518 184	84 222
	2008	14 630	13 173 485	583 782	272 545	650 851	88 321
	2009	13 803	13 212 648	416 590	371 075	640 999	84 754
	2010	**7 206**	**14 153 851**	**366 739**	**354 247**	**826 488**	**73 272**
北京	Beijing	47	37 491	…	1 336	1 253	651
天津	Tianjin	…	52 378	709	465	1 058	37
河北	Hebei	538	731 674	4 389	11 239	24 733	5 367
山西	Shanxi	1 080	154 929	5 758	2 127	16 800	5 440
内蒙古	Inner Mongolia	202	370 256	6 277	916	10 613	3 031
辽宁	Liaoning	58	355 816	18 944	6 158	32 034	14 826
吉林	Jilin	18	230 460	9 927	1 517	3 539	1 409
黑龙江	Heilongjiang	150	379 427	9 366	154 242	44 365	6 448
上海	Shanghai	44	260 219	1 597	14 538	7 319	778
江苏	Jiangsu	401	1 274 487	15 862	26 625	56 460	4 147
浙江	Zhejiang	1 320	1 495 056	9 935	25 607	77 097	485
安徽	Anhui	421	416 933	8 263	6 702	74 375	2 864
福建	Fujian	147	794 498	4 764	6 507	50 062	408
江西	Jiangxi	315	141 829	7 684	9 744	11 588	2 217
山东	Shandong	283	2 317 429	80 951	23 047	175 866	5 117
河南	Henan	141	1 049 373	12 736	11 768	26 468	10 666
湖北	Hubei	119	378 739	6 566	8 579	12 607	1 500
湖南	Hunan	175	459 033	14 628	3 047	14 957	1 552
广东	Guangdong	1 020	744 643	6 887	4 543	39 350	307
广西	Guangxi	238	882 018	55 212	1 350	8 423	685
海南	Hainan	…	89 942	39	154	1 178	0
重庆	Chongqing	18	126 030	3 354	886	6 289	1 035
四川	Sichuan	14	326 681	24 906	977	7 212	699
贵州	Guizhou	330	58 236	2 576	281	1 029	133
云南	Yunnan	19	311 131	37 247	21 853	17 378	543
西藏	Tibet		41		…	2	
陕西	Shaanxi	14	244 318	5 007	4 191	9 133	114
甘肃	Gansu	2	53 804	1 620	1 466	6 549	71
青海	Qinghai	…	6 159	24	654	59	…
宁夏	Ningxia	14	244 601	1 923	1 523	60 444	1 272
新疆	Xinjiang	76	166 220	9 589	2 206	28 251	1 470

各地区工业废气排放及处理情况（一）

Discharge and Treatment of Industrial Waste Gas by Region（1）

（2010）

年份 地区	Year Region	煤炭消费量（万吨）Total Amount of Coal Consumed (10 000 t)	燃料煤 Coal Used as Fuel	原料煤 Coal Used as Material	燃料油消费量（万吨）Total Amount of Fuel Oil Consumed (10 000 t)	工业废气排放总量（标态）（亿米³）Total Volume of Industrial Waste Gas Emission (10^8 m^3)	燃料燃烧中排放的 From Process of Fuel Burning	生产工艺中排放的 From Process of Production
	2005	204 423	143 627	60 796	3 447	268 988	155 238	113 749
	2006	230 076	162 089	67 987	2 666	330 990	181 636	149 354
	2007	266 457	187 815	78 642	3 207	388 169	209 922	178 247
	2008	280 169	199 862	80 307	2 335	403 866	229 535	174 331
	2009	295 221	210 140	85 081	2 016	436 064	241 201	194 862
	2010	**337 999**	**240 534**	**97 465**	**2 087**	**519 168**	**303 897**	**215 271**
北京	Beijing	2 035	1 587	448	65	4 750	1 857	2 894
天津	Tianjin	4 890	4 471	419	23	7 686	5 254	2 432
河北	Hebei	23 606	16 186	7 421	31	56 324	27 750	28 574
山西	Shanxi	38 694	15 925	22 769	10	35 190	22 669	12 520
内蒙古	Inner Mongolia	26 594	19 849	6 745	35	27 488	17 476	10 013
辽宁	Liaoning	14 389	10 572	3 816	225	26 955	12 949	14 006
吉林	Jilin	6 920	5 907	1 012	31	8 240	5 786	2 454
黑龙江	Heilongjiang	8 891	6 930	1 961	35	10 111	8 304	1 807
上海	Shanghai	5 763	4 257	1 505	68	12 969	5 089	7 880
江苏	Jiangsu	22 628	19 565	3 062	167	31 213	20 235	10 978
浙江	Zhejiang	13 504	12 262	1 242	81	20 434	13 480	6 954
安徽	Anhui	11 634	8 500	3 135	33	17 849	8 860	8 989
福建	Fujian	6 551	5 488	1 064	97	13 507	8 322	5 185
江西	Jiangxi	5 966	4 010	1 956	22	9 812	5 248	4 564
山东	Shandong	29 235	20 962	8 272	129	43 837	25 874	17 963
河南	Henan	20 041	14 756	5 285	41	22 709	12 929	9 780
湖北	Hubei	8 244	5 098	3 146	18	13 865	7 451	6 414
湖南	Hunan	7 136	4 703	2 432	23	14 673	5 011	9 662
广东	Guangdong	15 857	14 231	1 626	424	24 092	16 627	7 465
广西	Guangxi	5 908	4 105	1 803	21	14 520	8 584	5 936
海南	Hainan	650	521	129	2	1 360	1 010	350
重庆	Chongqing	3 572	2 770	803	3	10 943	7 756	3 187
四川	Sichuan	8 511	6 112	2 398	16	20 107	11 111	8 995
贵州	Guizhou	8 250	5 769	2 481	25	10 192	6 866	3 327
云南	Yunnan	8 523	4 947	3 575	11	10 978	5 235	5 743
西藏	Tibet	23	23		…	16	13	3
陕西	Shaanxi	10 396	7 140	3 256	8	13 510	8 695	4 815
甘肃	Gansu	4 566	3 851	715	14	6 252	3 401	2 852
青海	Qinghai	1 234	805	429	2	3 952	855	3 097
宁夏	Ningxia	7 003	4 488	2 516	3	16 324	13 491	2 834
新疆	Xinjiang	6 786	4 744	2 042	423	9 310	5 711	3 598

各地区工业废气排放及处理情况（二）

Discharge and Treatment of Industrial Waste Gas by Region（2）

单位：吨　　　　（2010）　　　　(t)

年份 地区	Year Region	工业二氧化硫排放量 Volume of Industrial Sulphur Dioxide Emission	燃料燃烧中排放的 From Process of Fuel Burning	生产工艺中排放的 From Process of Production	工业二氧化硫去除量 Volume of Industrial Sulphur Dioxide Removed	燃料燃烧中去除的 Removed in Process of Fuel Burning	生产工艺中去除的 Removed in Process of Production
	2005	21 684 245	18 430 775	3 253 470	10 904 393	4 149 311	6 755 083
	2006	22 376 204	18 930 863	3 445 341	14 390 241	6 536 083	7 854 007
	2007	21 399 805	18 000 463	3 349 982	19 426 461	10 373 063	9 053 397
	2008	19 913 702	16 789 095	3 109 565	22 863 896	12 960 273	9 903 603
	2009	18 659 424	15 487 953	3 171 471	28 898 607	17 388 954	11 509 653
	2010	**18 644 201**	**15 695 602**	**2 847 273**	**33 039 968**	**22 312 865**	**10 727 103**
北　京	Beijing	56 844	54 969	1 870	129 888	121 856	8 032
天　津	Tianjin	217 620	206 634	10 991	374 014	369 149	4 865
河　北	Hebei	994 177	808 190	185 700	1 705 707	1 567 198	138 508
山　西	Shanxi	1 147 115	996 597	150 518	1 946 219	1 647 676	298 543
内蒙古	Inner Mongolia	1 192 984	1 033 577	157 133	1 602 604	1 479 486	123 118
辽　宁	Liaoning	859 410	746 893	112 283	1 294 885	595 762	699 123
吉　林	Jilin	300 636	264 767	35 836	203 112	140 151	62 961
黑龙江	Heilongjiang	417 051	380 998	35 870	110 038	91 425	18 614
上　海	Shanghai	221 476	206 966	14 410	349 849	333 389	16 460
江　苏	Jiangsu	1 002 445	918 804	82 952	2 157 808	1 665 599	492 209
浙　江	Zhejiang	653 889	612 531	41 213	1 297 423	900 785	396 637
安　徽	Anhui	483 912	383 277	96 115	1 613 141	458 150	1 154 990
福　建	Fujian	391 231	347 355	43 974	404 895	351 448	53 447
江　西	Jiangxi	470 954	360 418	110 536	1 627 852	338 459	1289 393
山　东	Shandong	1 382 874	1 236 865	145 991	3 152 480	2 486 771	665 708
河　南	Henan	1 162 874	952 698	209 491	1 433 447	885 348	548 099
湖　北	Hubei	516 044	413 192	102 071	1 091 439	618 617	472 823
湖　南	Hunan	627 383	392 946	232 796	1 001 682	544 698	456 984
广　东	Guangdong	989 088	865 154	115 800	1 450 107	1 071 175	378 932
广　西	Guangxi	847 970	704 155	143 816	984 372	639 467	344 905
海　南	Hainan	28 167	23 216	4 951	87 811	87 482	330
重　庆	Chongqing	572 747	477 143	84 039	962 388	889 730	72 658
四　川	Sichuan	937 632	809 904	115 743	867 915	700 258	167 657
贵　州	Guizhou	637 779	573 635	62 625	2 158 774	1 999 486	159 288
云　南	Yunnan	439 554	350 674	87 085	1 483 285	548 446	934 839
西　藏	Tibet	956	956				
陕　西	Shaanxi	706 960	589 376	103 784	1 076 472	872 615	203 857
甘　肃	Gansu	452 464	265 204	147 539	1 735 040	280 204	1 454 836
青　海	Qinghai	133 149	74 341	58 669	27 925	27 925	…
宁　夏	Ningxia	280 380	234 686	44 516	530 260	470 765	59 495
新　疆	Xinjiang	518 438	409 481	108 955	179 141	129 349	49 793

各地区工业废气排放及处理情况（三）

Discharge and Treatment of Industrial Waste Gas by Region（3）

单位：吤　　（2010）　　(t)

年份 地区	Year Region	工业烟尘去除量 Volume of Industrial Soot Removed	工业烟尘排放量 Volume of Industrial Soot Emission	工业粉尘去除量 Volume of Industrial Dust Removed	工业粉尘排放量 Volume of Industrial Dust Emission	工业氮氧化物去除量 Volume of Industrial Nitrogen Oxide Removed	工业氮氧化物排放量 Volume of Industrial Nitrogen Oxide Emission
	2005	205 870 817	9 489 033	64 538 505	9 111 883	—	—
	2006	235 645 611	8 644 934	72 799 462	8 084 446	941 567	11 362 582
	2007	251 663 943	7 711 389	76 695 820	6 987 435	1 146 856	12 613 314
	2008	305 428 242	6 707 464	84 712 248	5 849 478	718 772	12 503 135
	2009	328 481 458	6 043 631	87 226 480	5 236 472	676 427	12 847 753
	2010	**389 413 777**	**6 032 466**	**95 017 184**	**4 486 723**	**686 964**	**14 656 306**
北京	Beijing	1 907 537	21 266	780 542	16 550	1 185	108 851
天津	Tianjin	4 844 544	53 831	548 424	7 970	40 950	236 078
河北	Hebei	28 066 951	322 624	5 267 200	320 924	66 163	958 265
山西	Shanxi	25 816 610	432 252	3 021 640	365 131	142 805	781 174
内蒙古	Inner Mongolia	26 456 764	475 902	2 596 855	160 385	11 216	1 137 174
辽宁	Liaoning	22 202 804	397 943	5 622 053	167 346	35 683	706 433
吉林	Jilin	9 679 528	209 630	4 359 963	53 042	5	514 841
黑龙江	Heilongjiang	13 061 736	296 834	1 086 197	57 099	10 777	385 069
上海	Shanghai	4 723 090	41 793	1 381 789	9 705	16 561	261 498
江苏	Jiangsu	23 458 858	299 138	3 445 437	151 078	40 132	961 959
浙江	Zhejiang	12 465 566	165 417	8 131 355	139 178	27 524	702 375
安徽	Anhui	17 165 114	207 446	3 567 587	263 705	53 975	513 206
福建	Fujian	7 407 524	99 954	5 704 721	140 088	7 624	320 427
江西	Jiangxi	8 349 036	138 955	4 639 038	223 628	915	220 914
山东	Shandong	35 960 425	291 201	7 672 001	189 417	1 318	1 167 190
河南	Henan	28 107 260	473 683	5 405 789	226 970	4 204	1 000 242
湖北	Hubei	10 695 836	145 112	4 761 237	146 496	4 227	427 245
湖南	Hunan	7 867 768	234 986	3 731 501	394 350	11 541	313 367
广东	Guangdong	14 993 447	253 339	4 760 240	104 291	132 297	830 039
广西	Guangxi	5 268 131	250 115	3 357 477	317 748	20 530	235 552
海南	Hainan	781 546	6 520	46 997	6 580	133	31 455
重庆	Chongqing	3 773 199	102 132	954 137	83 601	…	231 313
四川	Sichuan	12 756 791	259 651	3 187 881	141 409	14 747	595 554
贵州	Guizhou	13 257 940	113 184	1 631 282	86 355	17 883	191 756
云南	Yunnan	10 882 214	89 154	2 110 433	91 761	…	329 352
西藏	Tibet	833	721	130	845		240
陕西	Shaanxi	8 364 627	115 018	2 841 239	185 451	10 671	469 486
甘肃	Gansu	3 555 779	98 128	1 425 171	92 626	12 815	224 033
青海	Qinghai	1 176 004	51 896	823 156	97 520	45	77 630
宁夏	Ningxia	22 214 547	136 231	1 262 514	60 851	185	322 525
新疆	Xinjiang	4 151 767	248 410	893 199	184 624	854	401 063

各地区工业废气排放及处理情况（四）

Discharge and Treatment of Industrial Waste Gas by Region（4）

（2010）

年份 地区	Year Region	废气治理设施数（套） Facilities for Treatment of Waste Gas（set）	脱硫设施数 Desulfurization Facilities	废气治理设施处理能力（标态）（万米³/时） Capacity of Facilities for Treatment of Waste Gas（10^4 m^3/h）	脱硫设施脱硫能力（吨/时） Capacity of Desulfurization Facilities（t/h）
	2005	145 043	22 648	—	12 907
	2006	154 557	24 530	801 085	16 116
	2007	162 325	24 867	769 863	24 410
	2008	174 164	27 281	1 663 969	23 992
	2009	176 489	26 995	906 574	25 389
	2010	**187 401**	**26 611**	**965 618**	**31 358**
北　京	Beijing	2 468	1 018	8 177	196
天　津	Tianjin	3 126	1 651	15 179	2 419
河　北	Hebei	13 743	2 358	87 801	2 903
山　西	Shanxi	9 517	3 475	44 766	1 395
内蒙古	Inner Mongolia	5 183	741	54 112	428
辽　宁	Liaoning	9 641	1 797	52 291	760
吉　林	Jilin	3 144	158	14 607	103
黑龙江	Heilongjiang	4 396	238	33 246	2 359
上　海	Shanghai	4 319	620	19 377	226
江　苏	Jiangsu	11 631	1 625	62 769	1 469
浙　江	Zhejiang	21 702	2 029	52 054	1 947
安　徽	Anhui	4 933	379	24 793	181
福　建	Fujian	6 470	159	24 047	1 863
江　西	Jiangxi	4 141	318	24 316	778
山　东	Shandong	11 886	2 743	81 613	2 485
河　南	Henan	9 079	942	56 316	1162
湖　北	Hubei	5 478	351	28 534	227
湖　南	Hunan	5 154	661	22 733	511
广　东	Guangdong	12 789	1 783	52 502	1 334
广　西	Guangxi	6 017	569	33 572	390
海　南	Hainan	472	17	5574	291
重　庆	Chongqing	3 511	276	12 264	663
四　川	Sichuan	7 346	493	28 885	3 365
贵　州	Guizhou	2 910	462	26 198	294
云　南	Yunnan	5 649	396	20 305	773
西　藏	Tibet	46		10	
陕　西	Shaanxi	3 983	420	24 532	1 049
甘　肃	Gansu	2 769	561	13 305	278
青　海	Qinghai	981	8	3 740	4
宁　夏	Ningxia	1 370	127	12 935	464
新　疆	Xinjiang	3 547	236	25 062	1 041

各地区工业废气排放及处理情况（五）

Discharge and Treatment of Industrial Waste Gas by Region（5）

（2010）

年 份 地 区	Year Region	废气治理设施运行费用（万元） Annul Expenditure for Operation （10 000 yuan）	脱硫设施运行费用（万元） Annul Expenditure for Operation of Desulfurization Facilities （10 000 yuan）	废气污染物在线监测仪器套数（套） On-line Appartus for Waste Gas Mornitoring （set）
	2005	2 670 764.6	—	—
	2006	4 643 681.2	1 323 719.7	3 028
	2007	5 549 659.8	1 946 889.6	4 713
	2008	7 733 885.7	2 815 906.0	7 357
	2009	8 736 718.5	3 550 653.1	9 108
	2010	**10 545 255.8**	**4 349 929.9**	**10 227**
北 京	Beijing	90 590.0	46 441.6	206
天 津	Tianjin	185 777.4	92 891.8	149
河 北	Hebei	901 330.8	321 112.6	772
山 西	Shanxi	501 016.8	277 280.2	818
内蒙古	Inner Mongolia	439 299.4	197 992.0	418
辽 宁	Liaoning	426 449.4	177 194.7	383
吉 林	Jilin	88 058.8	32 524.7	142
黑龙江	Heilongjiang	283 755.9	27 304.4	150
上 海	Shanghai	288 712.8	104 257.3	202
江 苏	Jiangsu	931 674.7	372 567.3	820
浙 江	Zhejiang	598 701.2	301 777.9	710
安 徽	Anhui	230 684.4	129 320.5	318
福 建	Fujian	237 264.5	95 457.9	185
江 西	Jiangxi	243 782.7	103 282.7	187
山 东	Shandong	1 039 838.9	454 694.5	868
河 南	Henan	490 261.7	262 330.8	619
湖 北	Hubei	315 735.2	128 357.2	255
湖 南	Hunan	223 270.1	102 324.7	214
广 东	Guangdong	973 399.7	374 936.1	752
广 西	Guangxi	183 402.0	47 495.2	256
海 南	Hainan	26 068.7	11 900.7	40
重 庆	Chongqing	163 084.5	90 279.4	129
四 川	Sichuan	564 880.5	114 577.8	281
贵 州	Guizhou	218 127.6	123 858.8	235
云 南	Yunnan	277 463.5	98 791.4	251
西 藏	Tibet	1 711.5		
陕 西	Shaanxi	181 976.3	92 295.4	273
甘 肃	Gansu	131 993.9	79 236.2	164
青 海	Qinghai	55 216.8	5 179.0	85
宁 夏	Ningxia	129 330.4	40 323.6	88
新 疆	Xinjiang	122 395.7	43 943.5	257

各地区工业固体废物产生及处置利用情况（一）

Generation and Utilization of Industrial Solid Wastes by Region（1-1）

单位：万吨　　（2010）　　（10 000 t）

年 份 地 区	Year Region	工业固体废物产生量 Industrial Solid Wastes Generated	危险废物 Hazardous Wastes	冶炼废渣 Smelting Residue	粉煤灰 Coalburning Powder	炉渣 Slag
	2005	134 449	1 162	18 199	23 377	13 722
	2006	151 541	1 084	20 893	26 602	15 603
	2007	175 632	1 079	23 720	32 672	17 797
	2008	190 127	1 357	24 909	35 096	18 512
	2009	203 943	1 430	28 209	36 179	19 818
	2010	**240 944**	**1 587**	**32 215**	**42 907**	**21 899**
北 京	Beijing	1 269	11	216	170	120
天 津	Tianjin	1 862	10	434	589	432
河 北	Hebei	31 688	37	6 418	2 858	1 418
山 西	Shanxi	18 270	11	2 059	3 136	1 389
内蒙古	Inner Mongolia	16 996	35	987	3 174	1 083
辽 宁	Liaoning	17 273	95	2 549	2 240	1 558
吉 林	Jilin	4 642	78	446	1 079	579
黑龙江	Heilongjiang	5 405	16	251	1 347	634
上 海	Shanghai	2 448	51	940	490	133
江 苏	Jiangsu	9 064	133	1 995	2 767	1 572
浙 江	Zhejiang	4 268	64	419	1 509	1 031
安 徽	Anhui	9 158	12	998	2 028	663
福 建	Fujian	7 487	8	526	530	488
江 西	Jiangxi	9 407	9	768	830	483
山 东	Shandong	16 038	289	2 867	3 879	2 211
河 南	Henan	10 714	19	1 024	3 312	1 423
湖 北	Hubei	6 813	17	1 232	1 182	829
湖 南	Hunan	5 773	67	1 326	1 096	451
广 东	Guangdong	5 456	118	484	1 553	1 051
广 西	Guangxi	6 232	24	1 032	751	369
海 南	Hainan	212	…	3	60	31
重 庆	Chongqing	2 837	36	182	539	444
四 川	Sichuan	11 239	133	1 079	1 305	1 107
贵 州	Guizhou	8 188	52	521	1 933	499
云 南	Yunnan	9 392	62	1 458	1 102	601
西 藏	Tibet	11			…	…
陕 西	Shaanxi	6 892	26	466	1 271	451
甘 肃	Gansu	3 745	34	624	654	232
青 海	Qinghai	1 783	86	210	123	96
宁 夏	Ningxia	2 465	4	194	954	224
新 疆	Xinjiang	3 914	49	505	448	298

各地区工业固体废物产生及处置利用情况（一）（续表）

Generation and Utilization of Industrial Solid Wastes by Region（1-2）

单位：万吨　　　　（2010）　　　　（10 000 t）

年份 地区	Year Region	工业固体废物产生量 Industrial Solid Wastes Generated				
		煤矸石 Coal Stone	尾矿 Gangue	放射性废物 Radioactive Wastes	脱硫石膏 Gypsum for Desulfurization	其他 Others
	2005	16 158	38 519	19.6	—	13 383
	2006	17 693	42 738	10.4	872.8	16 559
	2007	17 339	51 252	18.5	2 055.1	18 316
	2008	20 232	52 113	21.4	3 336.0	22 144
	2009	22 094	55 563	21.6	4 370.6	36 258
	2010	**25 600**	**65 548**	**27.8**	**5 940.5**	**29 385**
北　京	Beijing	54	499	…	28.7	83
天　津	Tianjin	…	…	…	99.8	186
河　北	Hebei	1 304	12 058	…	483.7	3 112
山　西	Shanxi	7 389	2 155	…	412.3	638
内蒙古	Inner Mongolia	2 554	3 385	7.3	401.8	4 232
辽　宁	Liaoning	605	7 938	…	152.4	1 353
吉　林	Jilin	129	1 783	…	59.8	154
黑龙江	Heilongjiang	2 185	529	…	16.7	202
上　海	Shanghai	1	…	…	79.9	553
江　苏	Jiangsu	302	322	…	347.3	927
浙　江	Zhejiang	16	236	…	219.8	548
安　徽	Anhui	2 517	1 994	…	167.4	562
福　建	Fujian	184	4 046	…	88.4	953
江　西	Jiangxi	215	5 787	0.9	144.2	830
山　东	Shandong	1 449	2 351	…	553.5	1 537
河　南	Henan	851	2 377	…	310.4	823
湖　北	Hubei	23	1 330	…	360.9	1 325
湖　南	Hunan	387	1 464	0.9	136.5	460
广　东	Guangdong	7	705	18.4	190.1	950
广　西	Guangxi	24	1 911	…	155.7	1 562
海　南	Hainan	…	25	…	21.4	60
重　庆	Chongqing	606	90	…	187.5	505
四　川	Sichuan	713	3 989	…	205.7	1 994
贵　州	Guizhou	1 575	1 175	…	504.7	1 624
云　南	Yunnan	278	2 871	…	110.5	2 063
西　藏	Tibet		7			4
陕　西	Shaanxi	1 120	2 763	0.1	238.3	428
甘　肃	Gansu	162	1 441	…	79.9	383
青　海	Qinghai	117	820	…	7.8	318
宁　夏	Ningxia	519	50	…	142.8	261
新　疆	Xinjiang	313	1 447	…	32.6	755

各地区工业固体废物产生及处置利用情况（二）

Generation and Utilization of Industrial Solid Wastes by Region（2-1）

单位：万吨 （2010） （10 000 t）

年份 地区	Year Region	工业固体废物综合利用量 Industrial Solid Wastes Utilized	危险废物 Hazardous Wastes	冶炼废渣 Smelting Residue	粉煤灰 Coalburning Powder
	2005	76 993	496	16 243	16 916
	2006	92 601	566	18 588	20 245
	2007	110 311	650	19 514	23 850
	2008	123 482	819	22 900	25 634
	2009	138 186	831	26 257	28 306
	2010	**161 772**	**977**	**29 474**	**31 557**
北京	Beijing	835	5	216	170
天津	Tianjin	1 845	2	434	589
河北	Hebei	17 973	19	6 417	2 129
山西	Shanxi	12 059	9	1 966	1 671
内蒙古	Inner Mongolia	9 562	16	719	2 001
辽宁	Liaoning	8 210	54	2 542	1 307
吉林	Jilin	3 114	54	442	704
黑龙江	Heilongjiang	4 169	7	254	787
上海	Shanghai	2 367	28	940	468
江苏	Jiangsu	8 761	75	1 989	2 714
浙江	Zhejiang	4 033	28	401	1 393
安徽	Anhui	7 849	8	998	1 701
福建	Fujian	6 215	3	520	518
江西	Jiangxi	4 379	8	762	775
山东	Shandong	15 297	270	2 796	3 090
河南	Henan	8 380	19	1 001	2 549
湖北	Hubei	5 521	9	1 225	916
湖南	Hunan	4 797	59	1 141	821
广东	Guangdong	4 953	75	486	1 438
广西	Guangxi	4 231	21	766	698
海南	Hainan	178	0	3	60
重庆	Chongqing	2 317	26	53	459
四川	Sichuan	6 159	68	628	1 000
贵州	Guizhou	4 174	21	398	811
云南	Yunnan	4 798	8	1 098	482
西藏	Tibet	…			…
陕西	Shaanxi	3 753	15	424	1 000
甘肃	Gansu	1 783	15	194	536
青海	Qinghai	759	11	205	109
宁夏	Ningxia	1 422	3	121	302
新疆	Xinjiang	1 877	41	335	360

各地区工业固体废物产生及处置利用情况（二）（续表）

Generation and Utilization of Industrial Solid Wastes by Region（2-2）

单位：万吨　　　　（2010）　　　　（10 000 t）

年　份 地　区	Year Region	工业固体废物综合利用量 Industrial Solid Wastes Utilized				
		炉渣 Slag	煤矸石 Coal Stone	尾矿 Gangue	脱硫石膏 Gypsum for Desulfurization	其他 Others
	2005	12 137	10 911	7 951	—	9 599
	2006	14 084	12 161	8 416	594	11 640
	2007	16 138	12 133	11 637	1 628	14 219
	2008	16 646	15 021	12 353	2 624	17 027
	2009	17 899	17 628	15 270	3 657	28 338
	2010	**20 001**	**19 962**	**17 161**	**5 176**	**22 578**
北　京	Beijing	120	14	115	29	81
天　津	Tianjin	442	…	…	100	170
河　北	Hebei	1 367	1 225	820	507	3 074
山　西	Shanxi	972	4 827	868	171	586
内蒙古	Inner Mongolia	847	1 971	216	223	2 925
辽　宁	Liaoning	1 467	571	256	152	1 013
吉　林	Jilin	526	123	808	66	145
黑龙江	Heilongjiang	507	1 832	157	18	108
上　海	Shanghai	132	1	…	79	506
江　苏	Jiangsu	1 552	303	151	347	866
浙　江	Zhejiang	979	16	213	219	474
安　徽	Anhui	628	2 464	935	163	554
福　建	Fujian	452	147	3 205	88	707
江　西	Jiangxi	475	204	1 093	138	674
山　东	Shandong	2 153	1 531	1 951	648	1 407
河　南	Henan	1 388	894	624	305	592
湖　北	Hubei	776	21	584	282	1 043
湖　南	Hunan	443	377	994	229	341
广　东	Guangdong	1 017	6	469	191	852
广　西	Guangxi	341	15	428	147	1 474
海　南	Hainan	31	…	13	21	39
重　庆	Chongqing	470	509	60	184	358
四　川	Sichuan	1 006	441	368	193	1 745
贵　州	Guizhou	305	922	632	269	663
云　南	Yunnan	452	204	1 144	71	845
西　藏	Tibet	…				…
陕　西	Shaanxi	449	853	206	219	361
甘　肃	Gansu	205	45	462	51	209
青　海	Qinghai	94	28	196	5	106
宁　夏	Ningxia	150	186	50	39	188
新　疆	Xinjiang	253	229	142	21	472

各地区工业固体废物产生及处置利用情况（三）

Generation and Utilization of Industrial Solid Wastes by Region（3）

单位：万吨　　（2010）　　（10 000 t）

年份 地区	Year Region	工业固体废物贮存量 Stock of Industrial Solid Wastes	危险废物贮存量 Hazardous Wastes	工业固体废物处置量 Industrial Solid Wastes Disposed	危险废物处置量 Hazardous Wastes	工业固体废物排放量（吨） Industrial Solid Wastes Discharged（t）	危险废物排放量 Hazardous Wastes
	2005	27 876	337.27	31 259	339.00	16 546 848	5 966.96
	2006	22 398	266.81	42 883	289.34	13 020 916	199 663.10
	2007	24 119	153.94	41 350	345.64	11 967 191	735.53
	2008	21 883	196.19	48 291	389.33	7 817 522	717.84
	2009	20 929	218.91	47 488	428.00	7 104 521	1.50
	2010	**23 918**	**166.34**	**57 264**	**512.66**	**4 981 976**	**1.50**
北　京	Beijing	40	0	780	6.49	611	0
天　津	Tianjin	0	0	27	8.66	0	0
河　北	Hebei	1 831	0	12 007	18.04	44 506	0
山　西	Shanxi	997	0.07	5 273	1.49	953 817	0
内蒙古	Inner Mongolia	2 152	1.94	5 295	17.94	45 294	0
辽　宁	Liaoning	2 535	3.48	6 767	50.89	27 877	1.50
吉　林	Jilin	952	0	578	23.89	0	0
黑龙江	Heilongjiang	835	0.01	445	10.81	18 619	0
上　海	Shanghai	1	0.04	94	23.44	2	0
江　苏	Jiangsu	215	0.54	139	58.09	0	0
浙　江	Zhejiang	69	0.88	174	35.55	6 177	0
安　徽	Anhui	518	0.12	916	6.99	15	0
福　建	Fujian	108	0.04	1 181	4.98	35 163	0
江　西	Jiangxi	557	0.09	4 487	1.25	132 314	0
山　东	Shandong	384	0.11	475	19.63	110	0
河　南	Henan	722	0.20	1 770	3.51	2 150	0
湖　北	Hubei	238	0.49	1 097	8.76	41 693	0
湖　南	Hunan	769	2.95	448	5.08	163 714	0
广　东	Guangdong	177	0.17	350	43.13	141 609	0
广　西	Guangxi	439	0.54	1 563	1.89	91 391	0
海　南	Hainan	33	0	0	0.24	10	0
重　庆	Chongqing	288	0.03	155	10.76	1 336 308	0
四　川	Sichuan	1 323	28.06	3 758	39.71	32 240	0
贵　州	Guizhou	1 473	36.47	2 498	25.84	597 582	0
云　南	Yunnan	1 845	3.25	2 911	49.99	363 098	0
西　藏	Tibet	7				41 243	
陕　西	Shaanxi	1 030	0.02	2 177	10.79	133 926	0
甘　肃	Gansu	1 012	8.62	1 150	13.83	113 758	0
青　海	Qinghai	1 035	75.51	1	0.16	21 652	0
宁　夏	Ningxia	564	0	490	1.47	9 220	0
新　疆	Xinjiang	1 771	2.68	256	9.34	627 877	0

各地区汇总工业企业概况（一）

Summarization of Industrial Enterprises Investigated by Region（1）

（2010）

年 份 地 区	Year Region	汇总工业企业数（个） Number of Industrial Enterprises Investigated（unit）	工业总产值（现价）（万元） Gross Industrial Output Value（current rate）（10 000 yuan）	企业专职环保人员数（人） Number of Professional Environmental Protection Employee（person）	工业炉窑数（台） Number of Industrial Furnaces（unit）	烟尘排放达标的 Furnaces with Soot Discharged Meeting Standard	二氧化硫排放达标的 Furnaces with Sulphur Dioxide Discharged Meeting Standard
	2005	70 612	1 120 864 276.0	188 051	81 117	60 525	53 545
	2006	76 185	1 426 002 200.0	268 473	82 814	64 599	57 450
	2007	106 457	1 833 178 647.4	311 624	82 576	64 188	58 564
	2008	110 373	2 202 623 978.2	312 665	85 468	70 296	64 939
	2009	110 905	2 293 049 566.3	282 648	80 831	68 163	63 442
	2010	**112 799**	**2 814 551 937.9**	**316 359**	**83 295**	**70 878**	**66 943**
北 京	Beijing	718	52 207 963.3	1 758	345	328	324
天 津	Tianjin	1 954	87 719 001.3	3 532	713	713	697
河 北	Hebei	5 890	138 039 030.9	15 615	4 257	3 715	3 372
山 西	Shanxi	4 020	94 188 727.4	15 250	4 845	4 569	4 265
内蒙古	Inner Mongolia	2 346	52 867 460.4	4 848	3 726	3 365	3 306
辽 宁	Liaoning	4 624	145 152 476.6	17 847	5 418	4 650	4 384
吉 林	Jilin	1 000	60 902 583.0	3 402	565	479	506
黑龙江	Heilongjiang	1 556	61 730 844.7	4 604	3 431	3 344	3 310
上 海	Shanghai	1 871	154 663 582.8	9 180	1 509	1 419	1 413
江 苏	Jiangsu	8 087	302 800 896.5	25 245	3 842	3 733	3 592
浙 江	Zhejiang	10 895	194 406 986.2	23 291	3 462	3 211	2 808
安 徽	Anhui	4 162	79 540 962.8	7 301	2 482	1 723	1 578
福 建	Fujian	6 080	82 835 183.6	5 032	2 438	1 863	1 815
江 西	Jiangxi	3 323	45 978 880.9	5 639	2 142	1 251	1 130
山 东	Shandong	6 307	326 546 004.1	25 277	3 977	3 855	3 762
河 南	Henan	4 423	96 606 207.8	13 282	5 512	5 181	4 738
湖 北	Hubei	2 643	84 308 535.0	6 152	2 187	2 069	1 878
湖 南	Hunan	3 929	65 211 574.1	6 555	3 165	2 631	2 540
广 东	Guangdong	12 034	258 625 905.9	33 129	4 493	3 399	3 117
广 西	Guangxi	4 443	47 072 984.3	6 623	3 360	2 929	2 792
海 南	Hainan	332	10 155 075.3	853	190	185	155
重 庆	Chongqing	2 489	51 744 138.7	4 459	1 804	1 537	1 421
四 川	Sichuan	6 615	85 387 548.4	24 869	5 178	4 552	4 315
贵 州	Guizhou	3 470	28 704 305.7	10 924	2 891	2 185	2 086
云 南	Yunnan	2 049	43 869 637.0	4 888	2 327	1 944	1 864
西 藏	Tibet	33	222 831.4	5 837	24	24	
陕 西	Shaanxi	2 629	65 425 641.3	7 938	2 155	1 441	1 279
甘 肃	Gansu	1 461	35 047 508.7	3 234	2 297	1 835	1 941
青 海	Qinghai	570	10 132 015.7	2 316	725	213	173
宁 夏	Ningxia	666	21 762 824.8	10 998	1 884	1 726	1 540
新 疆	Xinjiang	2 180	30 694 619.3	6 481	1 951	809	842

各地区汇总工业企业概况（二）

Summarization of Industrial Enterprises Investigated by Region（2）

（2010）

年份 地区	Year Region	工业锅炉数（台）Number of Industrial Boilers（unit）	烟尘排放达标的 Boilers with Soot Discharged Meeting Standard	二氧化硫排放达标的 Boilers with Sulphur Dioxide Discharged Meeting Standard	工业锅炉蒸吨数（蒸吨）Capacity of Industrial Boilers（steam. t）	烟尘排放达标的 Boilers with Soot Discharged Meeting Standard	二氧化硫排放达标的 Boilers with Sulphur Dioxide Discharged Meeting Standard
	2005	85 324	76 328	65 220	1 702 889	1 593 056	1 443 641
	2006	80 355	73 497	64 628	1 666 515	1 617 654	1 522 552
	2007	108 492	78 793	69 267	1 814 625	1 729 757	1 651 240
	2008	88 139	80 040	71 734	2 060 614	1 995 158	1 905 868
	2009	86 389	80 439	73 385	2 207 530	2 135 129	2 066 964
	2010	**85 487**	**79 300**	**73 626**	**2 294 520**	**2 232 872**	**2 165 812**
北京	Beijing	1 628	1 621	1 617	41 431	41 403	41 357
天津	Tianjin	2 173	2 167	2 169	65 851	64 751	65 191
河北	Hebei	4 320	4 143	3 654	94 535	94 161	91 909
山西	Shanxi	4 422	4 051	3 948	127 701	122 478	121 974
内蒙古	Inner Mongolia	3 256	2 864	2 686	213 385	203 046	193 546
辽宁	Liaoning	6 381	5 892	4 880	152 880	143 505	128 862
吉林	Jilin	2 270	2 179	2 045	68 739	67 940	65 450
黑龙江	Heilongjiang	4 646	4 288	4 106	77 709	76 231	71 967
上海	Shanghai	2 144	2 084	2 019	70 045	69 153	69 698
江苏	Jiangsu	5 631	5 527	5 467	172 061	168 980	168 615
浙江	Zhejiang	6 342	6 177	5 769	122 959	122 470	121 119
安徽	Anhui	1 818	1 744	1 571	59 461	56 581	55 906
福建	Fujian	3 238	2 908	2 680	60 395	59 716	58 012
江西	Jiangxi	1 579	1 208	1 073	29 571	28 339	27 843
山东	Shandong	6 029	5 945	5 713	205 617	202 525	199 971
河南	Henan	3 784	3 709	3 355	140 642	140 069	138 380
湖北	Hubei	1 982	1 899	1 799	64 984	64 675	64 213
湖南	Hunan	1 841	1 663	1 526	40 323	39 695	37 487
广东	Guangdong	6 721	6 120	5 603	130 256	126 417	124 114
广西	Guangxi	1 688	1 588	1 418	36 133	35 615	33 331
海南	Hainan	184	180	137	3 730	3 730	3 158
重庆	Chongqing	1 064	943	865	18 284	17 706	17 242
四川	Sichuan	2 731	2 527	2 369	36 028	34 737	30 754
贵州	Guizhou	1 095	604	603	26 136	25 190	25 329
云南	Yunnan	1 174	992	953	46 868	43 967	43 376
西藏	Tibet	11	11		31	31	
陕西	Shaanxi	2 074	1 969	1 815	35 735	35 546	32 966
甘肃	Gansu	1 421	1 213	1 169	40 075	39 111	38 740
青海	Qinghai	339	247	192	8 953	7 764	7 473
宁夏	Ningxia	767	659	557	29 021	28 547	26 025
新疆	Xinjiang	2 734	2 178	1 868	74 983	68 793	61 808

各地区汇总工业企业概况（三）

Summarization of Industrial Enterprises Investigated by Region（3）

（2010）

年份 地区	Year Region	工业用水总量（万吨）Quantity of Water Used by Industry（10 000 t）	新鲜水量 Fresh Water	重复用水量 Recycled Water	污水排放口数（个）Number of Waste Water Outlets（unit）	直接排海的污水排放口数 Direct Discharge into Sea	“三废”综合利用产品产值（万元）Output Value of Products Made from “Three Waste”（10 000 yuan）
	2005	31 992 327	7 965 406	24 026 774	64 583	1 141	7 555 064.3
	2006	34 653 012	7 059 207	27 593 815	67 074	1 156	10 267 926.0
	2007	37 837 725	6 819 681	31 018 041	75 736	1 660	13 512 691.9
	2008	37 295 361	6 037 238	31 258 117	77 580	1 696	16 213 662.0
	2009	38 739 629	5 825 629	32 914 000	78 542	1 396	16 082 440.4
	2010	**41 564 673**	**5 958 478**	**35 606 193**	**77 242**	**1 404**	**17 785 033.7**
北京	Beijing	513 901	18 996	494 905	563	0	34 365.8
天津	Tianjin	1 128 661	70 297	1 058 363	1 595	5	192 650.4
河北	Hebei	4 271 483	230 418	4 041 066	3 069	31	1 071 801.0
山西	Shanxi	2 309 499	111 859	2 197 640	1 256	0	426 371.8
内蒙古	Inner Mongolia	1 401 142	88 209	1 312 932	1 155	0	272 375.4
辽宁	Liaoning	2 252 790	236 088	2 016 701	2 756	96	328 090.2
吉林	Jilin	862 909	94 955	767 953	887	0	391 663.3
黑龙江	Heilongjiang	1 026 307	172 154	854 152	1 141	0	323 471.4
上海	Shanghai	1 448 485	400 514	1 047 971	1 842	19	170 379.1
江苏	Jiangsu	3 793 699	483 713	3 309 986	7 127	46	2 189 748.9
浙江	Zhejiang	2 092 871	891 970	1 200 902	9 570	692	2 863 867.2
安徽	Anhui	1 369 233	152 150	1 217 083	2 058	0	566 921.6
福建	Fujian	793 284	280 458	512 826	4 159	134	375 028.8
江西	Jiangxi	666 813	154 529	512 284	2 387	0	593 473.1
山东	Shandong	3 843 977	331 023	3 512 954	4 467	144	1 871 897.8
河南	Henan	2 730 575	256 951	2 473 624	3 035	0	743 908.8
湖北	Hubei	1 478 919	169 016	1 309 903	2 360	0	822 835.7
湖南	Hunan	1 018 088	145 894	872 195	3 458	0	901 206.8
广东	Guangdong	1 723 509	581 883	1 141 627	9 965	190	624 265.3
广西	Guangxi	1 250 248	388 317	861 932	2 195	28	510 233.4
海南	Hainan	181 132	11 852	169 279	221	19	31 623.2
重庆	Chongqing	732 290	197 167	535 123	2 092	0	291 326.6
四川	Sichuan	888 735	138 721	750 013	4 228	0	457 846.5
贵州	Guizhou	758 003	39 046	718 956	1 426	0	179 142.5
云南	Yunnan	839 211	71 290	767 921	862	0	654 554.6
西藏	Tibet	1 162	941	221	1		238.7
陕西	Shaanxi	601 007	73 309	527 698	1 452	0	293 499.6
甘肃	Gansu	477 003	37 260	439 742	665	0	224 120.8
青海	Qinghai	93 093	16 137	76 956	142	0	55 187.8
宁夏	Ningxia	233 631	43 195	190 436	239	0	100 750.3
新疆	Xinjiang	783 011	70 165	712 847	869	0	222 187.3

各地区工业污染治理项目建设情况（一）

Treatment Projects for Industrial Pollution by Region（1）

单位：个　　　　（2010）　　　　（unit）

年 份 地 区	Year Region	汇总工业企业数 Number of Industrial Enterprises Collected	本年施工项目总数 Number of Projects under Construction	治理废水 Treatment of Waste Water	治理废气 Treatment of Waste Gas	治理固体废物 Treatment of Solid Wastes	治理噪声 Treatment of Noise Pollution	治理其他 Treatment of Other Pollution
	2005	9 871	13 330	5 545	5 472	820	415	928
	2006	9 605	13 101	5 717	5 272	750	298	1 064
	2007	10 233	13 664	5 898	5 570	668	272	1 232
	2008	9 292	12 434	5 101	5 354	625	223	1 128
	2009	6 929	9 122	3 864	3 863	386	179	830
	2010	**5 231**	**6 597**	**2 942**	**2 519**	**372**	**143**	**601**
北 京	Beijing	27	45	5	32	0	2	6
天 津	Tianjin	98	124	45	48	9	4	18
河 北	Hebei	132	159	36	112	3	4	4
山 西	Shanxi	336	674	146	360	71	11	86
内蒙古	Inner Mongolia	114	172	60	77	8	1	26
辽 宁	Liaoning	115	106	34	45	11	0	16
吉 林	Jilin	76	98	45	41	4	2	6
黑龙江	Heilongjiang	40	57	18	27	0	0	12
上 海	Shanghai	129	173	65	79	4	11	14
江 苏	Jiangsu	346	388	185	139	13	9	42
浙 江	Zhejiang	610	633	446	142	15	4	26
安 徽	Anhui	59	69	35	21	0	3	10
福 建	Fujian	575	697	375	236	30	9	47
江 西	Jiangxi	119	136	78	47	3	0	8
山 东	Shandong	361	456	202	153	48	18	35
河 南	Henan	148	194	68	93	6	6	21
湖 北	Hubei	159	226	98	74	13	12	11
湖 南	Hunan	154	190	67	67	25	11	20
广 东	Guangdong	559	657	322	239	11	11	72
广 西	Guangxi	126	175	109	36	22	1	7
海 南	Hainan	21	17	13	1	1	0	2
重 庆	Chongqing	98	116	63	39	3	1	10
四 川	Sichuan	159	170	88	59	8	4	11
贵 州	Guizhou	87	116	53	43	10	6	4
云 南	Yunnan	247	364	134	168	29	5	28
西 藏	Tibet							
陕 西	Shaanxi	141	163	53	53	8	7	42
甘 肃	Gansu	101	133	58	45	16	1	13
青 海	Qinghai	12	12	3	9	0	0	0
宁 夏	Ningxia	16	21	7	11	0	0	3
新 疆	Xinjiang	66	56	31	23	1	0	1

各地区工业污染治理项目建设情况（二）

Treatment Projects for Industrial Pollution by Region（2）

单位：个　　（2010）　　(unit)

年　份 地　区	Year Region	本年竣工项目数 Number of Projects Completed	治理废水 Treatment of Waste Water	治理废气 Treatment of Waste Gas	治理固体废物 Treatment of Solid Wastes	治理噪声 Treatment of Noise Pollution	治理其他 Treatment of Other Pollution
	2005	11 158	4 710	4 613	708	353	731
	2006	11 822	5 047	4 843	679	274	979
	2007	12 199	5 169	5 038	582	253	1 157
	2008	11 184	4 477	4 887	556	215	1 049
	2009	8 235	3 415	3 540	338	167	775
	2010	**5 866**	**2 575**	**2 294**	**330**	**129**	**519**
北　京	Beijing	46	5	33	0	2	6
天　津	Tianjin	99	34	40	9	1	15
河　北	Hebei	150	34	107	3	4	2
山　西	Shanxi	599	114	338	63	10	74
内蒙古	Inner Mongolia	156	52	74	7	1	22
辽　宁	Liaoning	84	24	37	11	0	12
吉　林	Jilin	89	39	39	3	2	6
黑龙江	Heilongjiang	49	14	24	0	0	11
上　海	Shanghai	156	56	74	4	11	11
江　苏	Jiangsu	361	174	128	12	9	38
浙　江	Zhejiang	577	404	132	15	4	22
安　徽	Anhui	55	29	18	0	1	7
福　建	Fujian	628	349	206	24	8	41
江　西	Jiangxi	103	53	43	2	0	5
山　东	Shandong	399	177	130	47	17	28
河　南	Henan	169	56	83	6	5	19
湖　北	Hubei	195	84	64	10	10	10
湖　南	Hunan	167	58	57	23	11	18
广　东	Guangdong	613	296	225	11	11	68
广　西	Guangxi	166	104	34	20	1	7
海　南	Hainan	14	10	1	1	0	2
重　庆	Chongqing	89	46	37	1	1	4
四　川	Sichuan	154	82	53	6	3	10
贵　州	Guizhou	97	45	34	8	6	4
云　南	Yunnan	334	119	158	24	5	28
西　藏	Tibet						
陕　西	Shaanxi	124	32	45	7	5	35
甘　肃	Gansu	120	52	42	13	1	12
青　海	Qinghai	11	3	8	0	0	0
宁　夏	Ningxia	14	4	9	0	0	1
新　疆	Xinjiang	48	26	21	0	0	1

各地区工业污染治理项目建设情况（三）

Treatment Projects for Industrial Pollution by Region（3）

单位：万元　　（2010）　　（10 000 yuan）

年份 地区	Year Region	污染治理项目本年完成投资合计 Investment Completed in the Treatment of Industrial Pollution This Year	治理废水 Treatment of Waste Water	治理废气 Treatment of Waste Gas	治理固体废物 Treatment of Solid Wastes	治理噪声 Treatment of Noise Pollution	治理其他 Treatment of Other Pollution
	2005	4 581 908.7	1 337 146.9	2 129 571.3	274 181.3	30 613.3	810 395.9
	2006	4 839 485.1	1 511 164.5	2 332 697.1	182 630.5	30 145.1	782 847.9
	2007	5 523 909.4	1 960 721.8	2 752 642.2	182 531.9	18 278.6	606 837.9
	2008	5 426 403.8	1 945 977.4	2 656 986.8	196 850.6	28 382.9	598 206.1
	2009	4 424 521.9	1 494 606.0	2 324 616.0	218 535.7	14 100.0	372 664.2
	2010	**3 969 768.2**	**1 301 148.7**	**1 888 456.5**	**142 692.2**	**15 193.2**	**621 777.6**
北京	Beijing	19 339.5	1 761.6	17 264.4	0.0	11.2	302.3
天津	Tianjin	164 683.5	47 138.8	34 856.0	1 225.5	155.0	81 308.2
河北	Hebei	108 587.8	24 677.9	82 053.7	1.8	992.6	861.8
山西	Shanxi	279 573.8	54 819.9	149 810.1	30 135.3	918.5	43 890.0
内蒙古	Inner Mongolia	132 399.5	46 323.3	68 782.6	1 986.6	201.2	15 105.8
辽宁	Liaoning	147 707.9	47 213.5	94 551.4	803.8	0.0	5 139.2
吉林	Jilin	63 365.5	26 793.1	31 281.4	4 870.5	311.5	109.0
黑龙江	Heilongjiang	49 494.4	25 338.7	16 883.8	0.0	0.0	7 271.9
上海	Shanghai	94 106.5	6 308.1	34 780.6	26.6	679.3	52 311.9
江苏	Jiangsu	185 994.5	74 386.3	71 869.9	8 964.3	1 973.5	28 800.5
浙江	Zhejiang	119 568.2	71 261.7	39 981.9	2 746.0	211.5	5 367.1
安徽	Anhui	58 895.4	14 249.8	30 931.9	0.0	107.8	13 605.9
福建	Fujian	153 295.5	70 938.9	49 844.2	7 703.5	611.0	24 197.9
江西	Jiangxi	63 775.4	19 864.9	39 593.8	48.0	0.0	4 268.7
山东	Shandong	456 759.3	162 780.0	186 422.8	15 288.2	784.5	91 483.8
河南	Henan	125 120.3	44 300.6	75 616.4	856.4	445.0	3 901.9
湖北	Hubei	277 416.4	41 256.9	203 866.8	9 579.6	1 866.6	20 846.5
湖南	Hunan	137 948.6	59 247.5	63 845.3	11 221.7	500.5	3 133.6
广东	Guangdong	310 584.0	150 271.5	54 176.2	4 622.0	341.9	100 672.4
广西	Guangxi	92 845.1	47 387.8	27 249.5	17 024.1	80.0	1 103.7
海南	Hainan	4 353.5	3 923.3	140.0	52.0	0.0	238.2
重庆	Chongqing	77 501.9	37 760.6	27 368.6	3 189.0	500.0	8 683.7
四川	Sichuan	71 626.5	37 748.9	30 889.4	1 405.0	487.5	1 095.7
贵州	Guizhou	68 079.5	19 383.2	39 460.6	6 297.7	2 497.0	441.0
云南	Yunnan	106 271.5	24 183.6	72 910.3	6 301.7	239.6	2 636.3
西藏	Tibet						
陕西	Shaanxi	336 535.0	71 478.5	178 851.0	1 950.0	1 258.5	82 997.0
甘肃	Gansu	146 483.2	26 166.1	104 551.6	5 592.9	19.0	10 153.6
青海	Qinghai	9 747.2	1 018.6	8 728.6	0.0	0.0	0.0
宁夏	Ningxia	40 896.0	23 478.0	5 618.0	0.0	0.0	11 800.0
新疆	Xinjiang	66 812.8	19 687.1	46 275.7	800.0	0.0	50.0

各地区工业污染治理项目建设情况（四）

Treatment Projects for Industrial Pollution by Region（4）

单位：万元　　（2010）　　（10 000 yuan）

年份 地区	Year Region	污染治理项目本年投资来源 Source of Current Investment in the Treatment of Industrial Pollution				
		合计 Total	排污费补助 Subsidy of Pollution Emission Expenditure	政府其他补助 Other Funds	企业自筹 Self-financing	国内贷款 Domestic Loans
	2005	4 581 908.7	206 000.5	77 765.8	4 298 142.4	389 917.7
	2006	4 839 485.1	142 810.8	155 157.4	4 541 517.0	301 030.0
	2007	5 523 909.4	107 990.8	156 563.7	5 259 355.3	382 803.0
	2008	5 426 403.8	88 172.1	136 126.5	5 200 689.3	306 685.2
	2009	4 424 743.9	70 015.3	140 988.0	4 213 739.9	472 564.2
	2010	**3 969 768.2**	**49 122.4**	**151 422.4**	**3 769 222.2**	**313 548.2**
北京	Beijing	19 339.5	0.0	327.8	19 011.4	5 688.0
天津	Tianjin	164 683.5	202.5	16 684.2	147 796.8	2 189.7
河北	Hebei	108 587.8	1 555.0	3 925.7	103 107.1	20.0
山西	Shanxi	279 573.8	6 036.9	7 590.5	265 945.9	2 116.0
内蒙古	Inner Mongolia	132 399.5	1 300.0	1 126.0	129 973.3	1 731.0
辽宁	Liaoning	147 707.9	2 642.0	3 926.2	141 139.7	5 169.2
吉林	Jilin	63 365.5	1 845.0	3 040.6	58 479.9	1 648.0
黑龙江	Heilongjiang	49 494.4	40.0	3 865.0	45 589.4	0.0
上海	Shanghai	94 106.5	10.0	1 110.9	92 985.6	1 716.0
江苏	Jiangsu	185 994.5	4 501.0	3 724.5	177 769.0	5 705.0
浙江	Zhejiang	119 568.2	3 261.1	2 618.0	113 689.1	2 182.7
安徽	Anhui	58 895.4	893.4	1 173.2	56 828.8	133.0
福建	Fujian	153 295.5	367.0	598.9	152 329.3	5 870.0
江西	Jiangxi	63 775.4	4 292.0	1 380.0	58 103.4	14 750.0
山东	Shandong	456 759.3	1 802.5	8 805.0	446 151.8	7 499.3
河南	Henan	125 120.3	405.0	14 010.4	110 704.9	10 196.0
湖北	Hubei	277 416.4	1 894.0	3 891.6	271 630.8	162 940.0
湖南	Hunan	137 948.6	6 402.0	3 956.2	127 590.4	20 344.0
广东	Guangdong	310 584.0	2 161.3	1 335.3	307 087.5	574.0
广西	Guangxi	92 845.1	770.0	388.0	91 687.1	30.0
海南	Hainan	4 353.5		375.0	3 978.5	
重庆	Chongqing	77 501.9	1 530.0	7 739.8	68 232.1	857.0
四川	Sichuan	71 626.5	1 839.9	3 549.3	66 237.3	1 060.6
贵州	Guizhou	68 079.5	254.5	2 876.0	64 949.0	1 000.0
云南	Yunnan	106 271.5	1 325.0	1 397.0	103 549.5	4 754.0
西藏	Tibet					
陕西	Shaanxi	336 535.0	847.3	40 531.3	295 156.4	4 468.7
甘肃	Gansu	146 483.2	2 530.0	10 138.0	133 815.2	50 896.0
青海	Qinghai	9 747.2	145.0	169.0	9 433.2	0.0
宁夏	Ningxia	40 896.0	0.0	0.0	40 896.0	0.0
新疆	Xinjiang	66 812.8	270.0	1 169.0	65 373.8	10.0

各地区工业污染治理项目建设情况（五）

Treatment Projects for Industrial Pollution by Region（5）

（2010）

年 份 地 区	Year Region	本年竣工项目新增设计处理能力 Capacity for Treament of Industrial Pollution New-added		
		治理废水（吨/日）Treatment of Waste Water（t/d）	治理废气（标态）（万米 ^{3}/时）Treatment of Waste Gas（10^4 m^3/h）	治理固体废物（吨/日）Treatment of Solid Wastes（t/d）
	2005	25 602 344	25 573	2 769 710
	2006	13 490 365	94 949	15 960 737
	2007	28 214 561	87 519	906 026
	2008	17 924 808	36 996	1 747 620
	2009	21 782 264	29 467	2 180 759
	2010	**9 071 786**	**29 028**	**247 204**
北 京	Beijing	2 200	10	0
天 津	Tianjin	69 117	462	0
河 北	Hebei	79 566	1 484	0
山 西	Shanxi	191 362	4 261	36 366
内蒙古	Inner Mongolia	46 405	394	805
辽 宁	Liaoning	103 516	910	100
吉 林	Jilin	253 888	388	500
黑龙江	Heilongjiang	62 774	309	0
上 海	Shanghai	16 378	327	0
江 苏	Jiangsu	672 351	603	35 050
浙 江	Zhejiang	945 063	197	657
安 徽	Anhui	41 744	2 195	0
福 建	Fujian	891 546	10 741	1 309
江 西	Jiangxi	253 525	258	3
山 东	Shandong	1 574 177	1 417	133 457
河 南	Henan	133 305	717	5
湖 北	Hubei	146 819	445	22 766
湖 南	Hunan	453 199	615	610
广 东	Guangdong	1 765 652	451	2 204
广 西	Guangxi	342 043	850	2 214
海 南	Hainan	7 490	2	1
重 庆	Chongqing	330 753	257	0
四 川	Sichuan	152 799	645	2 224
贵 州	Guizhou	16 090	268	0
云 南	Yunnan	53 693	251	5 953
西 藏	Tibet			
陕 西	Shaanxi	86 502	194	425
甘 肃	Gansu	84 735	110	2 355
青 海	Qinghai	3 824	101	0
宁 夏	Ningxia	122 860	10	0
新 疆	Xinjiang	168 410	57	200

各地区工业企业“三废”治理效率（一）

Efficiency of Industrial Pollution Treatment by Region（1）

单位：%　　　　　　　　　　　（2010）　　　　　　　　　　　（%）

年份 地区	Year Region	二氧化硫排放达标率 Ratio of Industrial Sulphur Dioxide Meeting Discharge Standards	烟尘排放达标率 Ratio of Industrial Soot Meeting Discharge Standards	工业粉尘排放达标率 Ratio of Industrial Dust Meeting Discharge Standards	工业氮氧化物排放达标率 Ratio of Industrial Nirogen Oxide Meeting Discharge Standards
	2005	79.4	82.9	75.1	—
	2006	81.9	87.0	82.9	79.6
	2007	86.3	88.2	88.1	77.5
	2008	88.8	89.6	89.3	85.3
	2009	91.0	90.3	89.9	88.7
	2010	**92.1**	**90.6**	**91.4**	**87.9**
北京	Beijing	99.5	98.7	100.0	97.3
天津	Tianjin	99.9	100.0	100.0	99.9
河北	Hebei	98.9	99.3	99.3	81.9
山西	Shanxi	98.9	96.6	93.2	84.0
内蒙古	Inner Mongolia	90.6	81.1	91.4	88.2
辽宁	Liaoning	88.1	88.6	91.2	81.8
吉林	Jilin	93.5	96.0	90.5	84.2
黑龙江	Heilongjiang	96.2	93.7	81.2	84.1
上海	Shanghai	98.8	96.9	98.8	99.0
江苏	Jiangsu	95.2	98.3	97.6	96.6
浙江	Zhejiang	97.5	97.8	97.3	96.8
安徽	Anhui	94.4	95.8	97.5	80.0
福建	Fujian	98.0	95.2	98.2	97.9
江西	Jiangxi	94.9	95.3	96.2	92.4
山东	Shandong	97.8	98.4	96.9	84.4
河南	Henan	96.8	97.8	96.6	93.8
湖北	Hubei	89.8	96.0	98.0	96.0
湖南	Hunan	91.9	91.5	90.5	93.3
广东	Guangdong	84.2	91.1	70.1	96.7
广西	Guangxi	91.4	96.4	97.6	78.3
海南	Hainan	98.4	96.4	99.4	79.2
重庆	Chongqing	81.2	80.0	83.0	83.9
四川	Sichuan	93.2	87.1	95.7	92.2
贵州	Guizhou	83.3	67.9	70.2	76.6
云南	Yunnan	96.5	88.8	89.5	92.7
西藏	Tibet		5.1	5.4	45.0
陕西	Shaanxi	95.8	98.2	98.6	66.8
甘肃	Gansu	81.0	74.1	91.7	90.4
青海	Qinghai	75.7	38.9	58.1	51.4
宁夏	Ningxia	84.6	94.8	94.3	87.9
新疆	Xinjiang	62.3	52.8	53.4	71.2

各地区工业企业“三废”治理效率（二）

Efficiency of Industrial Pollution Treatment by Region（2）

单位：%（2010）（%）

年 份 地 区	Year Region	工业废水排放达标率 Ratio of Industrial Waste Water Meeting Discharge Standards	工业用水重复利用率 Ratio of Recycled Water Used in Industry	工业固体废物	
				综合利用率 Ratio of Industrial Solid Wastes Utilized	处置率 Ratio of Industrial Solid Wastes Disposed
	2005	91.2	75.1	57.5	23.2
	2006	90.7	79.6	60.2	27.4
	2007	91.7	82.0	62.1	23.4
	2008	92.4	83.8	64.3	25.2
	2009	94.2	85.0	67.0	23.2
	2010	**95.3**	**85.7**	**66.7**	**23.7**
北 京	Beijing	98.8	96.3	65.8	47.1
天 津	Tianjin	100.0	93.8	98.6	1.5
河 北	Hebei	98.6	94.6	56.6	37.9
山 西	Shanxi	94.7	95.2	65.5	28.9
内蒙古	Inner Mongolia	90.2	93.7	56.3	31.2
辽 宁	Liaoning	92.6	89.5	46.9	39.2
吉 林	Jilin	89.0	89.0	67.1	12.4
黑龙江	Heilongjiang	92.7	83.2	76.5	8.2
上 海	Shanghai	98.0	72.4	96.2	3.8
江 苏	Jiangsu	98.1	87.3	96.1	1.5
浙 江	Zhejiang	96.2	57.4	94.3	4.1
安 徽	Anhui	98.0	88.9	84.6	10.0
福 建	Fujian	98.7	64.7	82.9	15.8
江 西	Jiangxi	94.2	76.8	46.5	47.7
山 东	Shandong	98.4	91.4	94.7	3.0
河 南	Henan	97.4	90.6	77.1	16.5
湖 北	Hubei	96.8	88.6	80.5	16.1
湖 南	Hunan	93.7	85.7	81.0	7.8
广 东	Guangdong	93.1	66.2	90.2	6.4
广 西	Guangxi	96.9	68.9	67.8	25.1
海 南	Hainan	97.8	93.5	84.1	0.2
重 庆	Chongqing	94.7	73.1	80.2	5.5
四 川	Sichuan	96.5	84.4	54.8	33.4
贵 州	Guizhou	77.3	94.9	50.9	30.5
云 南	Yunnan	91.8	91.5	50.8	31.0
西 藏	Tibet	29.5	19.0	1.8	
陕 西	Shaanxi	97.5	87.8	54.4	31.6
甘 肃	Gansu	83.3	92.2	46.3	30.7
青 海	Qinghai	59.9	82.7	42.2	0.1
宁 夏	Ningxia	78.7	81.5	57.5	19.8
新 疆	Xinjiang	57.3	91.0	47.5	6.5

各地区危险废物集中处置情况（一）

Centralized Treatment of Hazardous Wastes by Region（1）

（2010）

年份 地区	Year Region	危险废物集中处置厂数（座） Number of Plants for Centralized Treatment of Hazardous Wastes（unit）	危险废物处置能力（吨/日） Capacity of Plants for Centralized Treatment of Hazardous Wastes（t/d）	焚烧处置能力 Treatment by Burning	填埋处理能力 Treatment by Landfilling
	2005	189	5 211	3 290	1 167
	2006	248	18 052	3 144	13 314
	2007	322	19 986	8 663	1 692
	2008	518	19 362	13 909	1 701
	2009	558	15 947	11 443	2 537
	2010	**546**	**17 795**	**13 147**	**3 768**
北　京	Beijing	8	378	260	38
天　津	Tianjin	14	192	98	94
河　北	Hebei	14	174	162	8
山　西	Shanxi	13	122	121	0
内蒙古	Inner Mongolia	11	40	34	6
辽　宁	Liaoning	41	962	487	475
吉　林	Jilin	6	212	56	156
黑龙江	Heilongjiang	5	37	22	15
上　海	Shanghai	39	764	366	226
江　苏	Jiangsu	82	2 588	2 155	250
浙　江	Zhejiang	70	2 960	2 336	203
安　徽	Anhui	1	1 695	1 682	12
福　建	Fujian	7	106	60	46
江　西	Jiangxi	20	1 735	1 692	23
山　东	Shandong	15	302	234	68
河　南	Henan	15	114	76	38
湖　北	Hubei	10	715	514	201
湖　南	Hunan	9	73	37	36
广　东	Guangdong	81	3 259	1 891	1 368
广　西	Guangxi	17	348	342	6
海　南	Hainan	3	73	32	41
重　庆	Chongqing	6	383	199	184
四　川	Sichuan	13	75	72	4
贵　州	Guizhou	4	74	74	
云　南	Yunnan	2	9	8	1
陕　西	Shaanxi	11	314	64	251
甘　肃	Gansu	4	17	11	6
青　海	Qinghai	2	8	8	
宁　夏	Ningxia	1	3	3	
新　疆	Xinjiang	9	64	52	13

各地区危险废物集中处置情况（二）

Centralized Treatment of Hazardous Wastes by Region（2）

（2010）

年份 地区	Year Region	危险废物处置量（吨） Volume of Hazardous Wastes Disposed（t）	焚烧量 Volume of Hazardous Wastes Burned	填埋量 Volume of Hazardous Wastes Landfilled	危险废物综合利用量（吨） Volume of Hazardous Wastes Utilized（t）	本年运行费用（万元） Annual Expenditure for Operation（10 000 yuan）
	2005	521 978	316 524	198 839	214 316	71 496.0
	2006	885 908	507 979	352 856	337 338	110 234.2
	2007	1 142 897	781 817	290 192	909 933	172 466.4
	2008	1 300 129	989 154	165 548	1 228 529	324 558.9
	2009	1 975 046	1 555 481	337 933	1 550 895	321 344.7
	2010	**1 812 298**	**1 343 836**	**420 980**	**1 500 853**	**388 787.6**
北京	Beijing	85 351	70 268	11 531	21 601	17 167.6
天津	Tianjin	46 958	18 851	28 107	144 930	24 670.0
河北	Hebei	12 099	11 057	143	17	1 750.9
山西	Shanxi	12 616	12 592	24	…	1 740.3
内蒙古	Inner Mongolia	5 080	3 683	1 397		1 330.2
辽宁	Liaoning	118 453	47 008	71 445	301 774	26 208.9
吉林	Jilin	15 532	13 043	2 489		2 437.4
黑龙江	Heilongjiang	5 716	5 149	567	…	2 453.6
上海	Shanghai	139 891	82 595	40 050	104 910	46 916.5
江苏	Jiangsu	261 648	190 909	45 040	119 236	53 016.7
浙江	Zhejiang	274 170	264 457	9 712	173 353	33 388.1
安徽	Anhui	22 761	20 150	2 528	8 887	5 869.5
福建	Fujian	18 287	15 234	3 053		3 994.0
江西	Jiangxi	11 593	9 251	2 342	6 344	7 132.9
山东	Shandong	53 929	49 885	4 044	2 635	9 966.3
河南	Henan	22 377	15 305	7 071	1 594	7 981.0
湖北	Hubei	109 828	102 070	7 758	80 000	16 520.0
湖南	Hunan	11 260	7 552	3 708	…	3 439.4
广东	Guangdong	453 237	320 684	132 553	486 189	103 878.7
广西	Guangxi	17 819	15 569	2 250	9 185	2 824.9
海南	Hainan	3 705	3 705			1 363.0
重庆	Chongqing	23 470	15 181	8 289	3 500	3 506.7
四川	Sichuan	14 659	14 095	564	907	3 581.7
贵州	Guizhou	7 639	7 639		240	970.0
云南	Yunnan	1 637	1 612	25		653.7
陕西	Shaanxi	50 383	14 702	35 680	35 537	4 212.3
甘肃	Gansu	3 246	3 231	15	…	741.4
青海	Qinghai	2 026	2 026			254.2
宁夏	Ningxia	650	650			158.0
新疆	Xinjiang	6 277	5 683	594	13	659.7

各地区城市污水处理情况（一）
Urban Waste Water Treatment by Region（1）

（2010）

年 份 地 区	Year Region	城市污水处理厂数（座） Urban Waste Water Treatment Plants（unit）	污水处理厂设计处理能力（万吨/日） Treatment Capacity（10 000 t/d）	工业区污水集中处理装置数（座） Industrilal Park Waste Water Treatment Plants（unit）	集中处理装置处理能力（吨/日） Treatment Capacity（t/d）	本年运行费用（万元） Annul Expenditure for Operation（10 000 yuan）
	2005	764	5 220	46	742 445	736 514.5
	2006	939	6 370	74	1 627 310	1 016 240.3
	2007	1 258	7 579	120	2 414 000	1 295 906.7
	2008	1 692	9 079	119	2 108 367	1 785 350.9
	2009	2 192	10 477	195	3 659 098	2 193 741.2
	2010	**2 881**	**12 331**	**221**	**4 466 918**	**2 589 580.7**
北 京	Beijing	68	370	5	82 000	100 341.4
天 津	Tianjin	28	202	6	64 600	41 116.3
河 北	Hebei	186	716	13	123 000	121 432.6
山 西	Shanxi	100	237	2	32 880	47 164.8
内蒙古	Inner Mongolia	65	213	10	124 960	47 157.4
辽 宁	Liaoning	78	471	3	43 000	100 776.0
吉 林	Jilin	44	249	0	0	40 216.0
黑龙江	Heilongjiang	38	216	1	2 000	35 885.3
上 海	Shanghai	52	685	1	2 400	124 975.4
江 苏	Jiangsu	368	1 120	66	1 123 650	313 617.3
浙 江	Zhejiang	127	707	40	1 419 000	260 566.2
安 徽	Anhui	78	389	1	100 000	53 885.5
福 建	Fujian	76	310	21	422 300	63 976.6
江 西	Jiangxi	46	214	2	90 000	30 661.7
山 东	Shandong	205	956	21	196 478	231 981.2
河 南	Henan	145	656	3	70 000	109 965.3
湖 北	Hubei	103	520	4	80 000	85 769.4
湖 南	Hunan	118	477	1	45 000	78 050.0
广 东	Guangdong	295	1 624	10	290 800	324 213.5
广 西	Guangxi	97	329	4	58 600	58 387.3
海 南	Hainan	24	86	0	0	10 060.4
重 庆	Chongqing	79	231	2	6 250	83 471.5
四 川	Sichuan	166	408	5	90 000	94 356.4
贵 州	Guizhou	102	163	0	0	19 636.0
云 南	Yunnan	48	177	0	0	25 168.5
西 藏	Tibet	1	1	0		173.0
陕 西	Shaanxi	54	224	0	0	29 949.4
甘 肃	Gansu	24	108	0	0	17 164.2
青 海	Qinghai	10	33	0	0	5 007.5
宁 夏	Ningxia	15	69	0	0	10 023.0
新 疆	Xinjiang	41	171	0	0	24 431.6

各地区城市污水处理情况（二）

Urban Waste Water Treatment by Region（2）

（2010）

年 份 地 区	Year Region	污水处理量（万吨）Quantity of Waste Water Treated（10 000 t）	处理生活污水量 Household Waste Water Treated	污水再生利用量（万吨）Waste Water Recycled（10 000 t）	化学需氧量去除量（吨）Quantity of COD Removed（t）
	2005	1 287 059	1 084 444	55 174	3 745 933
	2006	1 631 339	1 304 230	75 062	4 830 318
	2007	1 903 890	1 524 408	80 244	5 620 200
	2008	2 373 468	1 893 619	113 503	6 491 598
	2009	2 812 731	2 248 605	137 430	7 596 131
	2010	**3 372 644**	**2 769 243**	**148 998**	**8 825 745**
北 京	Beijing	114 625	111 229	26 451	463 069
天 津	Tianjin	52 585	39 785	395	195 074
河 北	Hebei	171 325	124 735	14 576	553 550
山 西	Shanxi	56 577	51 999	9 770	177 010
内蒙古	Inner Mongolia	46 168	37 005	7 128	162 035
辽 宁	Liaoning	127 473	112 746	7 887	306 037
吉 林	Jilin	54 345	41 428	255	156 949
黑龙江	Heilongjiang	45 866	41 001	729	117 875
上 海	Shanghai	213 929	187 875	1 255	633 830
江 苏	Jiangsu	327 294	227 240	6 865	883 583
浙 江	Zhejiang	229 772	149 728	9 519	1 018 333
安 徽	Anhui	106 233	88 544	3 054	173 046
福 建	Fujian	93 648	79 589	181	300 038
江 西	Jiangxi	61 441	61 080	52	89 723
山 东	Shandong	267 296	198 721	10 578	890 089
河 南	Henan	180 617	160 460	14 902	470 683
湖 北	Hubei	139 369	130 299	890	202 632
湖 南	Hunan	111 254	108 703	2 582	165 000
广 东	Guangdong	516 341	391 447	2 789	769 439
广 西	Guangxi	56 440	53 501	881	87 290
海 南	Hainan	16 066	15 451	571	30 618
重 庆	Chongqing	678 99	669 26	724	183 041
四 川	Sichuan	109 163	101 034	39 89	225 971
贵 州	Guizhou	36 830	30 830	4 223	58 473
西 藏	Tibet	47 243	46 465	12 669	101 226
云 南	Yunnan	223	223	4	522
陕 西	Shaanxi	50 242	43 328	921	193 897
甘 肃	Gansu	21 326	20 600	1 173	49 837
青 海	Qinghai	5 580	5 580		14 160
宁 夏	Ningxia	14 721	13 220	1 464	48 580
新 疆	Xinjiang	30 753	28 472	2 521	104 136

各地区城市污水处理情况（三）

Urban Wastc Watcr Treatment by Region（3）

单位：吨　　　　（2010）　　　　（t）

年 份 地 区	Year Region	氨氮去除量 Quantity of Ammonia Nitrogen Removed	总磷去除量 Quantity of TP Removed	污泥产生量 Quantity of Sludge Generated	污泥排放量 Quantity of Sludge Discharged
	2005	214 482	46 972	11 130 226	85 605
	2006	288 951	48 807	11 038 162	304 155
	2007	347 945	52 338	15 174 266	964 356
	2008	419 332	67 467	14 524 944	217 434
	2009	518 774	94 114	16 549 011	69 804
	2010	**641 348**	**98 103**	**21 164 074**	**79 137**
北 京	Beijing	22 629	4 907	1 195 416	…
天 津	Tianjin	14 356	2 298	130 982	…
河 北	Hebei	44 079	5 729	1 370 513	9 148
山 西	Shanxi	13 302	2 308	345 144	6 981
内蒙古	Inner Mongolia	9 110	2 160	135 508	…
辽 宁	Liaoning	16 974	3 024	2 037 497	31 642
吉 林	Jilin	7 871	1 434	201 642	…
黑龙江	Heilongjiang	12 852	2 412	365 665	2 219
上 海	Shanghai	35 887	5 744	1 207 693	…
江 苏	Jiangsu	64 641	9 695	2 651 444	…
浙 江	Zhejiang	52 861	8 149	2 621 852	…
安 徽	Anhui	13 084	2 439	532 813	…
福 建	Fujian	17 633	2 104	502 182	…
江 西	Jiangxi	5 956	913	196 829	198
山 东	Shandong	78 772	9 136	1 465 193	69
河 南	Henan	39 544	5 020	1 063 522	98
湖 北	Hubei	17 885	2 935	434 068	17 965
湖 南	Hunan	15 245	1 652	402 781	298
广 东	Guangdong	68 717	12 740	2 167 288	4 566
广 西	Guangxi	8 657	1 207	158 805	…
海 南	Hainan	1 645	1 385	36 848	86
重 庆	Chongqing	18 944	2 263	316 034	…
四 川	Sichuan	23 252	2 321	501 791	233
贵 州	Guizhou	5 327	672	102 066	721
云 南	Yunnan	7 763	1 667	325 529	96
西 藏	Tibet	31	4	1 500	
陕 西	Shaanxi	14 565	2 142	382 773	…
甘 肃	Gansu	4 468	762	123 116	2 176
青 海	Qinghai	165	81	40 514	357
宁 夏	Ningxia	1 863	448	81 806	…
新 疆	Xinjiang	3 270	353	65 260	2 284

各地区生活及其他污染情况（一）
Discharge of Household Pollutants by Region（1）
（2010）

年 份 地 区	Year Region	城镇生活污水排放量（万吨） Amount of Household Waste Water Discharged（10 000 t）	城镇生活污水中化学需氧量排放量（吨） Amount of Household COD Discharged（t）	城镇生活污水中氨氮排放量（吨） Amount of Household Ammonia Nitrogen Discharged（t）	城镇生活污水处理率（%） Ratio of Household Waste Water Treated（%）
	2005	2 813 968	8 594 387	972 743	37.4
	2006	2 966 340	8 859 086	988 656	43.8
	2007	3 102 001	8 707 531	982 601	49.1
	2008	3 300 290	8 631 216	972 836	57.4
	2009	3 552 261	8 378 613	952 552	63.3
	2010	**3 797 830**	**8 032 914**	**930 123**	**72.9**
北 京	Beijing	128 217	87 115	11 734	86.8
天 津	Tianjin	48 516	109 751	16 577	82.0
河 北	Hebei	148 311	328 089	36 347	84.1
山 西	Shanxi	68 418	195 439	29 658	76.0
内蒙古	Inner Mongolia	53 012	183 810	33 684	69.8
辽 宁	Liaoning	146 668	339 846	47 022	76.9
吉 林	Jilin	75 775	221 087	26 787	54.7
黑龙江	Heilongjiang	79 654	331 954	38 107	51.5
上 海	Shanghai	211 554	198 225	24 320	88.8
江 苏	Jiangsu	291 740	531 748	48 006	77.9
浙 江	Zhejiang	177 402	242 694	25 689	84.4
安 徽	Anhui	113 729	296 274	32 247	77.9
福 建	Fujian	114 334	289 695	23 157	69.6
江 西	Jiangxi	88 135	313 299	25 893	69.3
山 东	Shandong	228 115	325 405	51 042	87.1
河 南	Henan	208 273	324 123	49 350	77.0
湖 北	Hubei	176 162	407 076	46 486	74.0
湖 南	Hunan	172 505	610 247	57 832	63.0
广 东	Guangdong	535 947	623 962	96 342	73.0
广 西	Guangxi	147 419	444 177	32 919	36.3
海 南	Hainan	30 907	83 108	7 203	50.0
重 庆	Chongqing	82 933	147 983	19 723	80.7
四 川	Sichuan	162 651	490 013	44 190	62.1
贵 州	Guizhou	46 693	191 899	15 773	66.0
云 南	Yunnan	61 066	179 004	16 967	76.1
西 藏	Tibet	3 089	27 664	1 770	7.2
陕 西	Shaanxi	70 186	186 260	24 953	61.7
甘 肃	Gansu	35 889	122 509	15 961	57.4
青 海	Qinghai	13 578	38 386	5 986	41.1
宁 夏	Ningxia	18 676	28 889	4 036	70.8
新 疆	Xinjiang	58 277	133 182	20 362	48.9

各地区生活及其他污染情况（二）

Discharge of Household Pollutants by Region（2）

（2010）

年份 地区	Year Region	生活及其他煤炭消费量（万吨）Amount of Household Coal Consumed（10 000 t）	生活及其他二氧化硫排放量（吨）Amount of Household Sulphur Dioxide Emission（t）	生活及其他烟尘排放量（吨）Amount of Household Soot Emission（t）	生活及其他氮氧化物排放量（吨）Amount of Household Nitrogen Oxide Emission（t）	公路交通氮氧化物排放量 Amount of Road Traffic Nitrogen Oxide Emission
	2005	21 741	3 808 817	2 335 820	—	—
	2006	20 376	3 511 888	2 226 552	3 877 733	2 843 207
	2007	18 920	3 281 084	2 154 937	3 820 482	2 767 297
	2008	19 328	3 298 598	2 308 793	3 739 633	2 821 862
	2009	19 466	3 484 594	2 432 899	4 078 877	3 170 497
	2010	**18 369**	**3 207 293**	**2 258 855**	**3 867 538**	**2 906 008**
北京	Beijing	403	58 206	27 491	114 187	90 000
天津	Tianjin	146	17 530	10 721	9 627	5 931
河北	Hebei	1 539	239 603	177 092	157 244	76 050
山西	Shanxi	1 621	102 086	188 540	133 688	85 853
内蒙古	Inner Mongolia	1 358	201 116	171 042	156 625	90 062
辽宁	Liaoning	1 011	162 797	235 343	104 656	53 466
吉林	Jilin	754	55 674	91 120	49 338	30 235
黑龙江	Heilongjiang	1 103	73 113	124 786	115 943	55 532
上海	Shanghai	37	136 624	60 348	185 078	109 867
江苏	Jiangsu	333	48 043	36 186	297 353	270 937
浙江	Zhejiang	160	24 453	9 044	244 105	229 456
安徽	Anhui	384	48 164	47 958	123 219	93 602
福建	Fujian	129	17 820	38 747	113 255	91 393
江西	Jiangxi	368	86 118	25 384	118 173	92 630
山东	Shandong	1 221	154 944	100 464	240 733	193 955
河南	Henan	1 311	175 827	72 831	212 153	144 653
湖北	Hubei	498	116 538	47 962	129 769	107 463
湖南	Hunan	683	173 928	77 271	94 436	63 745
广东	Guangdong	311	61 420	57 731	466 480	436 776
广西	Guangxi	197	55 856	11 431	110 062	96 906
海南	Hainan	14	643	1 582	24 077	22 891
重庆	Chongqing	351	146 657	105 570	68 917	48 126
四川	Sichuan	615	193 320	81 534	80 266	55 422
贵州	Guizhou	1 266	511 051	138 890	34 086	9 147
云南	Yunnan	439	61 148	49 063	105 333	83 729
西藏	Tibet	7	2 901	1 674	118	
陕西	Shaanxi	498	71 689	48 137	48 940	25 818
甘肃	Gansu	488	99 321	64 976	67 800	46 081
青海	Qinghai	123	10 282	24 633	32 727	22 578
宁夏	Ningxia	234	30 372	35 712	26 523	21 196
新疆	Xinjiang	768	70 049	95 593	202 627	152 508

各地区环境污染治理投资情况

Investment in the Treatment of Environmental Pollution by Region

（2010）

年份 地区	Year Region	环境污染治理投资总额（亿元） Total Investment in Treatment of Environmental Pollution （100 million yuan）	城市环境基础设施建设投资 Investment in Urban Environment Infrastructure Facilities	工业污染源治理投资 Investment in Treatment of Industrial Pollution Sources	建设项目“三同时”环保投资 Investment in Environment Components for New Construction Projects	环境污染治理投资总额占GDP比重（%） Percentage of Treatment Investment of Environmental Pollution in GDP （%）
	2005	2 388.0	1 289.7	458.2	640.1	1.31
	2006	2 566.0	1 314.9	483.9	767.2	1.22
	2007	3 387.6	1 467.8	552.4	1 367.4	1.36
	2008	4 490.3	1 801.0	542.6	2 146.7	1.49
	2009	4 525.2	2 512.0	442.5	1 570.7	1.35
	2010	**6 654.2**	**4 224.2**	**397.0**	**2 033.0**	**1.67**
国家级	State Level	386.0	0	0	386.0	
北 京	Beijing	231.4	173.0	1.9	56.5	1.68
天 津	Tianjin	109.7	65.8	16.5	27.4	1.20
河 北	Hebei	370.9	279.9	10.9	80.1	1.84
山 西	Shanxi	206.9	86.1	28.0	92.8	2.28
内蒙古	Inner Mongolia	238.9	175.6	13.2	50.1	2.05
辽 宁	Liaoning	206.5	141.6	14.8	50.1	1.13
吉 林	Jilin	124.2	70.8	6.3	47.1	1.45
黑龙江	Heilongjiang	131.3	72.8	4.9	53.5	1.28
上 海	Shanghai	134.0	87.1	9.4	37.4	0.79
江 苏	Jiangsu	466.4	300.0	18.6	147.8	1.14
浙 江	Zhejiang	333.7	95.2	12.0	226.5	1.23
安 徽	Anhui	179.9	132.4	5.9	41.6	1.47
福 建	Fujian	129.7	78.0	15.3	36.3	0.90
江 西	Jiangxi	156.5	125.5	6.4	24.6	1.66
山 东	Shandong	483.9	284.6	45.7	153.6	1.23
河 南	Henan	132.2	71.0	12.5	48.7	0.58
湖 北	Hubei	146.8	89.9	27.7	29.2	0.93
湖 南	Hunan	106.6	62.1	13.8	30.7	0.67
广 东	Guangdong	1 416.2	1 262.7	31.1	122.4	3.11
广 西	Guangxi	164.1	99.0	9.3	55.8	1.73
海 南	Hainan	23.6	11.6	0.4	11.6	1.15
重 庆	Chongqing	176.3	124.6	7.8	43.9	2.23
四 川	Sichuan	89.0	44.3	7.2	37.5	0.53
贵 州	Guizhou	30.0	6.7	6.8	16.5	0.65
云 南	Yunnan	106.2	59.4	10.6	36.1	1.47
西 藏	Tibet	0.3	0.3	0	0	0.06
陕 西	Shaanxi	179.2	109.1	33.7	36.5	1.79
甘 肃	Gansu	63.9	42.1	14.6	7.1	1.55
青 海	Qinghai	17.0	6.3	1.0	9.7	1.26
宁 夏	Ningxia	34.5	20.1	4.1	10.3	2.10
新 疆	Xinjiang	78.4	46.2	6.7	25.5	1.45

各地区城市环境基础设施建设投资情况

Investment in Urban Envirsonment Infrastructure by Region

单位：亿元　　　　　　　　（2010）　　　　　　　　（100 million yuan）

年份 地区	Year Region	投资总额 Total Investment	燃气 Gas Supply	集中供热 General Heating	排水 Sewerage Projects	园林绿化 Gardening & Greening	市容环境卫生 Sanitation
	2005	1 289.70	142.37	220.19	368.03	411.32	147.79
	2006	1 314.92	155.05	223.59	331.52	429.01	175.75
	2007	1 467.81	160.37	230.03	410.01	525.56	141.84
	2008	1 801.03	163.54	269.68	495.96	649.82	222.03
	2009	2 511.97	182.17	368.67	729.80	914.86	316.47
	2010	**4 224.2**	**290.8**	**433.2**	**901.6**	**2 297.0**	**301.6**
北　京	Beijing	173.0	18.3	49.7	17.3	65.5	22.3
天　津	Tianjin	65.8	13.3	7.0	24.9	14.0	6.4
河　北	Hebei	279.9	24.0	76.9	53.9	113.9	11.2
山　西	Shanxi	86.1	9.7	22.5	20.2	29.7	4.1
内蒙古	Inner Mongolia	175.6	8.4	37.5	39.9	78.5	11.3
辽　宁	Liaoning	141.6	12.4	71.8	13.6	28.9	15.0
吉　林	Jilin	70.8	10.8	29.5	11.4	12.1	7.1
黑龙江	Heilongjiang	72.8	5.3	24.0	22.8	17.3	3.4
上　海	Shanghai	87.1	18.8	0	34.3	28.2	5.8
江　苏	Jiangsu	300.0	20.9	0.2	84.8	179.2	14.9
浙　江	Zhejiang	95.2	11.9	0.5	34.4	38.9	9.5
安　徽	Anhui	132.4	8.3	1.2	24.1	90.9	8.0
福　建	Fujian	78.0	6.1	0	14.5	36.1	21.3
江　西	Jiangxi	125.5	6.7	0	17.8	94.9	6.2
山　东	Shandong	284.6	35.5	64.3	59.3	108.3	17.1
河　南	Henan	71.0	8.6	14.1	20.1	25.3	2.9
湖　北	Hubei	89.9	14.4	0.1	24.4	31.2	19.8
湖　南	Hunan	62.1	6.4	0	19.5	22.8	13.3
广　东	Guangdong	1 262.7	11.0	0	212.4	980.5	58.9
广　西	Guangxi	99.0	3.1	0	39.3	48.2	8.4
海　南	Hainan	11.6	0.8	0	6.3	2.7	1.8
重　庆	Chongqing	124.6	8.6	1.6	7.2	106.0	1.3
四　川	Sichuan	44.3	4.7	0	12.2	22.8	4.6
贵　州	Guizhou	6.7	0.6	0	2.6	2.2	1.4
云　南	Yunnan	59.4	1.1	0	43.6	10.7	4.0
西　藏	Tibet	0.3	0	0	0	0.3	0
陕　西	Shaanxi	109.1	3.4	7.9	15.0	75.8	6.8
甘　肃	Gansu	42.1	1.2	6.6	10.1	15.3	8.9
青　海	Qinghai	6.3	0.4	0	3.3	1.8	0.9
宁　夏	Ningxia	20.1	10.0	3.3	1.7	4.3	0.8
新　疆	Xinjiang	46.2	5.8	14.4	11.0	10.9	4.1

3

重点城市环境统计

ZHONGDIAN CHENGSHI HUANJING TONGJI

ANNUAL STATISTIC REPORT ON ENVIRONMENT IN CHINA

2010

重点城市工业废水排放及处理情况（一）

（2010）

城市名称	工业废水排放量（万吨）	直接排入海的	工业废水排放达标量（万吨）	废水治理设施数（套）	废水治理设施治理能力（吨/日）	废水治理设施运行费用（万元）
总计	**1 278 820**	**50 836**	**1 243 046**	**42 970**	**143 722 043.7**	**3 337 851.0**
北京	8 198	0	8 096	481	1 698 186.5	68 058.5
天津	19 680	469	19 671	912	2 730 061.0	69 321.1
石家庄	19 254	0	19 114	905	2 691 091.3	60 323.7
唐山	18 170	80	17 852	654	14 650 006.2	160 941.8
秦皇岛	5 608	177	5 608	207	716 842.6	12 845.3
邯郸	7 686	0	7 464	265	1 319 918.0	49 923.8
保定	17 866	0	17 549	517	928 772.9	18 830.1
太原	2 557	0	2 490	230	1 086 529.2	50 670.8
大同	9 024	0	8 724	189	1 146 292.4	12 809.6
阳泉	1 134	0	1 134	79	191 700.4	8 468.9
长治	5 241	0	5 240	382	715 500.8	20 328.4
临汾	3 068	0	2 879	145	978 284.6	27 130.7
呼和浩特	2 637	0	2 637	56	169 976.5	7 957.1
包头	4 833	0	4 749	101	2 477 655.0	27 928.1
赤峰	2 723	0	2 656	112	308 572.0	5 335.9
沈阳	6 141	5	5 913	341	232 010.4	18 178.9
大连	27 421	22 754	26 076	294	240 075.0	4 205.9
鞍山	5 548	0	5 319	144	5 793 177.6	52 166.2
抚顺	3 528	0	3 419	127	668 530.0	14 208.7
本溪	2 583		2 154	119	2 844 572.0	24 083.7
锦州	3 859	87	3 648	98	222 002.0	28 110.8
长春	5 815		5 597	116	201 202.5	5 809.1
吉林	15 584	0	15 104	129	910 384.2	27 977.3
哈尔滨	3 283	0	3 216	189	224 671.4	19 656.2
齐齐哈尔	6 090	0	5 464	105	421 199.5	5 201.9
大庆	8 788	0	8 787	258	2 649 467.1	91 312.8
牡丹江	2 793		2 762	122	139 613.4	3 512.4
上海	36 696	2 009	35 973	1 749	5 101 233.8	157 318.3
南京	33 784	0	32 164	708	3 536 001.5	97 166.0
无锡	35 843	0	35 585	919	3 291 807.5	59 992.7
徐州	12 178	0	11 371	278	1 318 140.0	20 971.4
常州	37 715	0	37 708	727	953 378.1	34 505.3
苏州	62 338	150	62 021	1 386	3 416 437.9	116 254.5
南通	15 708	0	15 606	833	726 680.9	33 795.6
连云港	3 538	351	3 473	203	583 608.0	14 468.7
扬州	9 059	…	8 908	277	285 011.4	19 788.4
杭州	80 468	1 240	77 957	1 161	5 380 572.5	92 510.4

重点城市工业废水排放及处理情况（一）（续表）

（2010）

城市名称	工业废水排放量（万吨）	直接排入海的	工业废水排放达标量（万吨）	废水治理设施数（套）	废水治理设施治理能力（吨/日）	废水治理设施运行费用（万元）
宁波	18 967	8 328	18 108	884	1 100 554.3	80 042.0
温州	17 008	239	16 311	1 133	838 052.0	28 737.3
嘉兴	19 812	493	19 533	796	1 540 308.5	64 347.9
湖州	10 888	0	10 459	514	472 759.9	21 427.4
绍兴	30 230	0	29 744	856	1 387 140.5	78 655.9
台州	5 709	678	5 153	1 144	352 944.6	18 742.5
合肥	3 290	0	3 167	196	471 511.5	9 428.6
芜湖	4 306	0	4 276	146	261 638.0	8 074.5
马鞍山	5 563	0	5 470	124	3 625 674.0	56 108.7
福州	4 933	155	4 684	410	1 428 494.0	21 698.0
厦门	4 457	228	4 456	303	271 040.6	10 767.5
泉州	19 544	2 534	19 513	662	1 178 791.0	26 097.5
南昌	10 534	0	9 928	219	725 206.3	13 041.5
九江	8 784	0	8 086	173	600 265.6	28 733.7
济南	5 594	0	5 576	230	1 612 956.0	25 586.6
青岛	10 800	2 714	10 591	504	827 313.4	33 013.4
淄博	21 212	0	21 212	599	1 414 992.0	64 994.0
枣庄	16 185	0	16 067	178	630 142.0	17 894.2
烟台	8 386	2 424	8 386	464	962 806.3	20 583.6
潍坊	21 496	24	21 210	583	3 271 995.7	54 722.9
济宁	16 212	0	16 015	331	1 378 919.4	35 607.5
泰安	4 579	0	4 492	246	1 539 129.6	7 888.1
威海	2 872	527	2 872	164	150 640.0	6 357.1
日照	9 977	3 693	9 977	199	394 125.5	12 480.4
郑州	13 484	0	13 234	368	809 321.0	16 147.2
开封	5 921	0	5 825	84	121 230.0	8 613.5
洛阳	8 036	0	8 035	277	926 689.5	9 352.0
平顶山	6 107	0	5 926	167	615 281.0	13 792.9
安阳	15 585		15 305	313	1 203 838.2	53 472.1
焦作	21 807		21 383	260	1 247 711.0	19 703.3
武汉	22 465	0	22 286	251	3 274 091.6	31 663.6
宜昌	14 484	0	14 445	375	2 069 331.0	14 048.8
荆州	6 104		5 811	113	221 256.1	6 334.2
长沙	4 343	24	3 961	266	161 022.5	4 411.0
株洲	7 900	0	7 605	483	594 600.9	13 422.6
湘潭	7 558	0	7 105	205	1 931 002.0	14 731.2
岳阳	12 245	0	11 864	217	726 986.0	9 632.1
常德	12 575	0	12 502	174	537 031.0	14 465.1

重点城市工业废水排放及处理情况（一）（续表）

（2010）

城市名称	工业废水排放量（万吨）	直接排入海的	工业废水排放达标量（万吨）	废水治理设施数（套）	废水治理设施治理能力（吨/日）	废水治理设施运行费用（万元）
张家界	429	0	411	13	7 717.0	117.7
广州	23 586	394	22 783	1 012	2 550 760.5	62 139.3
韶关	9 523	0	8 306	296	1 425 928.5	45 008.1
深圳	9 001	563	8 676	1 530	524 373.6	75 980.6
珠海	6 124	65	6 004	377	237 100.0	18 701.4
汕头	6 150	163	5 219	352	445 706.0	12 630.8
佛山	26 729	0	25 506	809	1 481 257.4	48 350.6
湛江	5 355	206	4 301	268	673 450.2	6 692.7
中山	11 374	0	11 034	414	697 919.9	25 820.3
南宁	12 426	0	11 866	351	1 803 212.4	18 092.6
柳州	26 009	0	25 772	301	6 713 280.0	47 071.8
桂林	3 527	0	3 381	263	399 633.0	4 661.4
北海	1 369	62	1 308	56	205 718.0	1 806.8
海口	513	0	513	33	53 886.0	1 509.1
三亚	23	0	23	7	3 061.0	876.6
重庆	45 180	0	42 798	1 498	2 206 841.4	56 491.5
成都	12 259	0	12 234	1 203	1 646 017.4	85 280.5
攀枝花	1 904	0	1 851	179	1 174 704.0	64 124.3
泸州	6 255	0	5 963	197	467 987.7	28 727.2
绵阳	9 492		9 436	248	395 699.6	7 315.7
宜宾	8 471	0	8 431	251	532 512.1	15 464.8
贵阳	2 380	0	2 280	298	943 588.6	11 220.1
遵义	2 431	0	2 265	155	166 492.6	3 652.1
昆明	3 225	0	3 225	517	1 352 827.3	21 949.8
曲靖	2 976	0	2 956	311	1 059 128.9	21 817.6
拉萨	492		217	13	7 551.0	158.2
西安	12 330	0	11 800	346	442 281.0	12 907.1
铜川	328	0	326	40	38 436.0	1 459.7
宝鸡	11 521	0	11 482	206	736 863.0	6 193.0
咸阳	6 882	0	6 882	205	661 315.2	9 245.2
延安	1 390	0	1 350	443	139 561.0	29 317.4
兰州	2 529	0	2 406	81	632 358.0	9 462.7
金昌	1 614	0	1 313	88	108 215.0	6 821.1
西宁	4 052		3 607	70	370 608.0	4 927.4
银川	5 894	0	5 864	82	235 883.7	12 738.3
石嘴山	1 864		938	154	544 331.6	7 585.2
乌鲁木齐	5 174	0	4 039	82	711 996.5	10 621.8
克拉玛依	1 649	0	1 616	93	97 882.0	24 021.1

重点城市工业废水排放及处理情况（二）

（2010）

单位：吨

城市名称	工业废水中污染物排放量					
	汞	镉	六价铬	铅	砷	挥发酚
总计	**0.468**	**5.029**	**38.391**	**24.886**	**14.576**	**219.467**
北京			0.022	0.016	0	0.110
天津	…	…	0.078	0.002	0.022	1.519
石家庄	…	…	0.218	…	…	0.409
唐山	…	…	0.263	0.063	…	1.987
秦皇岛			0.014			
邯郸	…	…	0.001	1.325	…	1.739
保定			0.085	0.291		0.020
太原	0.003	0.014	0.120	0.137	0.015	4.791
大同	0.010	0.001	0.055	0.045	0.009	2.251
阳泉	…	…	…	…	…	0
长治	0.001	…	0.023	…	…	0.607
临汾	…	…	0.001	…	…	2.619
呼和浩特	…	…	…	…	…	0.236
包头			0.334	0.475		1.845
赤峰	…	0.015	0.019	0.014	0.023	0.121
沈阳	…	0.004	0.081	0.110	…	0.360
大连	…	…	0.053	…	…	9.420
鞍山						0.336
抚顺		0.025	0.010	0.208	0.096	1.149
本溪						0.438
锦州			0.022			0.330
长春			0.014			0.496
吉林	…	0.004	0.571	…	…	45.553
哈尔滨	…	0.030	0.014	0.029	…	16.137
齐齐哈尔	…	…	0.207	…	…	1.125
大庆	0.001	0.026	0.022	0.102	0.223	2.164
牡丹江	…	…		…	0.003	0.224
上海	0.002	0.010	0.722	0.078	0.013	4.930
南京	0.001	0.032	0.916	0.849	0.017	4.505
无锡	…	…	0.163	0.001	…	0.044
徐州	…	…	…	0.001	…	0.336
常州	…	…	2.509	…	…	2.188
苏州	…	…	2.103	1.159	0.040	2.814
南通	0	0.007	0.486	0.006	…	4.116
连云港	…	…	0.007	…	…	0.023
扬州	…	…	0.188	0.263	…	2.388
杭州	…	…	0.270	0.130	0.008	0.136

重点城市工业废水排放及处理情况（二）（续表）

（2010）

单位：吨

城市名称	工业废水中污染物排放量					
	汞	镉	六价铬	铅	砷	挥发酚
宁　波	…	…	2.267	0.065	0.006	0.387
温　州	…	…	3.217	0.144	0.001	0.001
嘉　兴	…	…	0.039	0.030	…	0.023
湖　州		…	0.095			0.009
绍　兴	…	…	0.133	0.014	…	0.056
台　州		…	0.197	0.057		…
合　肥	…	…	0.004	…	…	0.057
芜　湖	0.002	…	0.026	0.104	0.023	0.788
马鞍山	…	…	…	…	…	2.821
福　州	…	…	0.193	0.120	0.004	0.071
厦　门	…	0.001	0.180	0.019	0.024	0.030
泉　州	…	0.006	6.762	0.083	0.166	8.070
南　昌	…	0.018	0.308	0.003	…	4.928
九　江	…	0.306	0.330	1.118	0.448	5.247
济　南	…	…	0.178	0	0	1.023
青　岛	…	…	0.022	0.017	…	0.304
淄　博			…	…	…	3.188
枣　庄			0			
烟　台	…	0.001	0.034	1.361	0.018	0.001
潍　坊	…	…	0.080	…	…	0.821
济　宁			0.003			0.082
泰　安	0.001	…	0.015	0.011	0.094	0.156
威　海			0.059	0.005		
日　照	…	…	0.020	…	…	0.018
郑　州			0.008		…	0.020
开　封		…	0	0.001	…	0.008
洛　阳	…	…	0.045	…	0.015	0.030
平顶山			0.026			1.022
安　阳			0.216			1.580
焦　作			3.108			13.090
武　汉	…	0.002	0.476	0.092	0.168	4.499
宜　昌	…	…	0.009	…	…	1.692
荆　州			1.073			6.266
长　沙	…	0.097	0.126	0.056	0.336	0.119
株　洲	0.078	0.919	0.049	1.064	7.264	0.437
湘　潭	0.290	0.790	1.079	2.197	0.930	1.626
岳　阳	0.027	0.002	0.008	0.168	0.380	1.689
常　德	…	…	0.020	0.002	…	3.006

重点城市工业废水排放及处理情况（二）（续表）

（2010）

单位：吨

城市名称	工业废水中污染物排放量					
	汞	镉	六价铬	铅	砷	挥发酚
张家界	…	…	…	…	…	…
广州	…	0.011	0.270	0.053	0.006	0.394
韶关	0.005	0.599	0.344	4.007	1.031	1.220
深圳	0.004	0.003	0.228	0.139	0.001	0.037
珠海	…	0.031	0.209	0.040	…	0.071
汕头	…	…	0.021	0.009	…	…
佛山	…	0.058	0.357	0.236	0.054	0.911
湛江	…	0.800	0.005	…	0.002	0.001
中山	…	…	0.517	0.019	…	0.234
南宁	…	…	0.309	…	0.001	0.358
柳州	0.002	0.117	0.065	0.910	0.071	0.175
桂林	…	0.021	0.065	0.260	0.037	0.109
北海						0.006
海口				0.025		0
三亚						
重庆	…	0.069	3.582	0.094	…	1.350
成都		0.013	0.342	0.884	0.001	0.092
攀枝花	…	0.001	0.015	0.002	0.030	0.100
泸州			0.230	0		0.127
绵阳		0.002	0.091	0.003		0.079
宜宾	0.005	…	0.003	0.001	0.012	0
贵阳	…	0.002	0.067	0.080	0.001	0.032
遵义	0		0.002	…	…	…
昆明	0.022	0.031	0.013	1.679	0.465	0.232
曲靖		0.038	0.007	0.258	0.044	0.280
拉萨						
西安	…	0.002	0.094	…	…	0.012
铜川						
宝鸡	…	0.538	0.148	0.987	1.649	0.097
咸阳	0.007	…	0.911	0.209	0.010	0.382
延安	0.003	0.009	0.033	0.253	0.032	0.857
兰州	…	0.043	0.115	0.046	0.120	0.170
金昌	0.001	0.132	0.116	0.254	0.485	0.181
西宁		0.119	0.020	0.192	0.111	0.394
银川		0.030	0.002	0.005	0.038	10.568
石嘴山						0.240
乌鲁木齐	0.005	0.053	0.217	2.208	0.030	12.451
克拉玛依						0.695

重点城市工业废水排放及处理情况（三）

（2010）

单位：吨

城市名称	工业废水中污染物排放量			
	氰化物	化学需氧量	石油类	氨氮
总　计	**93.7**	**1 906 848.1**	**6 128.3**	**125 586.4**
北　京	0.1	4 882.3	57.6	392.1
天　津	0.1	22 217.7	97.2	3 196.8
石家庄	0.3	40 855.2	10.3	4 017.6
唐　山	1.2	49 941.3	105.1	1 677.8
秦皇岛	…	6 687.7	16.4	304.3
邯　郸	0.9	11 239.0	123.1	631.6
保　定	…	18 322.7	1.6	1 270.0
太　原	0.6	4 721.6	17.4	582.2
大　同	…	24 972.4	45.0	3 658.9
阳　泉	…	1 040.5	…	6.3
长　治	1.9	10 119.2	16.8	692.3
临　汾	1.2	7 203.0	110.8	792.5
呼和浩特	…	2 853.4	9.0	368.9
包　头	1.4	4 608.9	40.3	2 022.3
赤　峰	…	5 885.0	0.7	65.4
沈　阳	0.2	10 732.2	290.5	1 043.0
大　连	0.4	14 992.5	243.2	544.1
鞍　山	1.8	4 948.3	19.3	314.1
抚　顺	0.4	2 548.9	41.9	191.1
本　溪	0.4	13 546.1	45.6	163.3
锦　州	0	32 357.8	64.7	167.9
长　春	0.5	31 815.0	207.5	485.5
吉　林	4.1	26 317.0	306.2	833.1
哈尔滨	…	10 914.0	26.0	344.2
齐齐哈尔	…	10 211.3	91.0	762.1
大　庆	0.4	10 963.1	152.0	947.3
牡丹江	1.0	18 239.1	3.1	373.3
上　海	4.7	21 574.9	415.5	3 192.4
南　京	2.5	26 119.8	252.6	1 233.2
无　锡	0.1	28 943.8	71.7	1 277.6
徐　州	0.7	14 533.8	65.7	651.4
常　州	0.6	24 826.9	55.5	864.0
苏　州	2.3	59 366.9	92.8	3 074.6
南　通	0.7	23 309.1	59.7	2 147.4
连云港	…	4 206.1	0.4	299.1
扬　州	0.2	17 003.9	33.4	709.8
杭　州	0.1	87 312.0	18.8	1 558.8

重点城市工业废水排放及处理情况（三）（续表）

（2010）

单位：吨

城市名称	工业废水中污染物排放量			
	氰化物	化学需氧量	石油类	氨氮
宁　波	3.1	20 664.1	52.9	1 366.4
温　州	3.1	26 249.0	9.4	1 285.3
嘉　兴	…	17 957.7	0.5	2 691.6
湖　州	…	9 078.3	0.2	554.9
绍　兴	…	33 547.0	6.7	2 845.4
台　州	0.1	9 742.9	6.7	893.6
合　肥	…	4 008.7	12.2	251.3
芜　湖	…	13 694.2	2.3	119.7
马鞍山	1.4	5 590.6	63.1	319.9
福　州	0.1	4 510.8	31.9	513.2
厦　门	0.2	2 316.9	12.9	268.7
泉　州	1.2	25 725.6	181.3	1 902.6
南　昌	2.4	24 141.7	58.3	1 341.8
九　江	0.1	7 339.4	71.1	651.3
济　南	0.3	8 816.4	40.0	428.2
青　岛	…	14 758.3	38.3	786.3
淄　博	…	27 839.2	121.1	2 206.8
枣　庄		17 153.1	0.2	801.9
烟　台	…	15 254.5	5.2	761.6
潍　坊	0.1	22 481.1	27.8	1 793.2
济　宁	0.1	12 773.0	14.7	326.9
泰　安	0.2	6 607.8	7.3	232.5
威　海		3 699.1	0.1	180.8
日　照	…	12 015.0	5.3	603.6
郑　州	…	13 251.9	30.5	431.7
开　封		8 272.4	1.2	1 530.5
洛　阳	…	5 939.0	37.2	472.2
平顶山	0.8	9 177.8	29.6	709.2
安　阳	0.9	42 276.2	229.2	772.3
焦　作	5.4	37 906.0	49.0	3 015.0
武　汉	9.0	19 645.0	179.5	1 299.1
宜　昌	1.5	13 631.0	34.3	1 960.1
荆　州	0.2	23 610.5	41.3	895.2
长　沙	…	4 903.9	21.3	256.7
株　洲	2.5	9 865.9	112.5	1 197.7
湘　潭	1.4	19 606.5	50.3	1 847.7
岳　阳	8.6	19 964.1	57.6	2 317.9
常　德	0.6	30 564.4	8.8	932.3

重点城市工业废水排放及处理情况（三）（续表）

（2010）

单位：吨

城市名称	工业废水中污染物排放量			
	氰化物	化学需氧量	石油类	氨氮
张家界	…	914.9	0.4	54.3
广州	0.4	21 026.8	49.9	1 006.3
韶关	1.9	9 460.7	6.0	484.7
深圳	2.4	4 295.8	5.4	402.1
珠海	0.1	7 125.0	32.1	541.7
汕头	…	9 982.5	2.6	533.0
佛山	0.6	23 910.9	81.6	1 032.7
湛江	…	13 018.1	6.8	529.8
中山	0.7	11 276.9	9.9	484.0
南宁	…	66 507.1	6.1	1 183.8
柳州	5.9	52 893.4	135.1	2 958.1
桂林	…	10 041.0	2.2	268.2
北海		14 631.8	0.1	704.0
海口		350.4	0.8	11.1
三亚		14.2	0.0	1.9
重庆	0.4	86 549.0	221.2	5 440.1
成都	0.4	47 468.1	39.5	9 306.9
攀枝花	…	6 054.0	2.5	72.5
泸州		14 834.1	40.5	677.9
绵阳	0.1	9 263.5	120.5	668.8
宜宾	0.3	37 572.5	22.5	551.4
贵阳	…	1 707.0	9.1	82.3
遵义	…	3 900.5	33.9	124.2
昆明	0.2	3 127.3	18.4	165.1
曲靖	0.4	5 568.4	3.0	1 135.4
拉萨		411.1	…	2.7
西安	…	35 526.0	67.3	2 412.9
铜川		725.5	22.4	7.6
宝鸡	…	21 066.6	16.9	898.8
咸阳	…	16 207.2	76.7	622.0
延安	0.1	3 527.5	30.0	105.3
兰州	0.1	3 203.4	28.4	209.8
金昌	…	4 874.3	16.5	5 476.1
西宁	…	15 934.1	29.8	1 108.1
银川	0.5	14 834.3	25.5	1 844.5
石嘴山	0.3	5 257.3	18.9	1 592.6
乌鲁木齐	6.5	7 198.7	40.3	1 203.8
克拉玛依		1 676.7	45.0	60.6

重点城市工业废水排放及处理情况（四）

（2010）

单位：吨

城市名称	工业废水中污染物去除量				
	氰化物	化学需氧量	石油类	氨氮	挥发酚
总　计	**5 132.8**	**7 447 552.6**	**285 464.9**	**462 512.5**	**47 351.3**
北　京	47.0	37 491.1	1 336.1	1 252.6	651.3
天　津	0.2	52 377.8	464.7	1 057.5	37.5
石家庄	68.1	179 168.9	14.6	11 313.2	0.6
唐　山	63.0	123 200.4	10 089.2	1 769.7	2 835.4
秦皇岛	…	52 489.5	99.6	171.9	
邯　郸	249.0	19 114.5	27.4	1 435.1	1 842.8
保　定	…	65 893.1	5.0	1 830.8	…
太　原	62.7	28 688.2	224.1	1 033.4	911.3
大　同	2.8	9 434.2	111.0	272.0	8.5
阳　泉	…	892.6	0.0	0.1	…
长　治	8.5	29 900.0	62.4	1 014.5	79.2
临　汾	950.4	32 054.0	923.6	9 482.7	2 638.9
呼和浩特	…	46 059.0	61.0	1 736.2	17.0
包　头	192.6	23 715.1	603.2	7 128.4	2 921.8
赤　峰		6 256.0	11.3	4.7	
沈　阳	3.1	51 345.1	33.4	322.4	0.6
大　连	…	48 493.6	437.2	20 621.0	…
鞍　山		18 629.6	27.0	66.6	…
抚　顺	3.6	7 161.1	495.6	96.2	94.0
本　溪	24.6	15 434.1	30.0	244.6	852.0
锦　州	0.5	44 742.0	2 687.8	200.2	7.4
长　春	9.4	23 878.0	356.1	453.5	329.8
吉　林	2.0	80 398.0	880.6	786.0	202.7
哈尔滨	0.2	127 163.9	243.2	32 929.7	3 741.6
齐齐哈尔	0	13 900.6	88.5	968.0	290.8
大　庆	74.3	86 728.6	153 163.8	4 261.8	466.0
牡丹江	1.2	13 300.1	26.6	3 632.3	91.8
上　海	44.4	260 218.8	14 537.9	7 318.8	777.7
南　京	93.8	100 796.8	20 396.9	12 732.4	2 649.9
无　锡	1.1	150 420.6	4 111.1	5 309.2	78.0
徐　州	244.3	135 691.2	515.6	3 681.8	888.1
常　州	4.1	100 061.4	123.7	5 533.0	7.8
苏　州	36.0	430 120.6	250.4	12 613.0	145.4
南　通	0.5	68 179.7	135.2	1 770.5	2.6
连云港	…	33 948.8	4.9	609.4	125.6
扬　州	10.7	54 529.2	622.5	485.0	86.3
杭　州	3.1	357 995.0	157.9	2 508.1	189.1

重点城市工业废水排放及处理情况（四）（续表）

（2010）

单位：吨

城市名称	工业废水中污染物去除量				
	氰化物	化学需氧量	石油类	氨氮	挥发酚
宁　波	82.4	252 560.6	25 271.5	34 467.6	99.3
温　州	687.6	177 203.2	48.2	4 770.8	…
嘉　兴	0.1	189 204.0	0.2	3 773.9	33.9
湖　州	1.0	67 541.4	1.3	2 085.0	…
绍　兴	0	243 304.3	9.3	9 446.1	2.2
台　州	6.9	50 253.9	28.2	15 674.4	160.0
合　肥	3.8	13 733.9	977.6	485.4	102.0
芜　湖	28.8	12 017.2	30.9	71.9	140.5
马鞍山	288.6	67 418.8	3 645.5	1 900.3	1 933.4
福　州	35.7	83 839.3	790.3	1 446.4	153.9
厦　门	24.1	52 579.2	258.0	699.1	…
泉　州	39.5	289 293.9	4 530.5	16 373.2	190.5
南　昌	145.9	16 448.3	4 653.6	2 134.0	215.0
九　江	0.0	19 184.5	800.7	472.8	303.2
济　南	9.9	42 527.2	1 596.1	4 992.8	1 625.3
青　岛	2.3	82 483.5	944.7	5 514.1	127.4
淄　博	0.2	107 032.1	738.1	6 842.5	243.3
枣　庄		68 220.5		3 088.0	
烟　台	245.9	88 985.4	52.5	2 382.3	0.2
潍　坊	0.7	183 961.6	418.3	4 614.1	34.1
济　宁	4.8	207 478.5	11.5	40 255.5	1 770.8
泰　安	0.5	56 951.1	29.3	850.1	23.2
威　海	6.2	4 195.4	…	279.0	
日　照	…	163 561.8	83.2	2 002.6	…
郑　州	…	49 872.0	18.2	1 035.3	0.1
开　封		6 020.4	1.2	1 770.2	
洛　阳	…	7 406.0	10 398.4	704.3	67.2
平顶山	18.6	25 711.2	253.2	1 888.2	537.4
安　阳	43.6	61 426.2	849.5	1 956.9	9 665.1
焦　作		120 340.6	0.1	3 187.1	
武　汉	15.7	133 358.3	627.7	1 729.7	1 231.2
宜　昌	15.0	44 237.3	27.5	2 445.2	82.3
荆　州		23 392.8	0.9	1 202.8	…
长　沙	0.1	8 085.1	23.6	174.8	…
株　洲	31.6	16 941.3	386.2	4 273.6	0.7
湘　潭	9.5	9 204.6	37.2	607.1	535.9
岳　阳	2.0	46 701.9	2 076.3	1 437.3	258.9
常　德	…	38 739.1	25.9	93.5	…

重点城市工业废水排放及处理情况（四）（续表）

（2010）

单位：吨

城市名称	工业废水中污染物去除量				
	氰化物	化学需氧量	石油类	氨氮	挥发酚
张家界	…	57.6	…	1.3	…
广州	114.0	112 218.0	888.5	16 027.8	101.1
韶关	0.7	7 832.6	3.8	75.8	2.7
深圳	242.0	38 265.5	301.6	1 465.5	0.3
珠海	1.0	47 403.8	119.0	633.5	2.6
汕头	0.7	51 582.2	0.2	197.3	…
佛山	88.6	119 995.3	506.1	2 616.3	11.7
湛江	…	18 782.8	16.3	408.3	…
中山	43.2	44 509.6	27.8	1 516.9	0.3
南宁	0.2	209 403.1	7.0	4 522.3	3.4
柳州	230.7	22 485.9	901.2	582.0	544.1
桂林	0.2	25 571.9	5.3	241.3	0.2
北海		9 178.3	97.5	379.4	28.0
海口		4 047.5	15.1	71.1	
三亚		119.8	…	2.5	
重庆	17.8	126 030.3	885.9	6 289.4	1 034.8
成都	0.2	65 363.5	257.2	2 263.7	3.2
攀枝花	…	5 287.1	61.7	110.7	…
泸州		16 321.3	12.2	167.9	
绵阳		28 521.3	221.6	619.8	2.4
宜宾	…	30 229.6	0.4	697.0	…
贵阳	329.6	14 518.5	11.3	171.4	131.8
遵义	0.1	33 657.0	0.8	643.5	
昆明	11.0	21 264.8	446.1	504.5	425.9
曲靖	1.2	28 694.7	2.3	53.2	18.6
拉萨		36.9	0.2	0.9	
西安	0.9	116 680.5	193.2	1 697.0	14.6
铜川		321.4	2.0	6.7	
宝鸡	…	17 684.3	39.0	955.5	0.5
咸阳	0.7	20 826.9	896.8	686.9	19.1
延安	9.3	8 912.4	2 930.2	420.4	80.0
兰州	…	3 452.3	556.5	327.3	28.2
金昌	…	1 666.4	6.0	4 243.6	…
西宁		1 785.4	1.7	37.1	
银川	1.5	72 086.0	741.2	50 263.3	237.2
石嘴山	12.7	22 978.8	679.4	8 450.5	1 032.3
乌鲁木齐	74.2	31 026.3	822.4	11 247.3	1 119.7
克拉玛依		7 465.3	802.9	138.8	

重点城市工业废气排放及处理情况（一）

（2010）

城市名称	燃料煤消费量（万吨）	原料煤消费量（万吨）	燃料油消费量（万吨）	工业废气排放总量（标态）（亿米^3）	燃料燃烧废气排放	生产过程废气排放
总　计	**129 877**	**49 615**	**1 543.86**	**291 003**	**161 571**	**129 432**
北　京	1 587	448	65.04	4 750	1 857	2 894
天　津	4 471	419	22.52	7 686	5 254	2 432
石家庄	2 921	410	3.73	4 509	2 545	1 964
唐　山	3 802	3 372	1.24	20 941	6 601	14 340
秦皇岛	789	121	16.37	3 227	898	2 329
邯　郸	2 401	1 590	1.30	7 481	2 359	5 122
保　定	1 066	299	0.23	1 466	977	489
太　原	1 315	2 826	0.97	4 243	1 480	2 763
大　同	2 327	236	0.23	9 402	8 784	618
阳　泉	1 013	55	0.81	1 812	1 532	280
长　治	1 742	3 779	0.67	3 035	1 529	1 506
临　汾	726	3 442	0.36	1 756	725	1 031
呼和浩特	2 440	130	2.26	1 902	1 690	212
包　头	2 035	1 335	2.09	5 887	2 365	3 522
赤　峰	1 465	43	0.48	1 967	1 632	334
沈　阳	1 376	57	63.45	1 301	1 117	185
大　连	1 438	159	34.07	2 618	1 573	1 045
鞍　山	545	1 206	6.08	4 842	1 492	3 350
抚　顺	1 057	146	14.91	1 831	1 037	793
本　溪	234	1 270	0.02	3 754	948	2 806
锦　州	858	5	41.85	921	735	186
长　春	1 557	104	0.59	1 508	1 234	273
吉　林	1 289	109	25.24	2 378	1 588	790
哈尔滨	1 258	169	2.30	1 181	876	305
齐齐哈尔	1 021	152	0.30	940	709	232
大　庆	1 003	3	29.58	1 503	1 318	185
牡丹江	689	130	0.07	852	783	69
上　海	4 257	1 505	68.47	12 969	5 089	7 880
南　京	2 214	976	40.62	5 738	2 770	2 968
无　锡	2 468	182	6.56	3 868	2 367	1 500
徐　州	2 429	559	21.50	3 083	2 416	666
常　州	952	31	1.98	1 149	1 001	148
苏　州	4 575	739	59.54	8 236	5 369	2 867
南　通	1 503	30	5.46	2 024	1 447	577
连云港	388	40	0.06	517	334	183
扬　州	1 323	0	14.11	1 153	1 075	78
杭　州	1 300	235	16.77	4 071	2 511	1 560

重点城市工业废气排放及处理情况（一）（续表）

（2010）

城市名称	燃料煤消费量（万吨）	原料煤消费量（万吨）	燃料油消费量（万吨）	工业废气排放总量（标态）（亿米³）	燃料燃烧废气排放	生产过程废气排放
宁 波	3 892	160	22.02	5 349	4 014	1 335
温 州	961	5	10.90	1 010	965	45
嘉 兴	1 461	37	8.88	2 088	1 420	668
湖 州	667	207	5.66	1 648	846	802
绍 兴	975	83	4.24	1 490	942	548
台 州	1 404	2	2.63	1 396	1 218	178
合 肥	565	107	0.23	702	474	228
芜 湖	461	466	0.06	1 636	592	1 044
马鞍山	763	892	18.22	4 393	879	3 514
福 州	1 349	108	12.73	3 034	2 185	849
厦 门	423		15.43	883	529	355
泉 州	940	29	40.97	2 517	1 967	550
南 昌	457	192	2.23	910	445	464
九 江	502	163	3.92	1 760	947	812
济 南	1 061	760	3.81	3 258	1 344	1 914
青 岛	1 404	272	9.51	2 558	1 529	1 029
淄 博	2 021	752	5.63	3 501	2 448	1 053
枣 庄	875	933	4.96	2 491	937	1 553
烟 台	1 467	214	1.03	2 187	1 309	879
潍 坊	1 653	592	7.72	4 275	3 062	1 213
济 宁	3 322	705	1.23	3 593	3 119	475
泰 安	841	481	0.42	1 877	978	899
威 海	453	5	10.78	546	507	39
日 照	713	336	5.03	3 612	784	2 828
郑 州	2 052	41	10.72	3 165	2 185	980
开 封	405	57	0.19	296	262	34
洛 阳	2 293	123	6.57	2 681	1 569	1 112
平顶山	1 251	1 058	0.65	2 070	1 162	908
安 阳	538	1 832	0.11	2 631	704	1 927
焦 作	996	191	1.50	1 816	831	985
武 汉	946	1 283	12.48	4 721	2 103	2 618
宜 昌	403	439	0.37	1 044	601	442
荆 州	244	19	0.10	425	329	97
长 沙	350	185	1.68	627	368	259
株 洲	328	104	9.73	1 109	516	593
湘 潭	424	558	0.15	1 786	406	1 380
岳 阳	478	71	5.20	749	618	131
常 德	725	124	1.05	1 541	741	800

重点城市工业废气排放及处理情况（一）（续表）

（2010）

城市名称	燃料煤消费量（万吨）	原料煤消费量（万吨）	燃料油消费量（万吨）	工业废气排放总量（标态）（亿米³）		
					燃料燃烧废气排放	生产过程废气排放
张家界	25	11	…	131	39	92
广州	2 087	52	42.00	4 427	3 424	1 003
韶关	728	364	1.29	1 446	498	948
深圳	459	0	61.34	1 684	1 224	459
珠海	704	23	23.31	1 337	786	551
汕头	947	25	1.51	860	772	89
佛山	1 171	31	74.48	1 516	1 250	266
湛江	405	52	3.66	661	468	193
中山	199	…	30.50	398	296	102
南宁	263	102	7.06	844	647	196
柳州	397	763	0.58	2 221	784	1 437
桂林	320	146	1.01	915	481	434
北海	227	14	0.48	345	297	49
海口	1		0.12	15	8	7
三亚	…	…	0.16	30	30	
重庆	2 770	803	3.39	10 943	7 756	3 187
成都	694	102	0.59	2 225	1 142	1 083
攀枝花	680	350	0.93	3 138	1 497	1 640
泸州	430	74	7.30	2 179	881	1 298
绵阳	414	179	0.03	1 142	720	423
宜宾	330	91	0.08	793	640	153
贵阳	522	316	12.43	1 516	462	1 054
遵义	571	44	3.38	986	567	419
昆明	981	676	7.06	2 828	901	1 928
曲靖	2 214	1 887	1.45	3 038	2 004	1 034
拉萨	13		0.06	10	8	2
西安	708	30	1.71	792	437	355
铜川	307	115	0.02	1 052	240	813
宝鸡	771	266	0.19	2 892	2 298	594
咸阳	1 057	60	1.17	1 272	1 045	228
延安	127	…	2.95	134	113	21
兰州	1 027	169	2.99	1 805	641	1 164
金昌	392	72	6.46	632	384	248
西宁	504	178	0.21	2 478	631	1 847
银川	509	60	1.69	1 562	1 355	207
石嘴山	1 368	2 107	1.00	1 712	1 195	516
乌鲁木齐	1 311	553	5.57	2 225	963	1 262
克拉玛依	349	…	414.60	943	834	109

重点城市工业废气排放及处理情况（二）

（2010）

单位：吨

城市名称	工业二氧化硫去除量	燃料燃烧过程中去除的	生产工艺过程中去除的	工业二氧化硫排放量	燃料燃烧过程中排放的	生产工艺过程中排放的
总　计	**16 972 480**	**12 189 919**	**4 782 562**	**9 413 743**	**8 021 935**	**1 348 676**
北　京	129 888	121 856	8 032	56 844	54 969	1 870
天　津	374 014	369 149	4 865	217 620	206 634	10 991
石家庄	371 446	364 816	6 630	137 974	129 702	8 142
唐　山	410 080	328 439	81 641	238 061	182 617	55 480
秦皇岛	63 450	62 740	710	44 737	34 249	10 489
邯　郸	326 731	305 019	21 712	161 805	95 631	66 133
保　定	107 765	107 405	360	53 984	49 907	3 892
太　原	254 968	202 348	52 619	94 233	76 219	18 015
大　同	280 820	277 236	3 584	127 416	125 562	1 855
阳　泉	194 151	189 575	4 576	109 377	108 721	656
长　治	162 378	134 808	27 570	121 103	84 224	36 878
临　汾	96 717	60 813	35 904	91 210	60 172	31 038
呼和浩特	224 259	224 259	…	82 813	81 243	1 171
包　头	274 369	211 845	62 524	174 718	89 796	84 922
赤　峰	140 334	111 171	29 163	150 719	147 978	2 749
沈　阳	87 775	87 439	335	77 333	75 578	1 757
大　连	275 718	95 456	180 262	88 477	79 874	8 598
鞍　山	26 724	26 324	400	83 141	64 365	18 776
抚　顺	93 590	63 241	30 348	57 224	47 843	9 381
本　溪	46 572	20 120	26 452	95 630	64 871	30 646
锦　州	68 406	50 924	17 482	61 739	58 376	3 363
长　春	35 162	34 412	750	60 528	54 683	5 845
吉　林	86 633	25 484	61 149	66 448	60 932	5 506
哈尔滨	29 844	20 343	9 500	53 941	51 962	1 805
齐齐哈尔	15 245	15 245	…	46 730	46 458	269
大　庆	14 876	9 603	5 272	60 517	58 881	1 637
牡丹江	3 035	2 999	37	42 691	42 495	196
上　海	349 849	333 389	16 460	221 476	206 966	14 410
南　京	606 632	157 809	448 823	115 507	94 237	21 270
无　锡	187 192	186 984	208	128 114	124 660	3 454
徐　州	216 577	210 592	5 985	127 063	122 489	4 582
常　州	59 710	59 710	…	48 000	47 655	345
苏　州	438 029	421 555	16 474	195 166	161 947	33 304
南　通	145 965	145 019	946	60 740	60 183	554
连云港	26 668	26 668	…	34 430	32 315	2 116
扬　州	141 527	141 526	1	65 994	65 283	711
杭　州	92 156	88 856	3 299	88 682	75 679	13 003

重点城市工业废气排放及处理情况（二）（续表）

（2010）

单位：吨

城市名称	工业二氧化硫去除量	燃料燃烧过程中去除的	生产工艺过程中去除的	工业二氧化硫排放量	燃料燃烧过程中排放的	生产工艺过程中排放的
宁波	721 536	357 754	363 781	109 840	105 711	4 129
温州	71 449	56 460	14 990	61 788	61 641	144
嘉兴	91 349	89 577	1 772	67 198	66 307	892
湖州	37 342	37 170	172	50 976	40 679	10 297
绍兴	40 140	37 851	2 289	54 882	51 432	3 450
台州	129 314	129 101	213	100 343	99 989	220
合肥	22 370	22 140	230	31 988	27 409	4 579
芜湖	37 409	25 411	11 998	40 765	19 722	16 525
马鞍山	57 924	42 309	15 615	60 543	33 029	27 514
福州	112 208	112 164	45	93 639	93 065	575
厦门	17 461	17 461	…	51 545	48 487	3 058
泉州	79 617	38 560	41 057	53 546	51 104	2 428
南昌	52 134	48 317	3 817	30 636	21 349	9 288
九江	79 520	41 274	38 246	66 817	62 090	4 727
济南	149 428	120 027	29 401	70 296	57 517	12 780
青岛	156 983	131 395	25 588	86 190	64 289	21 851
淄博	425 366	269 629	155 737	163 602	146 365	17 237
枣庄	128 533	123 584	4 949	75 952	65 630	10 321
烟台	189 421	125 519	63 901	88 047	78 352	9 695
潍坊	257 745	177 934	79 811	116 122	109 556	6 557
济宁	312 561	281 932	30 629	118 716	118 127	589
泰安	196 946	185 639	11 308	67 699	60 454	7 264
威海	24 774	24 605	169	31 785	31 756	29
日照	102 890	93 810	9 080	49 592	29 122	20 470
郑州	51 684	50 966	719	116 857	110 010	6 847
开封	1 267	1 267	…	31 436	30 598	837
洛阳	241 652	198 503	43 149	190 113	182 967	6 587
平顶山	80 745	66 639	14 106	114 674	78 001	36 672
安阳	98 355	48 582	49 772	93 059	31 474	61 551
焦作	55 711	37 582	18 129	78 217	59 262	18 828
武汉	153 925	131 129	22 797	87 256	68 339	18 918
宜昌	61 342	47 197	14 145	69 389	62 223	7 167
荆州	59 629	54 314	5 315	33 375	29 996	3 317
长沙	48 333	44 819	3 514	54 690	31 691	22 798
株洲	324 194	39 835	284 359	57 883	32 444	25 439
湘潭	77 972	71 702	6 270	66 704	26 627	40 063
岳阳	76 644	59 950	16 694	60 114	54 460	5 409
常德	86 355	75 066	11 289	47 728	40 131	7 595

重点城市工业废气排放及处理情况（二）（续表）

（2010）

单位：吨

城市名称	工业二氧化硫去除量	燃料燃烧过程中去除的	生产工艺过程中去除的	工业二氧化硫排放量	燃料燃烧过程中排放的	生产工艺过程中排放的
张家界	1 756	158	1 598	18 971	16 876	2 095
广州	392 842	234 635	158 207	85 871	77 180	7 925
韶关	36 832	33 932	2 900	55 691	40 872	14 819
深圳	38 554	38 060	494	30 941	30 862	79
珠海	61 720	61 720		34 839	26 658	8 016
汕头	65 338	65 338	…	25 744	25 578	1
佛山	97 327	92 704	4 623	102 125	90 534	11 399
湛江	17 189	16 074	1 116	46 720	43 415	3 425
中山	1 370	1 370	…	47 150	47 034	116
南宁	24 157	24 027	130	65 696	63 969	1 727
柳州	161 451	67 516	93 935	75 140	52 564	22 577
桂林	66 918	66 879	39	48 181	38 753	9 428
北海	12 561	12 561		34 185	30 671	3 514
海口				92	92	
三亚	…	…		12	12	
重庆	962 388	889 730	72 658	572 747	477 143	84 039
成都	61 402	56 477	4 926	61 928	53 157	7 830
攀枝花	24 771	23 007	1 765	100 568	97 892	2 676
泸州	15 100	14 762	339	91 274	88 414	2 776
绵阳	40 765	31 797	8 968	46 081	40 422	5 677
宜宾	112 227	111 654	573	96 786	82 061	5 286
贵阳	361 595	349 769	11 826	84 508	74 209	10 298
遵义	174 350	172 293	2 057	67 928	62 329	5 250
昆明	567 547	137 972	429 575	77 461	50 818	26 610
曲靖	369 341	162 544	206 797	172 819	159 485	11 854
拉萨				378	378	
西安	63 563	62 836	727	81 503	78 661	818
铜川	56 511	56 511	…	16 343	15 842	501
宝鸡	60 619	27 436	33 183	57 851	48 639	6 088
咸阳	149 524	144 912	4 612	86 304	78 840	1 509
延安	17 439	15 726	1 713	11 196	8 571	2 625
兰州	86 667	86 667	…	94 557	89 668	4 885
金昌	1 184 058	22 435	1 161 623	86 896	38 737	48 159
西宁	27 817	27 817	0	72 874	39 330	33 544
银川	35 800	31 303	4 497	24 150	22 787	1 362
石嘴山	177 479	175 510	1 969	112 839	103 349	9 491
乌鲁木齐	68 796	56 213	12 583	94 146	85 978	8 168
克拉玛依	5 224	5 224	…	45 058	33 403	11 655

重点城市工业废气排放及处理情况（三）

（2010）

单位：吨

城市名称	氮氧化物去除量	氮氧化物排放量	工业烟尘去除量	工业烟尘排放量	工业粉尘去除量	工业粉尘排放量
总计	**407 842**	**7 809 697**	**210 437 850**	**2 668 729**	**48 667 559**	**1 893 233**
北京	1 185	108 851	1 907 537	21 266	780 542	16 550
天津	40 950	236 078	4 844 544	53 831	548 424	7 970
石家庄	670	163 421	3 823 559	32 631	52 216	14 104
唐山	23 804	262 326	9 065 152	98 670	2 205 962	140 881
秦皇岛	3 361	41 140	1 621 884	11 365	580 140	32 972
邯郸	393	128 506	3 436 523	34 697	847 056	48 560
保定	17 264	51 957	1 384 577	14 652	78 803	8 444
太原	23 418	94 218	3 245 803	34 814	897 276	27 852
大同	36 428	140 934	4 310 467	57 185	119 986	58 666
阳泉	637	48 515	824 681	18 892	37 995	6 289
长治	…	96 169	3 861 715	64 376	494 690	54 273
临汾	23 854	37 304	1 018 745	57 346	427 789	82 261
呼和浩特	…	98 291	1 358 919	14 766	151 689	2 190
包头	45	106 962	3 925 911	32 563	764 278	71 108
赤峰	…	74 052	2 482 606	10 882	156 465	1 826
沈阳	505	95 060	2 343 065	60 364	6 008	4 542
大连	182	96 379	3 061 279	24 171	288 202	3 842
鞍山	…	73 384	2 580 049	30 205	1 856 474	25 782
抚顺	79	59 802	3 994 381	17 938	329 169	13 011
本溪	21	51 991	402 184	27 055	1 677 328	34 369
锦州	11 888	61 834	707 539	41 158	100 093	4 357
长春	…	143 598	1 577 652	94 173	849 045	25 323
吉林	…	145 563	3 390 524	33 507	1 498 951	4 078
哈尔滨	9 036	48 888	1 639 587	29 515	210 801	5 794
齐齐哈尔	813	47 861	1 196 086	25 809	129 789	4 489
大庆	47	71 335	4 225 285	29 717	5 157	1 700
牡丹江	2	50 043	349 417	29 471	57 711	2 767
上海	16 561	261 498	4 723 090	41 793	1 381 789	9 705
南京	20 129	113 842	3 008 848	33 788	870 054	37 643
无锡	3 147	144 865	3 506 772	38 839	599 251	38 008
徐州	243	160 900	5 640 254	31 495	685 995	16 710
常州	361	39 188	1 359 084	19 582	15 430	3 284
苏州	7 578	190 348	3 905 017	46 303	256 130	19 945
南通	4 875	68 037	1 635 979	38 457	28 538	8 337
连云港	313	32 132	752 553	7 356	127 228	1 870
扬州	1 274	48 175	90 981	7 797	2 174	2 408
杭州	2 939	103 575	1 061 174	30 860	926 747	13 381

重点城市工业废气排放及处理情况（三）（续表）

（2010）

单位：吨

城市名称	氮氧化物去除量	氮氧化物排放量	工业烟尘去除量	工业烟尘排放量	工业粉尘去除量	工业粉尘排放量
宁波	21 140	251 887	5 776 668	32 677	530 090	8 718
温州	48	33 147	865 893	5 666	281	31
嘉兴	1 546	70 754	1 121 435	18 900	321 872	7 548
湖州	544	51 227	789 284	10 648	1 964 978	24 082
绍兴	46	32 374	307 883	17 883	222 491	6 176
台州	198	61 704	1 147 402	5 856	974	309
合肥	106	41 117	1 435 799	10 604	128 718	5 186
芜湖	3 104	37 164	254 236	12 958	271 942	27 948
马鞍山	11 689	53 909	1 703 589	8 814	1 612 587	15 073
福州	14	98 993	1 373 417	8 594	1 913 584	8 325
厦门	3 487	25 592	617 166	2 252	6 537	296
泉州	2 382	41 864	1 033 656	23 452	538 832	24 058
南昌	266	19 118	127 397	6 264	7 988	9 773
九江	…	45 013	1 084 513	23 141	772 105	14 797
济南	6	60 999	2 186 420	19 709	693 352	28 631
青岛	786	61 633	2 178 880	12 967	183 104	5 074
淄博	155	120 433	3 240 587	38 271	314 955	10 139
枣庄	…	52 299	1 522 005	13 142	1 152 575	26 381
烟台	245	72 784	1 883 089	13 051	191 823	21 116
潍坊	99	92 711	3 068 977	26 081	235 087	14 333
济宁	27	127 487	4 482 982	22 090	569 537	5 548
泰安	…	68 930	1 808 673	18 248	272 776	6 021
威海	…	24 220	493 462	8 238	7 642	1 169
日照	…	47 976	1 197 323	7 427	97 749	6 832
郑州	435	145 424	3 555 305	45 011	679 244	31 418
开封	57	14 611	836 245	33 402	…	397
洛阳	…	136 232	3 348 998	91 037	406 430	21 728
平顶山	…	117 837	4 280 392	60 082	1 644 814	21 432
安阳	653	35 889	1 397 165	28 160	713 772	70 512
焦作	…	65 784	1 357 532	32 880	322 450	15 864
武汉	232	109 631	2 473 257	12 631	274 425	8 571
宜昌	…	23 507	360 634	9 427	951 816	4 747
荆州	1 503	10 566	317 730	13 316	71 521	858
长沙	1 020	26 435	103 169	24 746	126 159	105 217
株洲	1 083	22 133	399 923	5 983	698 441	5 821
湘潭	…	25 052	1 113 869	16 903	669 156	16 383
岳阳	66	35 315	1 096 192	15 864	105 592	7 735
常德	266	38 184	196 203	10 516	405 534	5 600

重点城市工业废气排放及处理情况（三）（续表）

（2010）

单位：吨

城市名称	氮氧化物去除量	氮氧化物排放量	工业烟尘去除量	工业烟尘排放量	工业粉尘去除量	工业粉尘排放量
张家界	…	2 474	57 707	3 513	59 723	11 347
广州	61 272	89 127	5 394 260	9 371	196 277	1 504
韶关	77	22 760	1 064 746	8 904	807 336	3 808
深圳	1 100	25 732	408 996	912	661	758
珠海	197	75 800	763 062	8 041	112 024	1 603
汕头	1 570	20 776	496 129	4 937	23 604	281
佛山	8 464	123 220	983 192	32 936	290 775	2 897
湛江	828	20 711	107 101	12 443	83 758	6 527
中山	317	19 158	213 270	21 748	1 507	222
南宁	299	21 000	475 427	24 506	373 381	6 972
柳州	4 323	17 747	440 852	17 059	1 207 074	48 484
桂林	…	38 442	591 446	10 514	78 353	7 586
北海	22	12 834	441 633	5 188		9 016
海口		65	629	93	15	1
三亚	…	249	…	34		…
重庆	…	231 313	3 773 199	102 132	954 137	83 601
成都	3 225	45 620	727 269	28 901	152 959	5 521
攀枝花	…	41 105	2 321 338	25 025	482 316	25 787
泸州	10	29 830	956 323	9 013	27 140	1 498
绵阳	20	27 772	1 072 225	7 485	435 131	4 299
宜宾	198	25 305	821 257	7 296	39 788	2 551
贵阳	3 376	16 004	1 042 180	12 601	412 766	14 931
遵义	166	17 062	2 150 337	10 416	392 255	14 213
昆明	…	37 156	1 801 015	7 842	534 556	6 103
曲靖	…	108 878	6 881 120	24 889	129 816	13 736
拉萨		135	507	247	8	283
西安	…	44 463	1 299 580	15 075	137 873	3 027
铜川	…	29 952	161 705	3 336	247 541	37 424
宝鸡	7 411	43 489	747 323	8 714	218 250	11 434
咸阳	67	38 664	1 036 462	10 724	612 971	6 155
延安	89	2 992	89 273	7 018	745	31
兰州	7 688	45 265	1 079 974	9 356	52 001	11 913
金昌	3 919	21 922	567 139	13 815	445 503	6 546
西宁	…	45 577	915 839	21 034	403 559	45 658
银川	…	25 544	534 338	8 329	173 717	2 133
石嘴山	1	108 240	11 245 517	72 071	178 581	37 653
乌鲁木齐	…	108 603	966 256	32 993	275 147	8 590
克拉玛依	…	27 757	8 450	6 038	…	…

重点城市工业废气排放及处理情况（四）

（2010）

城市名称	废气治理设施数（套）	脱硫设施数	废气治理设施处理能力(标态)(万米^3/时)	脱硫设施脱硫能力（吨/时）	废气治理设施运行费用（万元）	脱硫设施运行费用
总计	**111 164**	**16 777**	**563 332**	**16 485**	**6 604 599.3**	**2 685 621.3**
北京	2 468	1 018	8 177	196	90 590.0	46 441.6
天津	3 126	1 651	15 179	2 419	185 777.4	92 891.8
石家庄	1 540	368	9 201	256	73 573.6	40 710.0
唐山	2 659	274	29 506	779	374 266.6	77 187.6
秦皇岛	3 136	113	3 688	11	50 338.7	14 506.8
邯郸	1 318	227	19 973	363	148 067.7	47 044.8
保定	1 286	113	3 871	9	36 261.0	23 987.5
太原	1 116	431	8 584	44	88 754.7	45 046.4
大同	1 178	782	5 223	46	30 421.6	20 841.8
阳泉	445	173	1 262	175	16 024.3	11 549.7
长治	994	42	5 871	23	80 609.1	38 707.6
临汾	684	211	2 328	281	46 990.5	27 300.4
呼和浩特	585	19	4 479	27	46 403.6	11 156.0
包头	683	62	10 763	53	104 570.8	54 849.4
赤峰	516	112	5 105	61	23 046.8	17 200.3
沈阳	2 150	489	4 942	124	32 511.9	23 288.7
大连	953	229	2 682	99	33 809.4	21 504.8
鞍山	2 106	169	8 541	23	61 927.3	20 787.0
抚顺	679	43	2 218	10	23 998.0	10 474.9
本溪	661	56	4 380	262	41 178.9	2 438.3
锦州	417	195	7 025	52	25 870.8	17 780.8
长春	676	61	1 477	75	14 361.8	4 687.0
吉林	703	44	3 624	12	28 172.2	11 653.7
哈尔滨	833	60	8 035	496	17 153.7	8 234.6
齐齐哈尔	746	4	9 655	6	16 167.3	2 582.7
大庆	288	90	1 496	3	12 087.4	1 924.6
牡丹江	286	29	2 046	2	2 988.6	507.7
上海	4 319	620	19 377	226	288 712.8	104 257.3
南京	1 184	164	7 034	136	160 618.0	73 866.2
无锡	1 852	173	9 298	71	105 473.9	49 455.1
徐州	670	113	6 828	445	85 987.9	47 506.4
常州	1 093	60	2 857	161	29 935.9	15 048.8
苏州	2 986	568	18 083	273	346 260.9	75 716.9
南通	1 058	46	4 234	173	39 262.1	22 101.9
连云港	318	16	1 077	13	17 613.8	9 235.4
扬州	292	29	4 043	29	27 921.5	16 261.5
杭州	8 917	436	18 612	35	91 838.2	38 395.8

重点城市工业废气排放及处理情况（四）（续表）

（2010）

城 市 名 称	废气治理 设施数 （套）	脱 硫 设施数	废气治理设施 处理能力（标态） （万米^3/时）	脱硫设施脱硫能力 （吨/时）	废气治理设施 运行费用 （万元）	脱硫设施 运行费用
宁 波	2 265	264	8 694	1 087	203 169.1	113 470.5
温 州	1 324	137	3 066	460	37 456.3	23 954.4
嘉 兴	1 725	446	5 035	24	67 517.7	41 805.0
湖 州	1 288	186	2 705	16	35 005.0	7 685.2
绍 兴	2 176	173	3 765	100	29 399.4	4 335.6
台 州	1 020	196	3 692	84	49 928.9	34 332.9
合 肥	404	42	1 307	14	13 805.6	6 473.7
芜 湖	341	79	2 507	15	18 421.7	6 678.4
马鞍山	275	14	6 482	11	22 332.6	11 362.0
福 州	490	17	6 562	64	82 816.0	40 269.5
厦 门	658	24	1 987	157	26 863.0	11 202.6
泉 州	1 024	18	4 402	263	32 957.1	13 805.8
南 昌	517	43	1 944	75	28 074.2	12 299.6
九 江	625	62	3 399	450	39 491.8	10 097.4
济 南	1 021	189	13 723	77	129 326.5	27 908.1
青 岛	912	307	6 824	581	54 471.5	30 973.9
淄 博	1 439	448	6 584	420	154 916.3	55 153.8
枣 庄	1 100	51	3 701	19	35 323.9	13 462.0
烟 台	853	346	4 527	65	80 817.9	25 876.3
潍 坊	1 136	293	5 080	92	63 446.0	35 727.3
济 宁	946	192	6 794	185	81 209.3	56 661.9
泰 安	658	174	3 908	70	40 767.2	26 436.9
威 海	194	19	894	7	7 627.1	3 678.3
日 照	485	26	5 159	25	74 273.4	15 540.1
郑 州	1 320	90	10 677	86	56 865.8	27 044.0
开 封	114	18	979	467	4 810.4	3 435.5
洛 阳	1 212	200	6 220	70	84 984.5	53 211.1
平顶山	547	22	3 594	19	37 261.0	16 986.5
安 阳	1 009	48	6 135	44	49 029.6	13 955.8
焦 作	746	214	3 157	155	29 722.4	15 762.0
武 汉	649	40	7 894	22	71 994.5	37 596.8
宜 昌	634	95	2 407	17	16 087.2	6 879.1
荆 州	212	13	443	10	10 179.7	6 563.9
长 沙	342	52	683	14	6 639.8	1 207.9
株 洲	647	66	1 088	39	32 952.2	14 712.1
湘 潭	324	67	2 693	20	20 042.7	14 700.1
岳 阳	495	37	1 764	28	13 772.0	7 843.3
常 德	696	16	2 256	13	26 787.5	12 965.0

重点城市工业废气排放及处理情况（四）（续表）

（2010）

城市名称	废气治理设施数（套）	脱硫设施数	废气治理设施处理能力（标态）（万米^3/时）	脱硫设施脱硫能力（吨/时）	废气治理设施运行费用（万元）	脱硫设施运行费用
张家界	51	5	140	2	662.3	45.2
广州	1 508	253	6 547	434	425 816.5	120 257.9
韶关	611	26	7 123	164	73 582.6	14 051.7
深圳	769	46	1 929	7	27 243.2	15 324.6
珠海	844	39	2 609	9	44 927.2	15 523.2
汕头	487	7	2 075	13	29 024.4	22 197.7
佛山	1 358	645	6 981	25	53 548.6	29 711.1
湛江	377	19	1 424	7	19 692.1	11 775.0
中山	814	24	1 407	3	11 264.5	2 029.9
南宁	815	188	2 472	13	12 571.9	3 195.3
柳州	805	71	5 768	53	64 390.2	8 730.4
桂林	585	34	7 242	11	9 483.7	3 584.2
北海	53	6	349	2	6 810.9	4 141.6
海口	25		36		2 029.3	
三亚	1		1		1.0	
重庆	3 511	276	12 264	663	163 084.5	90 279.4
成都	1 616	64	3 991	70	33 820.4	17 767.0
攀枝花	430	52	6 316	32	62 551.1	8 608.2
泸州	212	39	809	60	33 396.4	29 434.5
绵阳	616	47	2 318	12	30 220.7	20 875.4
宜宾	299	55	973	92	181 687.7	11 021.7
贵阳	694	131	1 948	98	42 489.7	8 036.2
遵义	485	78	2 660	14	35 189.8	23 057.7
昆明	1 369	90	6 076	128	94 332.9	30 675.9
曲靖	783	61	4 185	113	57 410.9	30 341.8
拉萨	13		6		1 345.0	
西安	816	88	2 743	34	17 320.7	10 140.5
铜川	312	45	1 585	8	14 190.2	9 142.4
宝鸡	535	36	6 596	50	18 957.5	7 070.8
咸阳	663	85	1 906	466	23 343.3	11 406.1
延安	88	20	1 993	2	5 432.5	2 711.7
兰州	534	18	2 680	25	35 705.3	18 182.8
金昌	299	18	716	88	42 982.1	36 907.2
西宁	550	5	2 841	4	48 495.0	5 109.0
银川	253	49	1 679	18	9 286.6	3 171.5
石嘴山	506	51	3 096	415	39 971.2	21 565.9
乌鲁木齐	554	84	5 522	235	36 920.3	8 677.7
克拉玛依	179	8	820	280	25 321.6	21 684.0

重点城市工业固体废物产生及处置利用情况（一）

（2010）

单位：万吨

城市名称	工业固体废物产生量	危险废物	冶炼废渣	粉煤灰	炉渣	煤矸石	尾矿	放射性废物	脱硫石膏	其他废物
总计	**115 092**	**1 059.52**	**20 815**	**22 433**	**11 536**	**10 687**	**23 105**	**27**	**3 186**	**14 597**
北京	1 269	11.45	216	170	120	54	499	0	29	83
天津	1 862	10.21	434	589	432	…	…	0	100	186
石家庄	1 568	23.33	260	628	217	3	35	0	109	77
唐山	10 663	9.31	2 861	746	550	531	1 846	0	91	2 498
秦皇岛	1 482	1.24	487	142	63	0	639	0	0	40
邯郸	3 027	0.11	1 594	458	135	475	142	0	123	30
保定	447	1.29	4	116	93	…	123		25	39
太原	2 555	2.28	447	357	100	871	693	0	22	25
大同	1 586	0.03	22	488	114	603	97	0	78	35
阳泉	1 965	0.02	2	121	160	1 327			41	130
长治	2 348	0.11	216	405	165	1 217	45	0	39	183
临汾	1 466	3.50	390	120	84	543	141	0	32	31
呼和浩特	826	0.23	14	504	108	…	36	0	87	66
包头	2 254	5.87	671	417	61	11	658	7	49	43
赤峰	1 350	2.14	52	256	96	16	638		33	229
沈阳	895	7.73	2	516	176	114	2	0	18	38
大连	290	7.06	9	99	125	…	3	0	2	29
鞍山	5 176	1.29	707	57	93		4 043		6	54
抚顺	2 569	3.69	206	185	52	107	906		18	968
本溪	2 703	8.61	909	35	33	2	1 668	0	2	17
锦州	335	30.27	42	151	70	6	4		12	11
长春	474	1.26	…	245	177	1			10	16
吉林	1 455	72.97	51	303	157	20	768		0	19
哈尔滨	1 443	1.27	43	266	115	898	33	0	4	32
齐齐哈尔	302	2.00	17	132	72	…	…	0	3	71
大庆	313	12.45	…	156	38	…	…	0	2	77
牡丹江	266	0.01	…	166	58	…	3	0	1	4
上海	2 448	51.25	940	490	133	1	…	0	80	553
南京	1 657	22.53	508	287	96	14	200	0	41	275
无锡	922	41.12	200	333	200	4	…	0	36	59
徐州	1 473	0.71	216	541	185	271	…	0	46	123
常州	304	7.52	49	75	127	…	…	0	10	23
苏州	2 208	28.87	750	604	419	7	12	0	89	134
南通	396	9.81	18	181	94	…	…	0	31	50
连云港	293	0.51	70	58	58	…	5	0	7	80
扬州	358	10.04	4	194	56	…	5	0	33	21
杭州	707	8.21	190	155	180	…	15	0	16	91

重点城市工业固体废物产生及处置利用情况（一）（续表）

（2010）

单位：万吨

城市名称	工业固体废物产生量	危险废物	冶炼废渣	粉煤灰	炉渣	煤矸石	尾矿	放射性废物	脱硫石膏	其他废物
宁波	1 154	21.46	161	613	148	2	…	0	88	102
温州	218	5.93	3	97	65	…	3	0	24	11
嘉兴	378	2.99	4	155	125	…	…	0	13	54
湖州	211	5.75	2	92	50	…	16		11	14
绍兴	345	11.25	1	76	139	…	74	0	…	22
台州	246	2.42	…	121	36	…	7		38	33
合肥	339	0.39	78	125	41	…	8	0	5	31
芜湖	251	0.50	66	111	23	5	…	0	9	25
马鞍山	1 991	0.96	750	205	36	…	804	0	19	166
福州	693	1.50	222	137	109	3	2		55	139
厦门	135	1.89	2	61	26	…	…	0	…	36
泉州	674	1.98	52	43	182	19	117	0	7	184
南昌	217	0.44	38	79	26	…	…	0	18	27
九江	753	1.86	59	119	152	…	251	1	14	126
济南	1 012	4.13	431	209	102	23	4	0	35	114
青岛	909	2.55	178	256	101	…	175	0	3	115
淄博	1 518	22.52	239	450	228	27	224	0	53	147
枣庄	513	0	…	151	117	181	7	0	20	26
烟台	2 155	111.46	21	195	217	28	1 239	0	27	143
潍坊	835	6.99	69	312	166	14	5	0	47	193
济宁	2 053	124.21	…	451	480	672	113	0	50	80
泰安	1 042	2.28	98	188	155	353	52	0	63	117
威海	223	0.61	13	56	21		101		…	18
日照	821	0.15	619	112	36	…	…	0	8	41
郑州	959	0.25	25	340	130	129	204	0	18	20
开封	192	0.01	…	84	92	…	…		6	5
洛阳	1 796	1.07	11	335	124	21	1 125		71	28
平顶山	1 237	0.24	55	433	162	241	131		18	56
安阳	1 057	0.13	557	236	46	30	115		14	37
焦作	1 083	0.12	12	378	165	120	68	0	6	297
武汉	1 325	3.64	662	220	50	…	29	0	117	224
宜昌	1 090	1.10	52	88	109	2	309	0	94	316
荆州	143	0.03	2	62	37		…		21	22
长沙	149	0.15	2	60	25	23	8	0	1	15
株洲	272	1.98	46	70	33	15	11	0	9	51
湘潭	653	0.83	396	108	33	15	12		22	39
岳阳	324	17.93	2	94	58	…	96	0	9	45
常德	368	1.03	1	181	69	19	14	0	23	19

重点城市工业固体废物产生及处置利用情况（一）（续表）

（2010）

单位：万吨

城市名称	工业固体废物产生量	危险废物	冶炼废渣	粉煤灰	炉渣	煤矸石	尾矿	放射性废物	脱硫石膏	其他废物
张家界	24	…	2	6	8	3	1	0	…	1
广州	692	24.54	61	241	136	…	…	0	28	156
韶关	963	16.67	263	126	82	…	148	18	7	283
深圳	146	38.38	…	49	32	1	…	0	…	25
珠海	275	6.35	108	76	20				19	37
汕头	101	0.69	…	52	29	1	…	0	7	10
佛山	333	2.00	…	60	117	…	…	0	17	134
湛江	305	0.15	…	76	51	…	54	0	9	75
中山	106	5.07	…	19	65	…	…	0	…	6
南宁	408	0.04	3	46	28	…	86	0	6	194
柳州	990	4.89	434	38	34	…	21	0	12	313
桂林	263	0.13	71	53	31	…	13	0	26	37
北海	171	0.08		40	18		82		3	11
海口	4	0.10		…	…					4
三亚	…	…			…					…
重庆	2 837	36.42	182	539	444	606	90	0	188	505
成都	513	1.94	123	135	69	…	18		23	110
攀枝花	3 994	89.12	516	206	134	268	2 436	0	9	175
泸州	311	0.10	…	93	39	54	59		25	23
绵阳	291	7.64	31	107	33	1	16		16	55
宜宾	507	0.07	8	103	74	78	39	0	22	138
贵阳	1 168	0.91	84	161	56	9	130	0	53	570
遵义	406	0.38	24	185	35	42	4	0	60	43
昆明	2 284	1.30	380	211	87	…	218	0	51	1 211
曲靖	1 873	4.64	186	681	234	151	28		43	260
拉萨	7				…		6			1
西安	248	1.01	1	118	45	…	15	0	20	48
铜川	196	0	…	66	11	97	…		21	1
宝鸡	545	0.31	1	79	43	1	377	0	8	4
咸阳	476	0.37	…	199	107	115	…	0	30	25
延安	63	3.50	…	3	5	39	…	0	0	3
兰州	507	13.99	68	207	67	15	24	0	31	57
金昌	1 007	0	187	61	34	2	627	0	3	92
西宁	392	13.26	156	83	35	8	2		7	84
银川	198	0.74	7	59	44	4	5		7	65
石嘴山	695	1.53	48	248	56	150	23	0	57	108
乌鲁木齐	692	1.03	339	115	79	2	4	0	18	134
克拉玛依	107	19.10	…	44	27	…	…	0	…	16

重点城市工业固体废物产生及处置利用情况（二）

（2010）

单位：万吨

城市名称	工业固体废物综合利用量								
		危险废物	冶炼废渣	粉煤灰	炉渣	煤矸石	尾矿	脱硫石膏	其他废物
总　　计	**82 278**	**743.20**	**19 487**	**17 576**	**10 697**	**7 683**	**4 370**	**2 948**	**10 902**
北　　京	835	4.97	216	170	120	14	115	29	81
天　　津	1 845	1.64	434	589	442	0	0	100	170
石 家 庄	1 498	11.08	260	388	195	3	23	143	70
唐　　山	8 590	6.62	2 861	630	550	522	129	91	2 484
秦 皇 岛	883	0.12	487	140	63	0	134	0	40
邯　　郸	2 758	0	1 594	246	131	416	53	122	29
保　　定	316	1.17	4	116	93	0	10	25	38
太　　原	1 335	2.23	423	131	71	554	3	16	11
大　　同	953	…	20	367	104	290	12	81	5
阳　　泉	460	0.01	0	64	48	157		0	146
长　　治	1 667	…	165	199	116	771	40	23	184
临　　汾	1 214	2.63	424	59	70	403	97	2	31
呼和浩特	320	0.05	14	183	36	0	0	21	57
包　　头	1 068	4.91	407	338	55	11	10	49	39
赤　　峰	518	0.20	49	204	82	2	28	33	102
沈　　阳	857	6.41	4	416	165	95	2	18	32
大　　连	274	2.49	9	98	126	0	3	2	19
鞍　　山	989	0	707	17	90		34	6	11
抚　　顺	1 127	2.73	205	103	44	107	55	18	538
本　　溪	1 026	8.60	909	21	32	2	32	2	14
锦　　州	210	30.54	42	59	48	6		12	8
长　　春	472	0.70	…	245	177	…		10	16
吉　　林	623	49.70	51	170	133	20	137	…	15
哈 尔 滨	1 294	0.93	44	160	94	814	0	4	32
齐齐哈尔	226	3.45	20	87	53	0	0	3	55
大　　庆	237	2.27	0	166	36	0	0	2	3
牡 丹 江	256	…	0	57	58	…	3	1	4
上　　海	2 367	28.47	940	468	132	1	0	79	506
南　　京	1 471	10.20	508	287	95	14	62	41	265
无　　锡	896	23.45	200	333	200	4	0	36	54
徐　　州	1 503	0.23	216	472	185	273	0	46	121
常　　州	288	5.77	49	75	117	0	0	10	20
苏　　州	2 180	15.30	750	604	418	7	12	89	121
南　　通	389	4.08	18	181	94	0	0	31	48
连 云 港	283	0.18	70	71	55	…	5	7	60
扬　　州	348	8.07	4	194	56	0	5	33	16
杭　　州	666	6.13	190	155	178	0	9	16	62

重点城市工业固体废物产生及处置利用情况（二）（续表）

（2010）

单位：万吨

城市名称	工业固体废物综合利用量	危险废物	冶炼废渣	粉煤灰	炉渣	煤矸石	尾矿	脱硫石膏	其他废物
宁波	1 035	10.91	144	491	99	2	…	87	83
温州	208	1.69	3	97	64	…	…	24	10
嘉兴	369	0.35	4	153	125	…	…	13	51
湖州	211	2.16	2	100	50	…	16	11	11
绍兴	322	3.13	0	76	139	0	73	0	15
台州	240	1.07	0	121	36	…	6	38	30
合肥	335	0.17	78	125	41	0	5	5	31
芜湖	243	0.08	66	107	21	5	0	9	22
马鞍山	1 298	0.83	750	190	36	0	144	19	147
福州	558	1.14	222	136	78	3	2	55	38
厦门	118	0.74	0	58	19	0	0	0	32
泉州	648	0.45	52	44	179	15	103	7	182
南昌	203	0.04	38	72	23	0	0	18	25
九江	461	1.79	59	117	147	0	42	10	60
济南	987	2.43	431	204	95	21	4	35	106
青岛	907	1.21	178	155	98	0	173	3	123
淄博	1 374	17.09	239	439	226	27	107	53	140
枣庄	525	….	0	151	117	192	7	20	26
烟台	1 908	109.91	21	195	217	28	1 027	27	142
潍坊	753	5.79	69	212	165	14	5	47	113
济宁	1 940	121.09	0	305	439	667	104	145	80
泰安	1 057	0.21	98	89	155	389	32	63	117
威海	214	…	13	56	21	0	99	0	11
日照	820	0.06	619	112	36	0	0	8	41
郑州	796	…	25	240	130	113	62	18	14
开封	192	0.01	…	83	93	0	0	6	5
洛阳	983	1.03	10	318	124	10	406	67	17
平顶山	1 094	0.10	56	289	160	331	23	18	83
安阳	946	0.13	557	215	45	30	25	14	37
焦作	840	0.04	12	175	172	113	28	6	99
武汉	1 337	0.71	662	146	50	0	27	120	213
宜昌	516	0.70	52	88	109	2	10	16	179
荆州	143	…	2	62	37	…	0	21	22
长沙	149	0.14	2	60	25	22	8	1	16
株洲	247	0.83	59	60	31	12	3	7	43
湘潭	625	0.80	378	1	32	15	12	22	37
岳阳	312	17.58	1	94	58	…	88	9	42
常德	344	…	1	167	69	12	13	23	19

重点城市工业固体废物产生及处置利用情况（二）（续表）

（2010）

单位：万吨

城市名称	工业固体废物综合利用量	危险废物	冶炼废渣	粉煤灰	炉渣	煤矸石	尾矿	脱硫石膏	其他废物
张家界	23	…	1	6	8	2	1	…	1
广州	622	4.13	61	237	135	…	…	29	113
韶关	895	16.25	265	131	82		110	7	269
深圳	135	27.28	0	49	32	1	0	…	24
珠海	270	4.60	108	76	20			19	35
汕头	97	0.59	0	52	25	1	0	7	9
佛山	331	1.23	0	60	117	…	0	17	134
湛江	281	…	…	66	48	…	54	9	71
中山	91	2.37	…	18	58	…	0	…	4
南宁	384	0.03	3	46	28	…	66	4	195
柳州	895	4.39	403	44	16	…	10	12	288
桂林	239	0.13	69	53	31	…	4	25	28
北海	126	…		40	18		42	3	11
海口	4	0.01		0	0	…			4
三亚	0				0				0
重庆	2 317	25.71	53	459	470	509	60	184	358
成都	511	1.01	123	35	69		18	23	109
攀枝花	675	58.06	101	155	51	25	79	6	142
泸州	238	0.06	0	85	33	52	6	29	17
绵阳	258	8.57	31	107	33	…	8	12	34
宜宾	446	…	8	101	73	58	23	21	127
贵阳	659	0.89	84	128	48	11	44	57	251
遵义	270	0.18	15	121	25	42	3	22	31
昆明	944	1.11	317	210	87	…	69	33	181
曲靖	970	3.98	153	172	178	119	29	27	108
拉萨	0				0				
西安	244	0.51	1	118	51	…	13	20	39
铜川	161	0	0	66	11	63	0	21	1
宝鸡	146	0.31	1	79	43	1	0	8	3
咸阳	472	0.03	…	199	107	112	0	30	24
延安	55	0.25	0	3	5	38	0	…	3
兰州	413	2.32	59	179	52	…	24	20	57
金昌	199	0	35	88	35	…	11	6	23
西宁	339	10.77	154	92	36	13	2	5	25
银川	149	0.64	7	25	37	3	5	1	64
石嘴山	378	0.15	46	77	31	119	23	32	49
乌鲁木齐	472	0.80	189	102	62	2	0	15	102
克拉玛依	71	17.12	0	30	23	0	0	0	0

重点城市工业固体废物产生及处置利用情况（三）

（2010）

城市名称	工业固体废物贮存量（万吨）	危险废物贮存量	工业固体废物处置量（万吨）	危险废物处置量	工业固体废物排放量（万吨）	危险废物排放量
总　　计	**7 129**	**4.92**	**26 695**	**323.65**	**233**	**0.00 015**
北　　京	40	…	780	6.49	…	0
天　　津	…	…	27	8.66	0	0
石 家 庄	73	…	34	12.25	…	0
唐　　山	433	…	1 639	2.68	1	0
秦 皇 岛	102	…	498	1.12	0	0
邯　　郸	245	…	63	0.12	0	0
保　　定	…	…	131	0.12	…	0
太　　原	32	…	1 179	0.06	8	0
大　　同	47	…	575	0.04	10	0
阳　　泉	150	0.01	1 329	…	32	0
长　　治	232	…	555	0.11	3	0
临　　汾	139	…	132	0.87	21	0
呼和浩特	52	…	454	0.18	…	0
包　　头	427	…	759	0.96		
赤　　峰	95	…	739	2.14	0	
沈　　阳	20	…	19	1.56	…	0.00 015
大　　连	0	…	15	4.57	1	0
鞍　　山	303	…	3 885	1.29		
抚　　顺	847	0.03	595	0.93	0	
本　　溪	284		1 385	0	0	
锦　　州	1	…	125	0.46	0	
长　　春	1	…	1	0.56		
吉　　林	565	…	267	23.27	0	0
哈 尔 滨	24	…	124	0.34	…	0
齐齐哈尔	66	…	11	0.04	…	0
大　　庆	6	…	92	10.18	0	0
牡 丹 江	9	…	0	0.24	0	0
上　　海	1	0.04	94	23.44	…	0
南　　京	147	0.16	38	12.52	…	0
无　　锡	…	…	27	17.67	0	0
徐　　州	…	0.03	4	0.45	0	0
常　　州	…	…	16	1.75	0	0
苏　　州	…	…	29	13.93	0	0
南　　通	…	…	7	5.76	0	0
连 云 港	25	0.16	0	0.17	0	0
扬　　州	…	0.15	9	1.83	0	0
杭　　州	…	0.03	41	2.06	…	0

重点城市工业固体废物产生及处置利用情况（三）（续表）

（2010）

城　市 名　称	工业固体废物贮存量（万吨）	危险废物贮存量	工业固体废物处置量（万吨）	危险废物处置量	工业固体废物排放量（万吨）	危险废物排放量
宁　波	59	0.11	60	10.44	…	0
温　州	1	0.40	9	3.84	…	0
嘉　兴	…	…	9	2.65	…	0
湖　州	1	0.01	7	3.58	0	0
绍　兴	3	0.06	21	8.06	…	0
台　州	1	0.05	5	1.33	0	0
合　肥	4	…	1	0.22	0	0
芜　湖	…	…	9	3.30	…	0
马鞍山	46	0.02	647	0.11	0	0
福　州	21	…	123	0.36	…	0
厦　门	7	0.03	10	1.56	0	0
泉　州	11	…	19	1.53	0	0
南　昌	11	…	2	0.40	1	0
九　江	36	0.07	257	0.07	…	0
济　南	23	0.01	3	1.69	…	0
青　岛	7	0.02	11	1.36	0	0
淄　博	34	…	111	5.43	0	
枣　庄	…	…	0	…	0	0
烟　台	…	0.04	247	1.59	0	0
潍　坊	79	0.01	6	1.29	0	0
济　宁	108	…	7	3.11	0	0
泰　安	25	…	2	2.07	0	0
威　海	7	…	3	0.61		
日　照	0	…	0	0.09	0	0
郑　州	15	…	148	0.25	0	0
开　封			0	…	0	
洛　阳	23	…	790	0.05	…	
平顶山	6	…	254	0.14	0	
安　阳	22	…	90	0.01	0	
焦　作	34	…	216	0.08	…	0
武　汉	2	…	17	2.93	0	0
宜　昌	1	…	574	0.40	0	0
荆　州	0	…	0	0.03	0	
长　沙	0	…	0	0.01	…	0
株　洲	24	…	28	1.14	…	0
湘　潭	0	0	28	0.02	…	0
岳　阳	9	0.33	3	0.01	…	0
常　德	23	…	2	1.03	0	0

重点城市工业固体废物产生及处置利用情况（三）（续表）

（2010）

城市名称	工业固体废物贮存量（万吨）	危险废物贮存量	工业固体废物处置量（万吨）	危险废物处置量	工业固体废物排放量（万吨）	危险废物排放量
张家界	1	…	0	0	1	0
广州	5	…	67	20.41	…	0
韶关	95	0.05	5	0.06	1	0
深圳	…	…	11	11.10	…	0
珠海	…	…	5	1.74	…	0
汕头	4	…	1	0.10	…	0
佛山	0	0.09	1	0.69	…	0
湛江	5	…	21	0.15	1	0
中山	3	…	13	2.70	0	0
南宁	7	…	18	0.02	…	0
柳州	29	…	66	0.50	…	0
桂林	11	…	12	…	1	0
北海	…	…	44	0.08	1	
海口	…	…	0	0.09	0	
三亚			0	…		
重庆	288	0.03	155	10.76	134	0
成都	0	0.10	2	0.93	0	0
攀枝花	147	0.01	3 172	31.05	0	0
泸州	25	…	48	0.05	…	
绵阳	15	…	21	1.45	…	0
宜宾	29	…	32	0.07	…	0
贵阳	25	…	489	0.02	…	0
遵义	29	…	107	0.20	0	0
昆明	126	…	1 207	0.20	9	0
曲靖	448	…	491	0.65	0	
拉萨	6				1	
西安	2	…	3	0.52	0	0
铜川	18	…	16	…	1	0
宝鸡	…	…	398	…	…	0
咸阳	…	…	5	0.35	…	0
延安	0	…	8	3.25	0	0
兰州	82	0.04	29	13.28	0	0
金昌	292	…	583	0	…	0
西宁	66	2.33	0	0.16	1	0
银川	5	…	45	0.10	1	0
石嘴山	178		147	1.37	0	
乌鲁木齐	179	…	41	0.23	1	0
克拉玛依	…	0.44	36	1.71	…	0

重点城市汇总工业企业概况（一）

（2010）

城市名称	汇总工业企业数（个）	工业总产值（现价）（万元）	企业专职环保人员数（人）	工业炉窑数（台）		
					烟尘排放达标的	二氧化硫排放达标的
总　计	**57 596**	**1 940 768 608.3**	**168 975**	**41 787**	**38 901**	**37 110**
北　京	718	52 207 963.3	1 758	345	328	324
天　津	1 954	87 719 001.3	3 532	713	713	697
石家庄	581	17 423 709.4	3 103	387	361	352
唐　山	488	40 352 672.8	1 745	928	910	876
秦皇岛	255	7 940 093.1	860	104	98	95
邯　郸	414	23 786 331.4	1 354	468	403	388
保　定	880	8 436 588.2	2 139	616	467	370
太　原	250	18 403 956.2	1 257	395	395	382
大　同	333	5 330 028.8	1 257	307	262	221
阳　泉	279	3 377 932.1	690	347	292	177
长　治	384	14 664 875.8	1 563	454	418	340
临　汾	259	10 906 704.7	1 133	207	197	192
呼和浩特	263	6 634 838.6	416	94	93	91
包　头	81	12 092 578.6	487	484	456	461
赤　峰	218	2 510 855.6	344	114	96	88
沈　阳	1 472	22 626 949.5	2 269	246	180	162
大　连	445	28 144 392.5	1 082	434	402	353
鞍　山	260	10 308 060.8	815	1 358	1 354	1 341
抚　顺	308	12 932 317.3	638	457	338	325
本　溪	236	8 679 038.9	441	413	331	262
锦　州	178	6 549 006.0	538	305	305	303
长　春	143	35 747 728.9	520	233	166	233
吉　林	244	11 814 372.0	1 120	179	179	177
哈尔滨	289	13 516 711.6	803	94	91	90
齐齐哈尔	301	4 680 947.5	406	455	405	415
大　庆	143	29 986 571.7	1 323	2 596	2 576	2 576
牡丹江	81	1 648 709.8	220	24	23	22
上　海	1 871	154 663 582.8	9 180	1 509	1 419	1 413
南　京	727	54 750 012.5	1 962	553	529	449
无　锡	1 057	45 460 525.3	3 349	665	665	665
徐　州	364	14 884 132.4	1 156	164	153	131
常　州	875	17 509 777.6	2 045	264	264	264
苏　州	1 359	79 641 701.3	5 713	578	576	575
南　通	893	13 638 019.0	4 115	400	400	399
连云港	199	6 970 376.7	600	17	17	17
扬　州	485	13 777 680.5	1 115	158	158	158
杭　州	1 191	36 658 197.6	3 896	310	306	304

重点城市汇总工业企业概况（一）（续表）

（2010）

城市名称	汇总工业企业数（个）	工业总产值（现价）（万元）	企业专职环保人员数（人）	工业炉窑数（台）	烟尘排放达标的	二氧化硫排放达标的
宁　波	1 466	55 399 139.7	3 017	830	798	709
温　州	1 204	11 867 549.4	2 498	201	189	187
嘉　兴	921	20 796 513.0	2 631	162	161	156
湖　州	732	7 695 710.7	1 760	288	282	261
绍　兴	775	17 777 837.6	2 453	239	233	216
台　州	2 186	17 519 980.0	3 058	374	272	223
合　肥	355	15 789 307.8	792	142	137	137
芜　湖	321	7 821 449.3	489	216	179	164
马鞍山	84	9 485 840.9	429	173	171	172
福　州	539	17 625 434.1	593	274	266	267
厦　门	394	18 116 115.7	535	100	100	100
泉　州	1 475	17 884 396.8	1 025	678	646	653
南　昌	403	7 595 977.6	708	196	196	196
九　江	406	6 596 076.7	707	315	119	119
济　南	413	18 435 011.7	1 117	521	514	514
青　岛	766	40 039 088.4	2 126	263	263	263
淄　博	530	27 605 766.1	2 001	694	687	646
枣　庄	221	4 521 064.8	925	172	172	161
烟　台	518	27 325 070.4	1 670	204	188	202
潍　坊	652	32 126 719.4	2 501	190	189	172
济　宁	390	21 302 372.8	2 067	344	337	327
泰　安	253	7 792 510.9	767	126	126	125
威　海	205	9 695 426.9	2 225	31	28	28
日　照	105	9 587 092.2	469	113	113	113
郑　州	594	11 769 314.7	1 740	1 442	1 441	1 281
开　封	126	1 008 705.1	421	114	93	100
洛　阳	345	14 058 932.4	1 135	727	719	713
平顶山	135	8 352 019.6	582	253	248	233
安　阳	522	9 125 956.7	1 559	551	475	428
焦　作	302	5 830 850.1	1 232	309	290	240
武　汉	338	33 207 386.3	976	203	189	183
宜　昌	298	12 183 424.6	675	213	212	209
荆　州	319	3 448 470.0	795	85	78	70
长　沙	484	16 978 251.8	853	219	151	129
株　洲	489	7 712 978.9	819	546	546	546
湘　潭	296	6 884 600.6	593	249	202	190
岳　阳	588	6 827 227.1	631	365	329	330
常　德	183	6 428 160.3	462	63	63	63

重点城市汇总工业企业概况（一）（续表）

（2010）

城市名称	汇总工业企业数（个）	工业总产值（现价）（万元）	企业专职环保人员数（人）	工业炉窑数（台）	烟尘排放达标的	二氧化硫排放达标的
张家界	49	97 254.0	51	45	25	23
广　州	1 007	43 702 215.8	3 155	243	235	197
韶　关	307	5 230 228.8	859	231	214	177
深　圳	1 621	61 269 350.0	5 970	27	24	26
珠　海	538	21 199 285.3	1 025	76	76	76
汕　头	575	3 998 378.2	923	28	22	22
佛　山	955	24 329 874.5	2 497	1 182	1 094	1 065
湛　江	445	7 464 414.0	821	87	47	31
中　山	596	10 236 065.5	1 880	69	58	41
南　宁	680	3 997 400.1	1 225	660	582	551
柳　州	341	14 718 510.4	450	269	244	234
桂　林	369	3 197 417.2	578	233	233	232
北　海	172	945 471.3	282	110	2	4
海　口	31	1 265 272.1	138	5	5	5
三　亚	10	413 185.0	28			
重　庆	2 489	51 744 138.7	4 459	1 804	1 537	1 421
成　都	1 991	23 086 855.7	2 975	1 027	1 024	1 026
攀枝花	177	10 181 329.7	427	250	241	240
泸　州	691	3 858 078.3	589	258	211	157
绵　阳	281	9 946 534.5	551	336	336	336
宜　宾	241	5 671 933.3	661	129	91	86
贵　阳	245	10 330 940.3	782	1 116	1 080	1 077
遵　义	464	4 136 991.3	686	531	487	462
昆　明	438	14 330 291.2	1 181	436	436	436
曲　靖	208	7 289 639.5	550	361	344	349
拉　萨	13	120 670.0	5 805	15	15	
西　安	326	16 321 451.7	1 383	149	144	142
铜　川	96	1 418 395.2	203	43	41	40
宝　鸡	215	5 143 813.4	528	124	124	124
咸　阳	221	11 079 387.8	813	170	170	170
延　安	105	4 261 722.0	2 041	77	33	33
兰　州	151	16 202 472.3	573	892	851	857
金　昌	53	5 663 974.9	249	84	82	77
西　宁	128	5 202 470.3	375	235	171	147
银　川	107	3 226 152.0	456	81	81	81
石嘴山	199	6 243 223.4	9 294	555	537	376
乌鲁木齐	214	6 775 640.7	426	151	138	137
克拉玛依	36	5 873 506.7	106	138	138	138

重点城市汇总工业企业概况（二）

（2010）

城市名称	工业锅炉数（台）	烟尘排放达标的	二氧化硫排放达标的	工业锅炉蒸吨数（蒸吨）	烟尘排放达标的	二氧化硫排放达标的
总　　计	**48 619**	**47 095**	**44 487**	**1 359 278.0**	**1 340 922.8**	**1 311 242.4**
北　　京	1 628	1 621	1 617	41 431.4	41 403.1	41 357.1
天　　津	2 173	2 167	2 169	65 850.9	64 750.9	65 190.9
石 家 庄	508	503	502	27 273.0	27 253.5	27 247.0
唐　　山	302	302	295	15 538.8	15 538.8	15 498.8
秦 皇 岛	278	277	260	5 112.5	5 112.5	5 050.5
邯　　郸	290	279	266	8 710.5	8 708.5	8 699.5
保　　定	721	694	369	4 925.7	4 831.7	3 321.2
太　　原	366	365	364	19 376.9	18 349.9	18 343.9
大　　同	754	732	725	21 177.0	21 146.0	21 135.0
阳　　泉	245	231	222	9 242.7	9 242.7	9 237.2
长　　治	553	549	529	25 391.5	25 385.5	25 341.5
临　　汾	242	239	238	7 559.0	7 539.0	7 535.0
呼和浩特	453	453	453	31 919.0	31 919.0	31 919.0
包　　头	178	138	132	25 476.0	20 385.0	18 536.0
赤　　峰	264	224	215	12 031.5	11 421.0	11 387.8
沈　　阳	1 887	1 818	1 362	28 083.7	27 850.7	23 288.6
大　　连	723	648	597	13 608.6	11 879.5	10 836.3
鞍　　山	562	550	544	13 909.7	13 869.2	13 754.2
抚　　顺	426	415	344	12 796.4	12 769.5	12 429.1
本　　溪	398	378	182	7 258.8	7 162.3	2 920.5
锦　　州	369	369	369	10 679.7	10 679.7	10 679.7
长　　春	589	586	589	16 633.7	16 620.7	16 633.7
吉　　林	365	363	361	11 686.7	11 670.7	11 660.7
哈 尔 滨	492	492	437	10 842.3	10 842.3	7 886.3
齐齐哈尔	665	549	585	13 101.9	12 454.7	12 484.1
大　　庆	1 187	1 187	1 185	15 043.6	15 043.6	15 041.1
牡 丹 江	246	246	246	2 454.1	2 454.1	2 454.1
上　　海	2 144	2 084	2 019	70 044.6	69 152.8	69 697.5
南　　京	427	420	407	13 659.0	13 582.0	13 166.0
无　　锡	737	737	735	28 957.2	28 957.2	28 957.2
徐　　州	288	246	250	14 367.6	13 927.1	14 149.1
常　　州	564	564	564	5 856.4	5 856.4	5 856.4
苏　　州	1 229	1 224	1 224	44 207.5	43 184.5	43 407.5
南　　通	539	539	537	10 833.4	10 833.4	10 823.4
连 云 港	180	180	180	5 077.5	5 077.5	5 077.5
扬　　州	265	265	265	14 642.2	14 642.2	14 642.2
杭　　州	797	794	794	13 437.4	13 430.2	13 437.4

重点城市汇总工业企业概况（二）（续表）

（2010）

城市名称	工业锅炉数（台）	烟尘排放达标的	二氧化硫排放达标的	工业锅炉蒸吨数（蒸吨）	烟尘排放达标的	二氧化硫排放达标的
宁波	821	811	711	53 721.0	53 678.5	53 273.5
温州	948	936	838	7 805.3	7 780.3	7 630.7
嘉兴	782	780	778	16 385.7	16 375.7	16 371.7
湖州	459	455	433	7 894.4	7 779.4	7 482.9
绍兴	824	821	822	8 406.8	8 389.8	8 390.3
台州	549	457	379	6 249.8	6 110.6	5 899.2
合肥	163	162	160	6 495.9	6 495.7	6 491.7
芜湖	133	119	110	552.1	507.6	502.1
马鞍山	91	85	87	9 254.7	9 245.7	9 249.7
福州	377	376	376	12 825.4	12 811.4	12 811.4
厦门	200	200	200	6 620.4	6 620.4	6 620.4
泉州	875	874	872	11 816.0	11 816.0	11 714.0
南昌	264	264	264	4 044.1	4 044.1	4 044.1
九江	187	105	88	5 903.6	5 630.5	5 571.3
济南	257	255	250	7 397.1	7 393.1	7 306.1
青岛	581	577	569	21 526.4	21 509.9	21 396.9
淄博	445	444	420	18 145.0	18 141.0	17 882.0
枣庄	208	208	203	4 356.5	4 356.5	4 246.5
烟台	597	592	592	17 673.6	17 664.1	17 664.1
潍坊	680	677	609	25 242.3	25 236.8	23 942.3
济宁	425	421	377	14 863.5	14 550.2	14 317.7
泰安	272	272	263	13 741.9	13 741.9	13 707.9
威海	135	135	135	5 145.2	5 145.2	5 145.2
日照	160	160	160	12 268.0	12 268.0	12 268.0
郑州	570	570	570	24 754.4	24 754.4	24 754.4
开封	134	134	133	5 027.0	5 025.0	5 019.0
洛阳	215	215	215	12 445.1	12 445.1	12 445.1
平顶山	159	159	158	15 180.0	15 180.0	15 155.0
安阳	204	166	164	4 775.9	4 765.1	4 759.1
焦作	496	496	341	8 843.3	8 843.3	8 068.5
武汉	304	299	298	16 712.5	16 704.5	16 703.5
宜昌	232	231	232	6 323.3	6 317.3	6 323.3
荆州	200	198	176	2 424.0	2 413.5	2 326.0
长沙	287	259	242	832.5	793.5	697.8
株洲	178	178	178	3 384.2	3 384.2	3 384.2
湘潭	202	185	174	1 177.7	1 122.7	1 066.7
岳阳	211	188	167	7 294.6	7 243.0	5 545.5
常德	170	170	170	7 852.5	7 852.5	7 852.5

重点城市汇总工业企业概况（二）（续表）

（2010）

城市名称	工业锅炉数（台）	烟尘排放达标的	二氧化硫排放达标的	工业锅炉蒸吨数（蒸吨）	烟尘排放达标的	二氧化硫排放达标的
张家界	15	9	8	34.0	20.0	18.0
广州	764	750	704	12 794.4	11 686.0	12 575.5
韶关	131	122	110	6 345.3	6 313.5	6 205.3
深圳	309	286	276	11 733.3	11 679.9	11 682.8
珠海	358	356	356	11 211.5	11 203.5	11 203.5
汕头	526	495	492	13 560.4	13 017.4	12 951.9
佛山	711	667	658	8 626.6	8 464.1	8 475.1
湛江	181	144	95	6 000.7	5 906.2	5 126.8
中山	528	522	374	4 081.8	4 069.8	3 277.2
南宁	370	364	343	3 707.8	3 690.0	3 642.0
柳州	116	113	86	3 296.7	3 278.7	3 013.0
桂林	181	171	168	3 987.2	3 956.1	3 918.1
北海	71	70	41	838.7	837.7	482.7
海口	38	38	38	235.5	235.5	235.5
三亚	5	5	5	81.0	81.0	81.0
重庆	1 064	943	865	18 283.8	17 706.1	17 241.8
成都	908	908	897	9 899.3	9 899.3	8 199.3
攀枝花	63	61	62	2 442.4	2 442.4	2 442.4
泸州	182	152	122	5 543.0	5 188.0	3 235.6
绵阳	170	170	170	3 004.2	3 004.2	3 004.2
宜宾	197	173	171	3 428.3	2 963.2	2 960.2
贵阳	123	104	105	4 241.0	4 203.5	4 205.5
遵义	178	103	134	2 073.2	2 035.4	2 036.7
昆明	226	226	226	7 810.8	7 810.8	7 810.8
曲靖	163	162	162	24 763.5	24 753.5	24 623.5
拉萨	6	6		14.5	14.5	
西安	526	525	509	11 594.9	11 592.9	11 218.9
铜川	80	79	77	326.0	326.0	324.5
宝鸡	236	236	236	1 584.4	1 584.4	1 584.4
咸阳	228	228	228	10 074.9	10 074.9	10 074.9
延安	181	159	135	1 864.8	1 801.2	1 697.2
兰州	159	140	142	1 592.0	1 477.0	1 527.0
金昌	89	86	65	4 042.5	4 038.5	3 793.5
西宁	150	140	138	5 385.1	5 373.6	5 297.6
银川	133	133	133	2 220.2	2 214.2	2 214.2
石嘴山	353	322	276	2 673.5	2 626.2	2 130.7
乌鲁木齐	496	490	467	25 390.0	25 359.0	24 687.1
克拉玛依	375	375	367	10 930.7	10 930.7	10 930.7

重点城市汇总工业企业概况（三）

（2010）

城市名称	工业用水总量（万吨）	新鲜水量	重复用水量	污水排放口数（个）	直排海的污水排放口数	“三废”综合利用产品产值（万元）
总计	**26 973 184**	**3 781 523**	**23 191 661**	**44 208**	**1 015**	**10 869 440.8**
北京	513 901	18 996	494 905	563	0	34 365.8
天津	1 128 661	70 297	1 058 363	1 595	5	192 650.4
石家庄	703 479	32 652	670 827	351	0	60 884.4
唐山	1 559 921	59 862	1 500 059	210	5	435 115.4
秦皇岛	75 202	8 963	66 239	229	11	34 836.7
邯郸	570 221	23 261	546 960	149	1	302 147.0
保定	262 491	24 334	238 157	561	0	43 007.7
太原	326 366	10 165	316 201	73	0	134 332.8
大同	197 200	18 856	178 344	162	0	19 946.2
阳泉	142 119	5 122	136 997	88	0	22 580.9
长治	230 608	11 507	219 100	152	0	43 010.5
临汾	163 972	8 687	155 285	42	0	38 032.9
呼和浩特	134 695	9 322	125 374	170	0	22 043.4
包头	474 331	18 145	456 186	41		36 801.7
赤峰	162 867	7 779	155 087	122	0	23 893.9
沈阳	91 983	9 646	82 337	1 047	0	15 372.5
大连	139 275	48 074	91 202	355	73	39 366.6
鞍山	394 106	12 720	381 386	112		6 132.2
抚顺	381 773	13 186	368 587	190	0	85 193.0
本溪	223 003	6 801	216 203	114	0	7 486.2
锦州	131 617	8 055	123 561	98	6	14 590.4
长春	138 174	11 318	126 856	232		88 359.6
吉林	273 597	51 785	221 812	148	0	29 082.6
哈尔滨	239 521	9 735	229 785	246	0	67 588.6
齐齐哈尔	115 077	73 946	41 130	287	0	16 335.6
大庆	330 833	29 401	301 432	102	0	76 558.5
牡丹江	30 526	15 030	15 496	49		5 477.4
上海	1 448 485	400 514	1 047 971	1 842	19	170 379.1
南京	1 047 251	124 571	922 680	700	0	210 436.3
无锡	319 755	49 312	270 443	970	0	167 264.6
徐州	499 388	27 111	472 277	298	0	70 124.7
常州	217 611	43 322	174 288	785	0	464 245.0
苏州	897 145	91 373	805 772	1 187	0	639 327.2
南通	130 129	24 097	106 033	753	0	87 653.5
连云港	93 561	7 570	85 991	184	4	10 046.7
扬州	191 297	15 563	175 734	492	0	26 692.6
杭州	378 928	100 673	278 255	1 079	2	1 799 985.0

重点城市汇总工业企业概况（三）（续表）

（2010）

城市名称	工业用水总量（万吨）	新鲜水量	重复用水量	污水排放口数（个）	直排海的污水排放口数	“三废”综合利用产品产值（万元）
宁波	723 868	473 285	250 583	1 145	257	230 911.2
温州	193 440	168 671	24 770	1 089	23	40 082.7
嘉兴	124 020	30 451	93 569	830	13	158 982.2
湖州	145 149	16 337	128 812	556	0	87 422.6
绍兴	138 569	38 830	99 739	758	1	193 441.9
台州	41 907	9 029	32 878	2 157	242	145 916.5
合肥	176 005	7 537	168 468	207	0	37 332.1
芜湖	17 093	5 848	11 245	192	0	78 618.7
马鞍山	292 416	10 916	281 500	67	0	48 277.0
福州	43 451	12 843	30 608	337	5	21 566.0
厦门	76 924	7 363	69 562	279	23	4 521.2
泉州	179 219	30 472	148 747	1 363	20	28 968.9
南昌	59 028	14 457	44 571	312	0	27 475.6
九江	160 974	61 020	99 954	275	0	40 412.6
济南	316 217	12 458	303 759	227	0	146 485.2
青岛	196 752	21 407	175 345	589	46	143 531.8
淄博	535 819	35 763	500 055	280	0	391 379.6
枣庄	164 580	20 627	143 953	199	0	60 777.1
烟台	83 958	13 878	70 079	401	65	61 295.0
潍坊	222 571	33 964	188 608	501	0	92 419.0
济宁	659 478	30 176	629 302	273	0	126 354.3
泰安	174 088	10 093	163 996	174	0	124 004.4
威海	27 212	4 207	23 005	179	19	7 223.6
日照	175 069	13 869	161 200	117	12	151 020.8
郑州	237 221	29 469	207 752	302	0	64 147.8
开封	128 511	8 374	120 137	112	0	3 937.2
洛阳	317 610	15 664	301 946	284	0	86 569.7
平顶山	246 565	14 041	232 524	106	0	43 143.6
安阳	340 244	21 397	318 846	208		114 434.1
焦作	153 088	32 870	120 218	229	0	94 842.8
武汉	611 003	42 486	568 517	340	0	251 717.2
宜昌	252 600	21 238	231 362	282	0	63 121.5
荆州	16 217	8 066	8 151	274	0	62 721.2
长沙	80 279	6 080	74 199	413	0	84 562.9
株洲	65 453	9 451	56 002	416	0	78 999.6
湘潭	133 940	8 541	125 399	280		49 574.9
岳阳	180 666	14 980	165 686	358	0	29 344.3
常德	32 084	12 725	19 359	153	0	25 757.0

重点城市汇总工业企业概况（三）（续表）

（2010）

城市名称	工业用水总量（万吨）	新鲜水量	重复用水量	污水排放口数（个）	直排海的污水排放口数	“三废”综合利用产品产值（万元）
张家界	171	126	45	57		5 189.0
广州	544 351	267 877	276 475	998	3	116 687.5
韶关	189 705	12 927	176 778	214	1	34 806.9
深圳	27 271	13 926	13 345	1 615	75	30 022.1
珠海	27 703	10 510	17 194	460	12	12 987.3
汕头	13 327	10 184	3 143	549	17	5 188.8
佛山	127 098	107 606	19 492	495	0	58 763.9
湛江	44 289	6 895	37 394	209	21	76 831.7
中山	20 657	15 380	5 277	489	19	30 173.4
南宁	106 548	14 887	91 661	279	0	78 702.8
柳州	292 910	60 060	232 850	153	0	58 386.5
桂林	69 597	6 047	63 550	246	0	29 521.8
北海	8 139	2 009	6 130	66	13	9 623.0
海口	2 976	741	2 235	34	0	464.8
三亚	93	62	31	3	2	1.5
重庆	732 290	197 167	535 123	2 092	0	291 326.6
成都	151 650	20 323	131 327	1 538	0	63 134.3
攀枝花	134 532	10 551	123 981	47	0	38 882.8
泸州	128 905	10 189	118 716	381		6 034.5
绵阳	79 296	12 939	66 357	241	0	23 533.3
宜宾	33 545	10 924	22 622	179	0	19 212.6
贵阳	127 127	6 003	121 124	128	0	44 020.0
遵义	123 401	6 871	116 529	132		35 014.7
昆明	254 502	12 872	241 630	206	0	213 577.8
曲靖	225 889	15 222	210 667	51	0	131 580.1
拉萨	827	688	140	1		83.0
西安	87 420	17 089	70 331	332	0	11 068.9
铜川	18 765	912	17 854	56	0	11 594.6
宝鸡	151 578	14 954	136 624	184	0	24 061.3
咸阳	119 403	10 453	108 950	175	0	61 075.5
延安	22 396	4 181	18 214	21	0	18 668.6
兰州	135 208	7 753	127 455	85	0	77 965.6
金昌	62 977	7 270	55 707	37	0	72 406.6
西宁	79 291	6 994	72 297	88		31 234.3
银川	96 714	10 604	86 110	74	0	29 766.4
石嘴山	99 752	8 634	91 118	38		44 842.1
乌鲁木齐	362 069	22 707	339 362	90	0	26 721.5
克拉玛依	188 453	6 352	182 101	33	0	3 643.3

重点城市工业污染治理项目建设情况（一）

（2010）

单位：个

城市名称	汇总工业企业数	本年施工项目数	治理废水	治理废气	治理固体废物	治理噪声	治理其他
总计	**3 356**	**4 337**	**1 936**	**1 637**	**225**	**103**	**418**
北京	27	45	5	32	0	2	6
天津	98	124	45	48	9	4	18
石家庄	71	75	14	58	2	1	0
唐山	15	20	3	12	0	1	4
秦皇岛	2	6	0	5	0	1	0
邯郸	11	27	6	21	0	0	0
保定	8	8	1	7	0	0	0
太原	39	93	14	59	11	1	8
大同	48	150	30	76	30	1	13
阳泉	24	37	8	21	3	0	5
长治	44	79	26	17	2	2	32
临汾	17	53	17	13	7	6	10
呼和浩特	46	82	35	37	6	0	4
包头							
赤峰	5	16	5	7	1	1	2
沈阳	48	13	1	8	0	0	4
大连	23	27	12	9	0	0	6
鞍山	4	10	2	5	2	0	1
抚顺	7	9	6	3	0	0	0
本溪	1	3	0	2	0	0	1
锦州	10	11	4	4	2	0	1
长春	18	27	18	8	1	0	0
吉林	16	20	6	8	1	0	5
哈尔滨	1	0	0	0	0	0	0
齐齐哈尔	12	16	4	12	0	0	0
大庆	9	18	5	4	0	0	9
牡丹江	1	1	0	1	0	0	0
上海	129	173	65	79	4	11	14
南京	46	77	23	41	2	2	9
无锡	24	15	8	0	1	1	5
徐州	21	34	11	21	1	0	1
常州	18	20	11	8	0	1	0
苏州	125	94	55	22	6	3	8
南通	47	67	41	11	1	1	13
连云港	15	25	12	12	1	0	0
扬州	12	6	2	3	0	0	1
杭州	258	270	223	43	0	2	2

重点城市工业污染治理项目建设情况（一）（续表）

（2010）

单位：个

城市名称	汇总工业企业数	本年施工项目数	治理废水	治理废气	治理固体废物	治理噪声	治理其他
宁波	22	27	17	3	0	0	7
温州	19	20	19	1	0	0	0
嘉兴	20	24	16	7	0	0	1
湖州	52	22	19	3	0	0	0
绍兴	96	111	67	28	5	2	9
台州	62	56	37	16	2	0	1
合肥	6	6	2	1	0	1	2
芜湖	3	1	0	1	0	0	0
马鞍山	7	15	9	5	0	0	1
福州	79	53	27	21	3	1	1
厦门	35	59	32	14	2	1	10
泉州	283	439	213	165	22	7	32
南昌	18	14	11	3	0	0	0
九江	19	27	12	15	0	0	0
济南	24	37	13	21	2	0	1
青岛	23	25	14	9	0	1	1
淄博	49	74	34	34	0	1	5
枣庄	10	11	4	6	0	1	0
烟台	58	63	13	11	37	1	1
潍坊	37	39	21	10	0	1	7
济宁	22	31	17	7	1	1	5
泰安	15	20	2	17	0	0	1
威海	19	16	10	5	0	1	0
日照	11	21	7	6	2	1	5
郑州	35	46	15	23	0	1	7
开封	3	4	3	1	0	0	0
洛阳	19	25	9	9	3	1	3
平顶山	19	27	10	12	0	3	2
安阳	2	3	1	2	0	0	0
焦作	8	10	3	6	1	0	0
武汉	3	19	1	0	0	0	0
宜昌	22	28	15	9	1	2	1
荆州	20	21	12	5	1	2	1
长沙	14	15	7	7	1	0	0
株洲	18	33	8	18	0	6	1
湘潭	11	18	8	7	2	1	0
岳阳	12	6	3	2	0	0	1
常德	11	18	9	7	0	1	1

重点城市工业污染治理项目建设情况（一）（续表）

（2010）

单位：个

城市名称	汇总工业企业数	本年施工项目数	治理废水	治理废气	治理固体废物	治理噪声	治理其他
张家界							
广州	57	68	19	46	1	1	1
韶关	26	43	21	11	2	1	8
深圳	60	71	54	10	2	5	0
珠海	26	34	19	10	2	1	2
汕头	46	51	43	8	0	0	0
佛山	139	166	38	88	0	1	39
湛江	24	11	6	5	0	0	0
中山	19	24	10	14	0	0	0
南宁	30	46	29	2	14	0	1
柳州	6	8	4	4	0	0	0
桂林	6	5	1	4	0	0	0
北海	3	3	3	0	0	0	0
海口							
三亚							
重庆	98	116	63	39	3	1	10
成都	19	25	13	10	2	0	0
攀枝花	13	20	5	9	1	1	4
泸州	19	12	11	0	0	0	1
绵阳	17	13	6	2	2	2	1
宜宾	6	7	6	1	0	0	0
贵阳	23	40	17	17	2	2	2
遵义	7	6	4	2	0	0	0
昆明	86	140	63	43	10	4	20
曲靖	21	39	14	22	2	0	1
拉萨							
西安	27	32	9	17	2	1	3
铜川	8	9	2	0	0	0	7
宝鸡	23	2	1	0	0	1	0
咸阳	18	31	15	7	0	0	9
延安	13	36	8	6	2	4	16
兰州							
金昌	7	16	7	8	0	0	1
西宁	8	9	1	8	0	0	0
银川	14	19	6	10	0	0	3
石嘴山	1	0	0	0	0	0	0
乌鲁木齐							
克拉玛依							

重点城市工业污染治理项目建设情况（二）

（2010）

单位：万元

城市名称	污染治理项目本年完成投资合计	治理废水	治理废气	治理固体废物	治理噪声	治理其他
总　计	**2 276 873.3**	**753 826.8**	**999 257.7**	**69 141.8**	**11 957.7**	**442 689.3**
北　京	19 339.5	1 761.6	17 264.4	0	11.2	302.3
天　津	164 683.5	47 138.8	34 856.0	1 225.5	155.0	81 308.2
石家庄	10 947.2	8 063.7	2 807.7	1.8	69.0	5.0
唐　山	20 106.4	674.0	18 653.6	0	218.0	560.8
秦皇岛	905.6	0	827.0	0	78.6	0
邯　郸	56 365.0	3 986.0	52 379.0	0	0	0
保　定	365.0	40.0	325.0	0	0	0
太　原	41 307.5	6 792.7	22 789.5	7 439.4	15.8	4 270.1
大　同	31 594.2	7 019.7	13 081.5	3 158.0	3.0	8 332.0
阳　泉	39 892.0	5 801.4	31 190.6	2 107.0	0	793.0
长　治	49 509.1	15 229.8	17 856.4	1 066.0	678.1	14 678.8
临　汾	7 259.7	2 541.7	2 527.7	888.3	153.6	1 148.4
呼和浩特	33 268.1	13 694.9	16 924.8	1 827.4	0	821.0
包　头						
赤　峰	50 040.4	9 436.0	38 572.4	108.0	133.0	1 791.0
沈　阳	6 409.7	1 253.9	4 487.4	0	0	668.4
大　连	9 803.4	3 330.3	3 088.1	0	0	3 385.0
鞍　山	12 236.4	440.0	11 660.0	84.5	0	51.9
抚　顺	35 508.5	17 400.0	18 108.5	0	0	0
本　溪	17 042.0	0	16 677.0	0	0	365.0
锦　州	22 046.0	14 833.6	6 824.0	151.4	0	237.0
长　春	11 539.2	10 466.5	1 032.2	40.5	0	0
吉　林	3 494.0	1 765.0	1 129.0	500.0	0	100.0
哈尔滨	280.0	0.0	280.0	0	0	0
齐齐哈尔	21 103.5	13 138.0	7 965.5	0	0	0
大　庆	10 551.3	5 605.9	368.3	0	0	4 577.1
牡丹江	5 000.0	0	5 000.0	0	0	0
上　海	94 106.5	6 308.1	34 780.6	26.6	679.3	52 311.9
南　京	31 637.2	5 737.9	19 936.0	69.3	51.0	5 843.0
无　锡	8 425.2	4 475.2	1 905.0	590.0	12.0	1 443.0
徐　州	13 351.4	4 087.6	8 083.8	1 000.0	0	180.0
常　州	5 459.0	1 692.0	3 415.0	0	352.0	0
苏　州	75 478.4	30 174.5	18 531.2	7 260.0	1 548.0	17 964.7
南　通	11 359.7	6 514.7	1 664.0	15.0	1.5	3 164.5
连云港	15 076.1	9 098.0	5 968.1	10.0	0	0
扬　州	2 779.6	2 563.6	196.0	0	0	20.0
杭　州	37 946.1	28 179.8	9 515.8	0	159.5	91.0

重点城市工业污染治理项目建设情况（二）（续表）

（2010）

单位：万元

城 市 名 称	污染治理项目 本年完成投资合计	治理废水	治理废气	治 理 固体废物	治理噪声	治理其他
宁 波	5 109.6	4 443.8	280.0	0	0	385.8
温 州	809.0	549.0	260.0	0	0	0
嘉 兴	7 239.5	4 284.5	2 895.0	0	0	60.0
湖 州	6 823.7	4 573.7	2 181.0	29.0	0	40.0
绍 兴	24 128.6	14 686.6	6 289.0	88.0	52.0	3 013.0
台 州	5 862.0	3 489.0	2 178.0	50.0	0	145.0
合 肥	1 151.5	165.6	595.9	0	30.0	360.0
芜 湖	970.0	505.0	415.0	0	0	50.0
马鞍山	5 727.4	738.4	4 673.0	0	0	316.0
福 州	11 640.8	6 260.3	5 027.0	308.5	15.0	30.0
厦 门	11 854.3	7 747.0	3 419.1	11.9	20.0	656.3
泉 州	96 420.9	35 466.7	30 005.6	7 066.6	546.0	23 336.0
南 昌	6 564.8	6 351.8	213.0	0	0	0
九 江	11 053.0	3 956.7	7 096.3	0	0	0
济 南	27 392.8	8 463.3	18 870.5	57.0	0	2.0
青 岛	10 219.0	6 457.7	3 668.0	0	23.3	70.0
淄 博	39 179.0	21 830.9	15 679.3	0	200.0	1 468.8
枣 庄	8 851.3	2 737.3	6 094.0	0	20.0	0
烟 台	27 422.9	9 874.7	6 691.0	10 838.3	2.9	16.0
潍 坊	131 691.2	15 773.3	31 797.5	0	49.8	84 070.6
济 宁	8 136.1	3 061.3	4 411.2	170.0	38.0	455.6
泰 安	2 856.1	1 583.0	1 183.1	0	0	90.0
威 海	18 359.0	3 436.0	14 828.0	0	95.0	0
日 照	45 978.2	13 780.0	24 887.3	2 720.9	10.0	4 580.0
郑 州	10 689.1	5 277.0	4 955.3	0	20.0	436.8
开 封	1 020.0	1 020.0	0	0	0	0
洛 阳	2 418.0	956.0	636.0	760.0	10.0	56.0
平顶山	15 193.9	8 023.1	5 337.3	0	115.0	1 718.5
安 阳	21 220.0	200.0	21 020.0	0	0	0
焦 作	2 051.5	495.0	1 521.1	35.4	0	0
武 汉	14 968.5	5 637.5	6 574	0	1 000	1 757
宜 昌	21 908.9	7 784.2	7 838.0	121.7	118.0	6 047.0
荆 州	3 998.4	3 441.7	266.7	10.0	120.0	160.0
长 沙	2 911.5	895.1	2 008.0	8.4	0	0
株 洲	4 951.6	1 017.3	3 791.8	0	92.5	50.0
湘 潭	21 859.4	9 605.4	9 666.0	2 580.0	8.0	0
岳 阳	17 770.7	11 962.2	5 071.0	0	80.0	657.5
常 德	45 355.0	21 405.0	22 650.0	0	0	1 300.0

重点城市工业污染治理项目建设情况（二）（续表）

（2010）

单位：万元

城市名称	污染治理项目本年完成投资合计	治理废水	治理废气	治理固体废物	治理噪声	治理其他
张家界						
广州	15 996.1	3 860.4	11 988.2	15.0	45.0	87.5
韶关	40 635.7	10 838.1	13 031.3	3 446.0	3.0	13 317.3
深圳	10 464.3	9 353.8	593.6	403.0	113.9	0
珠海	4 246.4	2 944.9	857.5	280.0	120.0	44.0
汕头	5 984.5	5 786.3	198.2	0	0	0
佛山	67 931.7	4 374.7	15 956.4	0	20.0	47 580.6
湛江	2 703.0	1 373.0	1 330.0	0	0	0
中山	3 578.0	479.0	3 099.0	0	0	0
南宁	7 682.7	7 405.7	22.0	245.0	0	10.0
柳州	6 424.6	679.1	5 745.5	0	0	0
桂林	563.5	45.0	518.5	0	0	0
北海	580.0	580.0	0	0	0	0
海口						
三亚						
重庆	77 501.9	37 760.6	27 368.6	3 189.0	500.0	8 683.7
成都	5 081.9	2 430.3	2 524.6	127.0	0	0
攀枝花	21 123.5	6 871.4	13 723.0	23.0	220.5	285.6
泸州	10 276.0	8 571.5	1 555.0	19.5	0	130.0
绵阳	3 193.1	1 807.5	471.5	295.0	262.0	357.1
宜宾	1 114.2	1 091.9	22.3	0	0	0
贵阳	31 349.6	15 299.4	11 289.6	2 520.6	2 200.0	40.0
遵义	482.2	157.0	325.2	0	0	0
昆明	20 613.6	7 891.6	9 731.7	1 804.3	230.7	955.3
曲靖	59 363.4	2 131.2	53 762.2	3 400.0	0	70.0
拉萨						
西安	9 678.4	3 026.5	4 279.6	65.0	14.7	2 292.6
铜川	10 975.0	1 514.0	335.0	0	0	9 126.0
宝鸡	5 663.1	3 184.2	335.1	10.0	3.8	2 130.0
咸阳	26 805.5	3 360.0	21 104.5	0	0	2 341.0
延安	74 813.3	48 321.0	17 267.9	875.0	1 240.0	7 109.4
兰州						
金昌	38 808.9	2 113.7	36 107.0	0	0	588.2
西宁	6 627.6	195.0	6 432.6	0	0	0
银川	28 508.0	13 200.0	3 508.0	0	0	11 800.0
石嘴山	130.0	0	130.0	0	0	0
乌鲁木齐						
克拉玛依						

重点城市工业污染治理项目建设情况（三）

（2010）

单位：万元

城　市 名　称	施工项目本年 投资来源合计	排污费补助	政府其他补助	企业自筹	银行贷款
总　计	**2 276 873.3**	**34 452.1**	**73 629.6**	**2 168 791.2**	**60 738.0**
北　京	19 339.5	0	327.8	19 011.4	5 688.0
天　津	164 683.5	202.5	16 684.2	147 796.8	2 189.7
石家庄	10 947.2	25.0	2 318.0	8 604.2	0
唐　山	20 106.4	0	938.7	19 167.7	0
秦皇岛	905.6			905.6	
邯　郸	56 365.0	1 400.0	50.0	54 915.0	0
保　定	365.0	0	0	365.0	0
太　原	41 307.5	4 014.0	460.0	36 833.5	36.0
大　同	31 594.2	412.9	1 711.9	29 469.4	0
阳　泉	39 892.0		955.0	38 937.0	
长　治	49 509.1	820.0	2 103.6	46 585.5	280.0
临　汾	7 259.7		200.0	7 059.7	
呼和浩特	33 268.1	235.0	170.0	32 863.1	
包　头					
赤　峰	50 040.4	60.0	36.0	49 944.4	
沈　阳	6 409.7	12.0	2 385.0	4 012.7	0
大　连	9 803.4	960.0		8 843.4	0
鞍　山	12 236.4		620.0	11 616.4	0
抚　顺	35 508.5	0	0	35 508.5	0
本　溪	17 042.0	440.0		16 602.0	
锦　州	22 046.0	250.0	130.0	21 666.0	3 573.0
长　春	11 539.2	760.0	100.6	10 678.6	150.0
吉　林	3 494.0	0	360.0	3 134.0	0
哈尔滨	280.0	0	0	280.0	0
齐齐哈尔	21 103.5	40.0	2 000.0	19 063.5	0
大　庆	10 551.3			10 551.3	
牡丹江	5 000.0			5 000.0	
上　海	94 106.5	10.0	1 110.9	92 985.6	1 716.0
南　京	31 637.2	3 500.0	1 876.0	26 261.2	0
无　锡	8 425.2	0	0	8 425.2	0
徐　州	13 351.4	40.0	375.0	12 936.4	1 325.0
常　州	5 459.0	0	20.0	5 439.0	0
苏　州	75 478.4	75.0	354.5	75 048.9	1 430.0
南　通	11 359.7	668.0	30.0	10 661.7	0
连云港	15 076.1	80.0	350.0	14 646.1	2 700.0
扬　州	2 779.6	88.0	0	2 691.6	0
杭　州	37 946.1	2 911.5	825.0	34 209.6	1 284.7

重点城市工业污染治理项目建设情况（三）（续表）

（2010）

单位：万元

城市名称	施工项目本年投资来源合计	排污费补助	政府其他补助	企业自筹	银行贷款
宁　波	5 109.6	0	41.0	5 068.6	0
温　州	809.0	25.6	10.0	773.4	0
嘉　兴	7 239.5	0	0	7 239.5	0
湖　州	6 823.7	5.0		6 818.7	190.0
绍　兴	24 128.6	0	327.0	23 801.6	363.0
台　州	5 862.0	0	149.0	5 713.0	102.0
合　肥	1 151.5	0	125.0	1 026.5	0
芜　湖	970.0	0	160.0	810.0	
马鞍山	5 727.4		306.2	5 421.2	133.0
福　州	11 640.8	11.0	105.0	11 524.8	50.0
厦　门	11 854.3	0	0	11 854.0	300.0
泉　州	96 420.9	3.0		96 417.9	0
南　昌	6 564.8	802.0	500.0	5 262.8	0
九　江	11 053.0	770.0	0	10 283.0	0
济　南	27 392.8		1 250.0	26 142.8	1 350.0
青　岛	10 219.0	315.0	580.0	9 324.0	500.0
淄　博	39 179.0	0	0	39 179.0	100.0
枣　庄	8 851.3		650.0	8 201.3	1 000.0
烟　台	27 422.9	20.0	70.0	27 332.9	0
潍　坊	131 691.2	0	2 830.0	128 861.2	0
济　宁	8 136.1	0	0	8 136.1	50.0
泰　安	2 856.1		200.0	2 656.1	
威　海	18 359.0		900.0	17 459.0	
日　照	45 978.2	8.0	0	45 970.2	20.0
郑　州	10 689.1		80.0	10 609.1	
开　封	1 020.0			1 020.0	
洛　阳	2 418.0		168.0	2 250.0	196.0
平顶山	15 193.9	75.0	1 144.2	13 974.7	
安　阳	21 220.0	0	10 000.0	11 220.0	10 000.0
焦　作	2 051.5	310.0	0	1 741.5	0
武　汉	14 968.5	1 710.0	0.0	13 258.5	0
宜　昌	21 908.9	1.0	1 705.6	20 202.3	0
荆　州	3 998.4	0	0	3 998.4	0
长　沙	2 911.5		190.0	2 721.5	145.0
株　洲	4 951.6	0	0	4 951.6	443.0
湘　潭	21 859.4		989.2	20 870.2	1 636.0
岳　阳	17 770.7	0	0	17 770.7	60.0
常　德	45 355.0	6 000.0	0	39 355.0	13 800.0

重点城市工业污染治理项目建设情况（三）（续表）

（2010）

单位：万元

城市名称	施工项目本年投资来源合计	排污费补助	政府其他补助	企业自筹	银行贷款
张家界					
广州	15 996.1	1 250.3	868.3	13 877.6	100.0
韶关	40 635.7	0	0	40 635.7	263.0
深圳	10 464.3	68.0	344.0	10 052.3	0
珠海	4 246.4	0		4 246.4	1.0
汕头	5 984.5		43.0	5 941.5	0
佛山	67 931.7	0	0	67 931.7	210.0
湛江	2 703.0	0	80.0	2 623.0	0
中山	3 578.0	0	0	3 578.0	0
南宁	7 682.7	0	130.0	7 552.7	0
柳州	6 424.6			6 424.6	
桂林	563.5			563.5	
北海	580.0			580.0	
海口					
三亚					
重庆	77 501.9	1 530.0	7 739.8	68 232.1	857.0
成都	5 081.9	62.0	0	5 020.0	125.0
攀枝花	21 123.5	460.0	140.0	20 523.4	106.0
泸州	10 276.0	100.0	0	10 176.0	0
绵阳	3 193.1	45.0	205.0	2 943.1	829.6
宜宾	1 114.2	160.0	30.0	924.3	0
贵阳	31 349.6	0	1 530.0	29 819.6	1 000.0
遵义	482.2	0	0.0	482.2	0
昆明	20 613.6	650.0	140.0	19 823.6	3 100.0
曲靖	59 363.4		680.0	58 683.4	1 429.0
拉萨					
西安	9 678.4	679.3	2 297.1	6 702.0	507.0
铜川	10 975.0	40.0	0	10 935.0	0
宝鸡	5 663.1	48.0	300.0	5 315.1	1 400.0
咸阳	26 805.5	0	0	26 805.5	0
延安	74 813.3	0	0	74 813.3	0
兰州					
金昌	38 808.9	2 155.0	60.0	36 593.9	0
西宁	6 627.6	145.0	70.0	6 412.6	0
银川	28 508.0	0	0	28 508.0	0
石嘴山	130.0			130.0	
乌鲁木齐					
克拉玛依					

重点城市工业污染治理项目建设情况（四）

（2010）

城　市 名　称	本年竣工 项目数 （个）						本年竣工项目新增设计处理能力		
		治理 废水	治理 废气	治理 固废	治理 噪声	治理 其他	治理废水 （吨/日）	治理废气（标态） （米^3/时）	治理固废 （吨/日）
总　计	**3 854**	**1 709**	**1 476**	**205**	**92**	**355**	**5 849 937**	**120 569 589**	**214 912**
北　京	46	5	33	0	2	6	2 200	1 074 400	0
天　津	99	34	40	9	1	15	69 117	4 619 215	0
石家庄	71	13	55	2	1	0	3 310	612 070	0
唐　山	17	3	11	0	1	2	3 840	6 395 000	0
秦皇岛	6	0	5	0	1	0	0	0	0
邯　郸	27	6	21	0	0	0	4 400	6 702 000	0
保　定	8	1	7	0	0	0	700	75	0
太　原	86	11	57	10	1	7	23 998	12 676 191	0
大　同	147	30	75	28	1	13	67 140	11 035 698	16 915
阳　泉	35	8	20	2	0	5	0	13 000	0
长　治	48	12	11	1	1	23	33 612	1 477 730	350
临　汾	52	16	13	7	6	10	3 605	639 356	65
呼和浩特	82	35	37	6	0	4	798	649 100	505
包　头									
赤　峰	12	2	7	0	1	2	12 000	1 928 000	300
沈　阳	11	1	7	0	0	3	70	180 164	0
大　连	16	7	6	0	0	3	1 286	120 183	0
鞍　山	7	0	4	2	0	1	0	1 709 000	0
抚　顺	7	5	2	0	0	0	44 760	1 300 000	0
本　溪	3	0	2	0	0	1	0	8 900	0
锦　州	11	4	4	2	0	1	40 600	1 050 000	100
长　春	26	17	8	1	0	0	25 320	22 000	0
吉　林	20	6	8	1	0	5	82 900	2 410	500
哈尔滨	0	0	0	0	0	0	0	172 000	0
齐齐哈尔	14	2	12	0	0	0	23 774	2 816 800	0
大　庆	17	5	4	0	0	8	0	0	0
牡丹江	0	0	0	0	0	0	0	105 333	0
上　海	156	56	74	4	11	11	16 378	3 274 776	0
南　京	70	20	39	2	2	7	242 520	665 300	0
无　锡	15	8	0	1	1	5	34 990	208 000	0
徐　州	20	6	14	0	0	0	214 900	1 294 200	0
常　州	20	11	8	0	1	0	5 860	718 300	0
苏　州	87	52	19	6	3	7	104 290	1 146 590	35 000
南　通	67	41	11	1	1	13	12 520	11 831	0
连云港	25	12	12	1	0	0	7 880	864 100	0
扬　州	6	2	3	0	0	1	3 646	30 020	0
杭　州	250	203	43	0	2	2	796 772	144 850	0

重点城市工业污染治理项目建设情况（四）（续表）

（2010）

城市名称	本年竣工项目数（个）	治理废水	治理废气	治理固废	治理噪声	治理其他	本年竣工项目新增设计处理能力 治理废水（吨/日）	治理废气（标态）（米³/时）	治理固废（吨/日）
宁波	23	14	2	0	0	7	52 540	396 200	0
温州	20	19	1	0	0	0	600	0	0
嘉兴	20	12	7	0	0	1	18 644	330 498	0
湖州	20	17	3	0	0	0	3 850	3 100	5
绍兴	100	58	27	5	2	8	23 690	669 600	52
台州	49	35	12	2	0	0	10 510	54 025	0
合肥	4	1	1	0	0	2	200	308 730	0
芜湖	1	0	1	0	0	0	2 500	9 934	0
马鞍山	11	8	3	0	0	0	1 300	760 000	0
福州	47	26	16	3	1	1	7 122	5 838 960	55
厦门	47	28	11	2	0	6	82 239	551 000	0
泉州	392	193	145	16	7	31	347 055	12 010 410	1 003
南昌	8	6	2	0	0	0	51 450	201 522	0
九江	24	10	14	0	0	0	152 480	397 714	0
济南	30	11	17	2	0	0	15 120	5 975 370	100 000
青岛	19	11	7	0	1	0	13 000	850 000	0
淄博	71	33	34	0	0	4	154 060	804 170	0
枣庄	8	3	4	0	1	0	52 720	819 656	0
烟台	56	9	8	37	1	1	17 014	527 037	3 251
潍坊	31	19	8	0	1	3	60 765	252 500	0
济宁	28	16	5	1	1	5	28 140	470 000	986
泰安	20	2	17	0	0	1	3 000	509 500	0
威海	12	9	2	0	1	0	12 570	862 900	0
日照	17	6	3	2	1	5	169 400	100 000	29 200
郑州	41	14	20	0	1	6	12 690	593 167	0
开封	2	1	1	0	0	0	5 000	0	0
洛阳	25	9	9	3	1	3	615	78 000	0
平顶山	22	8	10	0	2	2	25 195	1 780 050	0
安阳	2	1	1	0	0	0	500	10 000	0
焦作	8	2	5	1	0	0	7 080	480 000	5
武汉	18	1	0	0	0	0	0	0	0
宜昌	24	13	7	1	2	1	14 185	13 596	22 000
荆州	20	12	4	1	2	1	32 798	1 174	10
长沙	11	7	4	0	0	0	100 515	120 000	0
株洲	33	8	18	0	6	1	2 790	2 724 080	0
湘潭	17	8	6	2	1	0	108 000	200 000	10
岳阳	2	1	1	0	0	0	66 730	60 712	0
常德	14	7	6	0	1	0	5 206	2 513 385	0

重点城市工业污染治理项目建设情况（四）（续表）

（2010）

城市名称	本年竣工项目数（个）						本年竣工项目新增设计处理能力		
		治理废水	治理废气	治理固废	治理噪声	治理其他	治理废水（吨/日）	治理废气（标态）（米³/时）	治理固废（吨/日）
张家界									
广州	63	17	43	1	1	1	752 955	957 035	0
韶关	36	15	11	2	1	7	833 357	680 200	200
深圳	71	54	10	2	5	0	13 428	238 000	2
珠海	34	19	10	2	1	2	13 658	155 700	2 002
汕头	48	41	7	0	0	0	25 514	41 384	0
佛山	156	37	79	0	1	39	16 684	1 357 150	0
湛江	10	5	5	0	0	0	17 100	445 000	0
中山	24	10	14	0	0	0	7 801	343 500	0
南宁	43	27	2	13	0	1	75 870	0	20
柳州	7	3	4	0	0	0	1 542	16 054	0
桂林	5	1	4	0	0	0	0	13 800	0
北海	3	3	0	0	0	0	1 350	0	0
海口									
三亚									
重庆	89	46	37	1	1	4	330 753	2 571 020	0
成都	21	11	8	2	0	0	2 102	1 223	248
攀枝花	18	5	8	1	1	3	18 500	2 926 200	0
泸州	12	11	0	0	0	1	2 350	1 040 000	0
绵阳	10	5	2	1	1	1	2 872	120 000	0
宜宾	7	6	1	0	0	0	2 171	2 000	0
贵阳	33	15	12	2	2	2	2 610	173 048	0
遵义	6	4	2	0	0	0	1 200	50	0
昆明	131	59	39	9	4	20	32 611	1 315 781	1 903
曲靖	34	12	19	2	0	1	1 373	389 970	0
拉萨									
西安	24	7	13	2	0	2	5 865	1 443 000	90
铜川	6	0	0	0	0	6	0	0	0
宝鸡	1	1	0	0	0	0	39 415	118 474	119
咸阳	24	10	7	0	0	7	17 682	63 094	0
延安	26	4	5	1	4	12	700	100 052	16
兰州									
金昌	10	4	5	0	0	1	1 475	24 272	0
西宁	8	1	7	0	0	0	1 680	1 010 000	0
银川	13	4	8	0	0	1	82 860	80 000	0
石嘴山	0	0	0	0	0	0	0		0
乌鲁木齐									
克拉玛依									

重点城市工业企业“三废”治理效率

（2010）

单位：%

城市名称	工业废水排放达标率	工业用水重复用水率	二氧化硫排放达标率	烟尘排放达标率	工业粉尘排放达标率	工业固体废物	
						综合利用率	处置率
总计	**95.7**	**81.4**	**95.4**	**95.5**	**96.2**	**80.0**	**14.1**
北京	98.8	96.3	99.5	98.7	100.0	65.8	47.1
天津	100.0	93.8	99.9	100.0	100.0	98.6	1.5
石家庄	99.3	95.4	99.3	100.0	99.8	93.4	2.2
唐山	98.3	96.2	99.6	99.9	99.1	80.6	15.4
秦皇岛	100.0	88.1	100.0	100.0	100.0	59.6	33.6
邯郸	97.1	95.9	99.2	97.6	99.4	90.0	2.1
保定	98.2	90.7	96.3	99.0	98.6	70.8	29.2
太原	97.4	96.9	99.1	98.5	98.7	52.3	46.2
大同	96.7	90.4	98.7	96.8	91.2	60.1	36.3
阳泉	100.0	96.4	100.0	100.0	100.0	23.4	67.6
长治	100.0	95.0	99.9	99.7	99.9	67.9	23.6
临汾	93.8	94.7	99.9	99.3	94.5	81.0	9.0
呼和浩特	100.0	93.1	98.6	100.0	100.0	38.7	55.0
包头	98.3	96.2	98.4	96.7	98.2	47.4	
赤峰	97.5	95.2	99.7	99.6	99.9	38.4	54.8
沈阳	96.3	89.5	94.9	93.8	98.4	95.7	2.1
大连	95.1	65.5	98.8	97.6	96.5	94.7	5.1
鞍山	95.9	96.8	96.8	96.8	96.6	19.1	
抚顺	96.9	96.6	91.0	82.2	68.3	43.9	
本溪	83.4	97.0	69.9	67.5	95.1	37.9	
锦州	94.6	93.9	99.9	99.4	99.4	62.5	37.3
长春	96.2	91.8	100.0	99.8	99.2	99.6	
吉林	96.9	81.1	100.0	96.7	98.9	42.8	18.4
哈尔滨	98.0	95.9	97.7	96.5	99.3	89.7	8.6
齐齐哈尔	89.7	35.7	93.8	85.5	77.0	74.5	3.8
大庆	100.0	91.1	100.0	100.0	100.0	70.8	29.4
牡丹江	98.9	50.8	100.0	99.7	100.0	96.4	0.1
上海	98.0	72.4	98.8	96.9	98.8	96.2	3.8
南京	95.2	88.1	96.9	96.3	97.4	88.8	2.3
无锡	99.3	84.6	99.9	99.8	99.7	97.1	2.9
徐州	93.4	94.6	74.5	96.1	88.9	99.8	0.3
常州	100.0	80.1	99.4	100.0	100.0	94.9	5.1
苏州	99.5	89.8	100.0	100.0	100.0	98.7	1.3
南通	99.4	81.5	100.0	100.0	100.0	98.2	1.8
连云港	98.1	91.9	99.3	99.3	98.8	91.9	0.1
扬州	98.3	91.9	99.8	98.9	100.0	97.4	2.5
杭州	96.9	73.4	99.2	99.6	99.9	94.1	5.8

重点城市工业企业“三废”治理效率（续表）

（2010）

单位：%

城市名称	工业废水排放达标率	工业用水重复用水率	二氧化硫排放达标率	烟尘排放达标率	工业粉尘排放达标率	工业固体废物	
						综合利用率	处置率
宁波	95.5	34.6	96.2	98.1	98.4	89.7	5.2
温州	95.9	12.8	96.9	98.2	96.7	95.4	4.3
嘉兴	98.6	75.5	99.8	99.5	99.4	97.6	2.4
湖州	96.1	88.7	93.1	97.2	95.6	96.5	3.4
绍兴	98.4	72.0	99.9	100.0	99.8	93.2	6.1
台州	90.3	78.5	97.8	90.5	95.3	97.6	2.0
合肥	96.3	95.7	98.4	99.1	99.7	98.8	0.2
芜湖	99.3	65.8	88.5	99.3	99.1	90.0	0.4
马鞍山	98.3	96.3	99.5	99.0	98.5	65.2	32.5
福州	95.0	70.4	99.0	99.8	99.9	80.4	17.5
厦门	100.0	90.4	100.0	100.0	100.0	87.3	7.2
泉州	99.8	83.0	99.9	100.0	100.0	95.6	2.7
南昌	94.3	75.5	93.8	95.7	99.6	93.6	1.0
九江	92.1	62.1	92.4	94.8	90.8	61.2	34.1
济南	99.7	96.1	99.0	99.2	99.9	97.5	0.3
青岛	98.1	89.1	99.1	95.1	99.9	98.3	1.2
淄博	100.0	93.3	100.0	100.0	100.0	90.5	7.3
枣庄	99.3	87.5	100.0	99.9	99.9	100.0	0.0
烟台	100.0	83.5	100.0	100.0	100.0	88.5	11.5
潍坊	98.7	84.7	86.1	92.1	92.2	89.9	0.7
济宁	98.8	95.4	98.0	99.0	100.0	94.4	0.3
泰安	98.1	94.2	97.9	99.5	95.3	97.5	0.2
威海	100.0	84.5	100.0	100.0	93.8	95.7	1.1
日照	100.0	92.1	100.0	100.0	100.0	100.0	0
郑州	98.2	87.6	100.0	100.0	100.0	83.0	15.4
开封	98.4	93.5	99.9	99.8	100.0	100.0	
洛阳	100.0	95.1	98.8	99.0	99.8	54.7	44.0
平顶山	97.1	94.3	99.5	99.9	99.1	80.9	20.5
安阳	98.2	93.7	94.6	98.4	93.0	89.5	8.5
焦作	98.1	78.5	97.5	99.5	99.7	77.3	20.0
武汉	99.2	93.1	99.5	99.1	100.0	98.6	1.3
宜昌	99.7	91.6	41.6	100.0	100.0	47.3	52.6
荆州	95.2	50.3	97.0	82.4	97.4	100.0	0
长沙	91.2	92.4	91.3	92.9	91.5	99.7	0.2
株洲	96.3	85.6	99.4	97.1	98.9	82.7	10.4
湘潭	94.0	93.6	92.9	93.0	92.4	95.6	4.3
岳阳	96.9	91.7	97.7	97.2	96.7	96.1	1.0
常德	99.4	60.3	99.8	100.0	100.0	93.4	0.4

重点城市工业企业“三废”治理效率（续表）

（2010）

单位：%

城市名称	工业废水排放达标率	工业用水重复用水率	二氧化硫排放达标率	烟尘排放达标率	工业粉尘排放达标率	工业固体废物	
						综合利用率	处置率
张家界	95.8	26.2	81.7	92.9	95.8	93.2	0
广州	96.6	50.8	84.4	94.3	99.9	89.8	9.6
韶关	87.2	93.2	90.8	72.9	92.5	89.9	0.5
深圳	96.4	48.9	99.3	99.9	100.0	92.0	7.8
珠海	98.0	62.1	99.5	99.9	99.3	98.2	1.7
汕头	84.9	23.6	80.9	84.4	95.0	94.9	0.9
佛山	95.4	15.3	95.0	97.5	93.5	99.5	0.4
湛江	80.3	84.4	75.6	77.0	77.2	91.3	6.8
中山	97.0	25.6	99.0	100.0	99.6	85.2	12.1
南宁	95.5	86.0	95.7	97.5	98.9	94.0	4.3
柳州	99.1	79.5	95.4	95.4	97.6	90.4	6.7
桂林	95.9	91.3	97.8	95.1	97.8	90.9	4.6
北海	95.6	75.3	89.5	93.8	93.0	73.9	25.8
海口	100.0	75.1		100.0	100.0	97.0	
三亚	100.0	33.2		100.0	100.0	99.5	
重庆	94.7	73.1	81.2	80.0	83.0	80.2	5.5
成都	99.8	86.6	93.2	100.0	100.0	99.6	0.4
攀枝花	97.2	92.2	97.2	96.5	100.0	16.9	79.4
泸州	95.3	92.1	97.0	95.4	97.1	76.5	15.3
绵阳	99.4	83.7	98.5	98.1	98.8	88.0	7.4
宜宾	99.5	67.4	87.2	99.4	99.9	88.0	6.2
贵阳	95.8	95.3	98.0	96.3	99.3	56.2	41.9
遵义	93.2	94.4	96.4	81.4	93.2	66.6	26.3
昆明	100.0	94.9	100.0	100.0	100.0	41.3	52.8
曲靖	99.3	93.3	98.1	98.4	94.0	51.4	26.2
拉萨	44.2	16.9		15.0	0.9	1.3	
西安	95.7	80.5	89.4	98.7	92.5	98.1	1.3
铜川	99.2	95.1	99.9	98.8	96.3	82.3	8.2
宝鸡	99.7	90.1	93.2	98.8	99.5	26.9	73.0
咸阳	100.0	91.3	93.1	100.0	100.0	99.0	1.0
延安	97.2	81.3	98.2	98.7	100.0	87.8	12.1
兰州	95.2	94.3	92.4	90.5	99.6	78.9	5.7
金昌	81.4	88.5	84.1	97.6	93.8	18.5	57.9
西宁	89.0	91.2	90.7	59.1	83.2	83.6	0.0
银川	99.5	89.0	99.9	100.0	99.9	75.4	22.6
石嘴山	50.3	91.3	87.5	95.3	93.6	53.7	21.2
乌鲁木齐	78.1	93.7	88.3	92.2	90.0	68.2	5.9
克拉玛依	98.0	96.6	99.6	97.1		66.3	33.4

重点城市污水处理情况（一）

（2010）

城市名称	城市污水处理厂数（座）	污水处理能力（吨/日）	污水处理量（万吨）	处理生活污水量	污水再用量（万吨）	本年运行费用（万元）
总　计	**1 644**	**85 705 853**	**2 415 813**	**1 984 678**	**118 438**	**1 861 171.6**
北　京	68	3 702 505	114 625	111 229	26 451	100 341.4
天　津	28	2 015 500	52 585	39 785	395	41 116.3
石家庄	24	1 646 666	42 932	29 940	1 088	20 465.8
唐　山	14	1 000 000	23 828	9 999	6 456	14 587.0
秦皇岛	9	450 000	12 696	10 388	720	8 399.7
邯　郸	21	786 000	15 339	14 126	757	13 684.4
保　定	29	1 002 000	26 480	19 070	489	23 096.6
太　原	12	523 900	14 480	13 872	2 691	11 077.9
大　同	11	224 000	5 936	5 532	1 045	4 501.5
阳　泉	4	150 000	4 073	3 764	772	3 324.0
长　治	8	205 000	5 835	5 435	1 839	4 247.3
临　汾	6	169 600	3 921	3 874	314	1 185.3
呼和浩特	5	268 500	7 792	7 306	484	4 874.1
包　头	6	285 000	7 414	6 801	1 959	8 951.4
赤　峰	11	445 000	8 741	7 595	…	3 831.8
沈　阳	16	1 421 200	44 969	40 540	2 904	33 253.8
大　连	15	797 000	25 758	23 942	2 313	17 393.7
鞍　山	3	260 000	5 848	5 848	…	6 997.0
抚　顺	2	310 000	8 929	7 965	326	2 631.0
本　溪	5	325 000	8 504	4 692		2 747.8
锦　州	8	350 000	5 424	5 398	131	4 462.9
长　春	12	887 500	19 484	15 055	120	6 514.5
吉　林	7	649 000	18 495	10 955	135	21 278.0
哈尔滨	6	770 000	18 149	16 009	25	10 226.3
齐齐哈尔	4	230 000	5 669	5 653	…	2 981.1
大　庆	8	445 000	6 977	6 932	639	13 023.7
牡丹江	3	140 000	4 630	4 374	…	3 628.0
上　海	52	6 853 677	213 929	187 875	1 255	124 975.4
南　京	38	1 609 000	58 597	47 125	933	29 652.2
无　锡	69	2 435 500	52 913	28 728	1 438	69 585.0
徐　州	14	564 503	20 184	18 383	1 740	15 754.7
常　州	34	861 000	27 974	14 554	54	30 582.5
苏　州	91	2 640 200	75 028	44 151	1 281	89 351.5
南　通	55	801 500	26 955	19 521	887	18 633.3
连云港	10	283 000	7 832	6 156	112	8 426.8
扬　州	9	493 500	15 211	11 196	133	10 234.0
杭　州	24	1 800 000	78 705	47 169	4 701	81 322.3

重点城市污水处理情况（一）（续表）

（2010）

城　市 名　称	城市污水 处理厂数 （座）	污　　水 处理能力 （吨/日）	污　水 处理量 （万吨）	处理生活 污水 量	污　水 再用量 （万吨）	本年运行 费　　用 （万元）
宁　波	14	1 055 000	31 756	24 541	2 359	28 005.9
温　州	11	622 000	19 785	15 582	6	14 562.3
嘉　兴	12	425 000	13 383	10 423	228	23 438.6
湖　州	22	558 000	13 380	8 709	670	14 067.8
绍　兴	4	1 337 500	34 341	9 963	1	56 580.4
台　州	7	440 000	13 362	11 899	732	14 911.9
合　肥	8	805 000	25 138	18 446	0	13 822.0
芜　湖	5	345 000	5 032	4 839		1 418.0
马鞍山	3	180 000	8 785	3 649	3 052	3 234.3
福　州	13	680 000	23 436	21 737	82	12 317.4
厦　门	10	833 400	19 927	16 838	26	14 261.5
泉　州	13	591 000	25 215	17 580	20	19 421.8
南　昌	4	940 000	28 254	27 543	0	7 184.2
九　江	8	170 500	5 528	5 526	7	4 413.5
济　南	15	732 000	21 871	21 080	1 374	20 297.8
青　岛	19	1 365 000	38 477	25 613	1 025	44 080.8
淄　博	14	768 000	23 299	13 075	1 560	11 493.6
枣　庄	8	310 000	9 338	6 903	81	7 643.6
烟　台	18	761 500	22 985	15 402	556	15 774.6
潍　坊	22	1 083 000	32 532	19 988	0	32 194.9
济　宁	15	720 000	18 406	15 916	1 789	17 006.2
泰　安	7	350 000	9 790	9 101	789	6 963.5
威　海	7	270 000	9 197	7 139	982	7 259.7
日　照	9	248 000	6 275	3 684	91	6 426.7
郑　州	11	1 015 000	33 020	31 860	2 663	26 373.8
开　封	7	370 000	6 603	6 202	300	3 591.0
洛　阳	12	605 000	14 856	13 795	955	6 954.2
平顶山	7	400 000	12 018	9 885	365	6 003.0
安　阳	8	443 000	10 348	6 005	3 397	8 212.3
焦　作	8	340 000	11 121	9 525	2 286	8 200.2
武　汉	16	1 865 000	57 514	54 708	99	29 009.9
宜　昌	22	484 600	11 680	11 425	2	9 384.3
荆　州	8	340 000	8 205	7 938	0	5 622.8
长　沙	12	1 456 800	32 110	31 244	114	25 267.7
株　洲	9	393 000	10 225	9 758	268	4 380.0
湘　潭	5	260 000	7 793	7 741	156	4 908.5
岳　阳	8	335 000	7 984	7 984	4	4 850.2
常　德	9	330 000	6 288	5 866	50	4 570.5

重点城市污水处理情况（一）（续表）

（2010）

城市名称	城市污水处理厂数（座）	污水处理能力（吨/日）	污水处理量（万吨）	处理生活污水量	污水再用量（万吨）	本年运行费用（万元）
张家界	6	115 000	1 391	1 390	0	1 258.4
广州	19	3 615 000	93 497	88 521	1 169	53 133.6
韶关	11	260 000	5 647	5 346	0	2 500.7
深圳	18	2 815 000	103 635	84 266	1 074	52 853.6
珠海	11	583 000	13 253	11 912	131	18 412.0
汕头	8	587 500	13 405	11 538	129	8 952.8
佛山	50	1 908 402	44 626	43 491	53	34 090.1
湛江	5	250 000	7 502	7 502	0	2 606.9
中山	18	675 000	20 331	20 336	0	15 613.9
南宁	4	452 000	14 610	14 578	0	9 249.2
柳州	9	585 000	10 862	8 786	196	12 433.9
桂林	19	438 000	8 395	7 921	351	9 788.0
北海	2	150 000	2 770	2 453		1 430.3
海口	4	540 000	10 499	9 897	37	7 671.3
三亚	5	109 800	3 156	3 156	471	466.4
重庆	79	2 314 000	67 899	66 926	724	83 471.5
成都	85	1 956 800	59 341	50 660	3 736	44 266.1
攀枝花	6	100 000	1 773	1 773	0	5 586.0
泸州	5	93 500	0	0	0	0.0
绵阳	4	267 000	6 808	5 742	3	6 296.2
宜宾	9	130 500	1 774	4 753	54	3 071.1
贵阳	14	621 500	18 252	11 467	2 979	6 829.8
遵义	19	213 500	4 682	4 945	1 060	2 720.2
昆明	15	946 000	27 249	26 812	11 655	9 828.4
曲靖	3	135 000	3 385	3 322	70	2 120.0
拉萨						
西安	11	1 100 000	23 115	22 212	407	6 433.9
铜川	3	67 000	763	749	0	809.7
宝鸡	8	246 000	6 316	3 880	6	5 145.6
咸阳	12	333 000	8 622	5 506	480	6 513.2
延安	5	65 800	1 803	1 765	4	1 922.0
兰州	3	455 000	11 011	10 636	66	6 678.0
金昌	1	80 000	1 254	1 254	1 064	1 162.0
西宁	5	255 000	4 372	4 372		3 416.8
银川	7	365 000	9 212	7 948	1 140	4 817.0
石嘴山	3	155 000	2 661	2 443	112	3 009.0
乌鲁木齐	5	565 000	12 743	12 393	591	11 779.4
克拉玛依	2	130 000	3 101	3 167	0	3 386.2

重点城市污水处理情况（二）

（2010）

单位：吨

城 市 名 称	化学需氧量 去 除 量	氨 氮 去除量	总 磷 去除量	污 泥 产生量	污 泥 排放量
总 计	**6 957 900**	**488 412**	**79 484**	**17 135 816**	**50 103**
北 京	463 069	22 629	4 907	1 195 416	0
天 津	195 074	14 356	2 298	130 982	0
石家庄	157 892	12 810	890	312 269	0
唐 山	100 820	5 995	1 117	222 852	0
秦皇岛	42 961	2 792	435	57 447	0
邯 郸	41 941	2 909	392	192 372	0
保 定	75 130	6 023	1 506	296 913	3 253
太 原	51 267	2 232	567	64 639	0
大 同	21 975	2 422	948	29 607	6 324
阳 泉	11 190	1 010	64	23 468	
长 治	18 009	1 508	212	29 217	
临 汾	11 110	516	40	38 634	0
呼和浩特	22 497	1 095	187	41 648	0
包 头	32 066	1 983	479	9 909	0
赤 峰	27 436		290	9 355	0
沈 阳	125 224	4 348	918	330 441	6 058
大 连	60 576	4 074	792	1 170 554	25 549
鞍 山	11 995	843	199	42 032	0
抚 顺	24 615	721	174	51 295	
本 溪	13 471	1 816	204	9 421	0
锦 州	13 215	453	81	45 588	
长 春	56 755	3 066	894	70 583	
吉 林	62 981	2 735	208	77 548	0
哈尔滨	46 683	6 122	1 731	173 103	219
齐齐哈尔	12 372	1 276	95	14 882	0
大 庆	25 022	2 474	193	57 281	0
牡丹江	10 528	887	182	54 272	0
上 海	633 830	35 887	5 744	1 207 693	0
南 京	100 941	13 525	1 056	141 664	0
无 锡	178 695	11 347	2 058	637 155	0
徐 州	41 107	5 267	666	94 391	0
常 州	92 486	5 653	816	328 749	0
苏 州	253 064	14 041	2 906	939 935	0
南 通	83 898	3 194	781	240 755	0
连云港	15 815	1 684	139	13 767	0
扬 州	32 619	2 912	431	54 895	0
杭 州	445 212	16 441	3 528	1 176 565	0

重点城市污水处理情况（二）（续表）

（2010）

单位：吨

城市名称	化学需氧量去除量	氨氮去除量	总磷去除量	污泥产生量	污泥排放量
宁　波	80 608	5 372	1 252	222 112	0
温　州	64 600	4 997	416	120 206	0
嘉　兴	54 549	2 597	330	179 573	0
湖　州	31 011	1 526	228	85 471	0
绍　兴	237 090	15 165	1 532	632 211	0
台　州	44 744	2 513	482	85 847	0
合　肥	40 168	3 530	1 086	149 474	0
芜　湖	6 937	208	108	32 612	
马鞍山	13 055	569	134	45 882	0
福　州	43 484	4 913	725	126 874	0
厦　门	55 370	5 319	687	114 535	0
泉　州	114 494	3 162	219	146 468	0
南　昌	41 226	1 718	504	66 890	0
九　江	6 663	620	41	15 570	55
济　南	62 628	6 024	860	107 271	0
青　岛	197 258	20 091	2 088	309 187	69
淄　博	79 642	7 181	1 515	221 240	0
枣　庄	22 133	1 848	183	53 635	0
烟　台	90 869	7 662	1 083	192 764	0
潍　坊	119 498	7 238	765	123 058	0
济　宁	44 890	3 827	421	136 503	0
泰　安	27 095	2 138	331	34 359	
威　海	31 640	2 161	253	60 136	0
日　照	15 428	1 281	127	13 393	0
郑　州	92 287	7 737	1 516	243 748	0
开　封	15 242	1 430	101	20 559	0
洛　阳	52 212	4 214	710	109 311	98
平顶山	27 811	1 562	241	66 263	0
安　阳	18 428	1 894	165	78 682	0
焦　作	26 279	1 865	266	67 793	0
武　汉	60 695	6 397	1 010	183 689	0
宜　昌	21 904	2 025	186	26 780	0
荆　州	14 885	539	173	34 254	3 810
长　沙	61 197	5 240	498	161 859	0
株　洲	11 181	1 248	157	43 517	0
湘　潭	9 112	889	77	26 243	0
岳　阳	10 323	1 604	157	7 748	0
常　德	6 565	596	70	65 077	0

重点城市污水处理情况（二）（续表）

（2010）

单位：吨

城市名称	化学需氧量去除量	氨氮去除量	总磷去除量	污泥产生量	污泥排放量
张家界	1 883	263	56	2 202	0
广州	194 473	17 777	2 807	455 146	0
韶关	5 508	373	70	1 961	0
深圳	218 888	20 800	4 429	692 828	0
珠海	23 435	1 427	461	77 401	
汕头	20 360	2 615	238	60 264	0
佛山	62 023	6 545	1 232	126 503	0
湛江	13 664	1 568	188	30 805	2 060
中山	34 187	3 074	486	78 717	2 032
南宁	23 179	2 627	300	49 832	0
柳州	17 498	1 830	167	15 598	0
桂林	15 283	1 040	311	14 993	0
北海	5 962	790	93	23 430	0
海口	21 262	808	1 296	28 059	0
三亚	5 858	529	50	7 827	
重庆	183 041	18 944	2 263	316 034	0
成都	120 572	13 197	1 317	216 279	0
攀枝花	5 299	452	52	101 594	0
泸州	0	0	0	0	0
绵阳	16 371	1 348	99	27 681	0
宜宾	3 947	248	42	9 143	0
贵阳	28 586	2 409	439	52 604	11
遵义	9 021	778	53	14 493	556
昆明	70 665	5 505	1 309	275 423	0
曲靖	4 015	232	75	5 599	0
拉萨					
西安	112 119	8 466	1 028	140 642	0
铜川	2 156	322	24	5 338	0
宝鸡	17 769	1 873	209	26 272	0
咸阳	25 830	1 747	371	58 513	0
延安	6 428	623	320	18 327	0
兰州	17 515	708	314	57 100	
金昌	3 801	420	58	3 678	0
西宁	11 670	143	71	33 853	9
银川	33 966	1 091	318	50 614	0
石嘴山	5 902	311	42	16 168	
乌鲁木齐	41 466	1 139	0	37 737	0
克拉玛依	7 559	443	75	5 067	

重点城市生活及其他污染情况

（2010）

城市名称	城镇生活污水排放量（万吨）	城镇生活污水中化学需氧量排放量（吨）	城镇生活污水中氨氮排放量（吨）	生活及其他二氧化硫排放量（吨）	生活及其他烟尘排放量（吨）	生活及其他氮氧化物排放量（吨）	城镇生活污水处理率（%）
总　计	**2 415 842**	**3 877 395**	**460 935**	**1 410 934**	**961 441**	**2 183 288**	**76.4**
北　京	128 217	87 115	11 734	58 206	27 491	114 187	86.8
天　津	48 516	109 751	16 577	17 530	10 721	9 627	82.0
石家庄	31 153	78 722	5 640	43 106	22 877	15 713	96.1
唐　山	14 045	26 438	3 712	11 292	20 857	25 203	71.2
秦皇岛	10 543	11 563	1 368	11 565	7 055	20 091	98.5
邯　郸	16 599	29 728	4 685	13 489	15 848	7 923	85.1
保　定	22 990	35 774	4 276	22 813	6 782	12 044	86.5
太　原	14 046	22 659	5 175	29 538	21 162	3 196	98.8
大　同	6 648	20 909	3 387	5 392	19 038	26 427	83.2
阳　泉	4 076	8 640	994	1 876	6 034	13 325	92.4
长　治	6 855	14 151	2 242	1 967	4 458	13 958	79.3
临　汾	7 166	26 053	4 072	2 540	34 874	23 888	54.1
呼和浩特	10 082	34 148	5 859	19 187	9 002	29 921	72.5
包　头	8 108	24 391	5 134	32 307	38 441	5 052	83.9
赤　峰	9 563	14 115	4 876	35 281	16 148	21 074	79.4
沈　阳	45 933	57 868	7 906	31 572	40 150	6 745	88.3
大　连	25 417	35 508	5 137	12 723	23 470	14 000	94.2
鞍　山	9 118	37 752	2 414	10 159	2 920	10 252	64.1
抚　顺	9 317	30 052	3 627	21 776	12 809	7 010	85.5
本　溪	5 888	21 554	2 291	7 670	2 569	563	79.7
锦　州	7 746	16 442	4 564	6 261	8 100	2 883	69.7
长　春	24 353	32 179	5 458	14 929	44 613	5 635	61.8
吉　林	14 280	25 680	3 381	3 486	5 998	12 173	76.7
哈尔滨	21 832	68 116	5 846	22 376	44 484	37 813	73.3
齐齐哈尔	8 123	30 099	4 232	8 210	8 951	6 371	69.6
大　庆	8 651	24 689	2 759	4 585	8 631	26 725	80.1
牡丹江	6 670	22 501	2 908	2 650	17 710	3 940	65.6
上　海	211 554	198 225	24 320	136 624	60 348	185 078	88.8
南　京	48 615	94 803	937	5 825	3 205	43 310	96.9
无　锡	29 614	20 388	3 176	4 599	3 149	39 447	97.0
徐　州	23 713	45 101	4 830	3 819	1 019	24 751	77.5
常　州	19 363	19 329	1 591	1 299	5 420	2 057	75.2
苏　州	54 702	51 407	4 946	1 265	262	64 191	80.7
南　通	23 622	56 954	6 214	1 555	1 135	12 341	82.6
连云港	10 920	29 693	3 497	5 839	653	42 299	56.4
扬　州	13 167	33 850	3 497	4 266	3 139	6 500	85.0
杭　州	49 143	33 900	5 080	3 658	1 883	18 940	96.0

重点城市生活及其他污染情况（续表）

（2010）

城市名称	城镇生活污水排放量（万吨）	城镇生活污水中化学需氧量排放量（吨）	城镇生活污水中氨氮排放量（吨）	生活及其他二氧化硫排放量（吨）	生活及其他烟尘排放量（吨）	生活及其他氮氧化物排放量（吨）	城镇生活污水处理率（%）
宁　波	29 013	20 348	1 974	1 080	153	25 212	84.6
温　州	23 290	77 376	4 953	873	95	38 416	66.9
嘉　兴	11 848	11 118	1 760	3 033	592	38 455	88.0
湖　州	10 189	8 869	1 727	2 565	847	67 052	85.5
绍　兴	11 475	20 062	2 249	7 124	1 773	19 087	86.8
台　州	13 802	18 383	1 730	758	726	5 134	86.2
合　肥	19 206	27 823	2 605	1 312	1 110	6 560	96.0
芜　湖	6 831	31 245	2 762	735	1 180	6 241	70.8
马鞍山	4 495	17 139	2 031	357	210	5 315	81.2
福　州	27 542	44 853	2 115	2 001	480	21 818	78.9
厦　门	19 908	44 433	4 268	529	40	18 582	84.6
泉　州	22 890	53 374	4 543	4 037	1 994	22 788	76.8
南　昌	27 685	32 007	3 928	10 464	661	44 139	99.5
九　江	7 528	35 154	2 857	2 700	2 274	13 568	73.4
济　南	22 973	38 537	4 351	11 305	7 130	28 967	91.8
青　岛	27 355	31 875	7 290	26 573	10 204	18 520	93.6
淄　博	13 523	8 565	2 400	4 319	4 447	30 129	96.7
枣　庄	8 623	20 234	2 388	6 279	3 131	11 144	80.1
烟　台	17 116	26 719	3 166	8 798	2 086	22 529	90.0
潍　坊	24 237	18 798	4 922	5 998	9 819	38 741	82.5
济　宁	17 281	25 244	3 559	1 105	6 389	8 854	92.1
泰　安	10 594	19 103	2 662	12 904	8 697	14 282	85.9
威　海	8 532	10 615	2 206	20 262	4 917	4 596	83.7
日　照	5 005	11 080	1 554	4 509	1 830	2 750	73.6
郑　州	39 498	36 174	7 021	15 743	14 026	4 329	80.7
开　封	7 512	16 098	1 869	3 507	2 999	3 677	82.6
洛　阳	15 515	33 668	2 683	21 796	6 101	2 756	88.9
平顶山	12 529	17 943	2 926	8 310	2 333	24 833	78.9
安　阳	10 760	21 986	2 995	5 541	3 654	10 974	55.8
焦　作	10 614	12 154	2 143	6 420	3 341	3 665	89.7
武　汉	55 912	125 422	10 843	5 529	1 076	64 704	97.9
宜　昌	13 813	15 289	2 289	18 562	6 503	7 216	82.7
荆　州	13 876	30 732	5 966	11 289	1 401	7 500	57.2
长　沙	34 138	42 874	6 922	8 215	10 560	20 322	91.5
株　洲	14 294	51 922	3 441	9 861	6 484	9 324	68.3
湘　潭	10 273	36 153	2 593	7 636	508	380	75.4
岳　阳	14 307	70 166	5 551	5 807	2 851	11 847	55.8
常　德	11 619	46 358	5 081	20 869	7 141	3 283	50.5

重点城市生活及其他污染情况（续表）

（2010）

城市名称	城镇生活污水排放量（万吨）	城镇生活污水中化学需氧量排放量（吨）	城镇生活污水中氨氮排放量（吨）	生活及其他二氧化硫排放量（吨）	生活及其他烟尘排放量（吨）	生活及其他氮氧化物排放量（吨）	城镇生活污水处理率（%）
张家界	3 706	14 060	1 276	2 181	1 528	5 724	37.5
广州	100 526	73 964	6 571	4 416	1 896	89 412	88.1
韶关	9 972	19 299	3 288	187	199	490	53.6
深圳	94 788	17 933	5 535	350	70	98 320	88.9
珠海	14 068	19 333	2 106	800	100	15 211	84.7
汕头	19 924	56 600	6 595	765	327	9 160	57.9
佛山	54 605	39 706	8 111	206	70	40 533	79.7
湛江	18 960	43 330	5 755	616	190	13 473	39.6
中山	23 908	22 567	2 641	100	1 560	855	85.1
南宁	22 629	57 928	3 662	10 141	1 733	19 209	64.4
柳州	15 778	41 105	2 747	7 010	753	7 153	55.7
桂林	13 962	37 450	3 033	5 359	3 246	31 766	56.7
北海	9 946	25 129	1 555	2 384	996	1 452	24.7
海口	10 239	11 284	1 724	378	600	21 000	96.7
三亚	4 203	3 599	207	21	10	93	75.1
重庆	82 933	147 983	19 723	146 657	105 570	68 917	80.7
成都	57 410	70 872	9 503	76 945	10 383	6 806	88.2
攀枝花	3 261	11 652	1 066	332	114	40	54.4
泸州	5 996	30 645	2 332	5 080	2 877	1 862	0
绵阳	11 146	28 221	2 476	15 717	12 358	7 043	51.5
宜宾	6 747	24 904	2 249	9 205	2 329	908	70.5
贵阳	13 452	47 893	3 376	86 823	16 435		85.2
遵义	8 370	33 003	4 128	56 420	14 858	8 115	59.1
昆明	27 988	11 683	2 267	4 887	4 955	10 946	95.8
曲靖	6 050	28 025	2 378	4 589	9 853	11 533	54.9
拉萨							
西安	29 784	61 211	8 169	3 494	8 591	26 207	74.6
铜川	910	6 949	745	1 147	1 412	522	82.4
宝鸡	6 831	19 509	2 509	13 588	7 484	3 510	56.8
咸阳	7 878	19 868	2 717	10 024	2 651	2 921	69.9
延安	2 870	11 115	1 439	7 288	6 191	4 130	61.5
兰州	15 148	43 597	5 336	1 843	6 290	11 805	70.2
金昌	1 638	2 726	331	3 104	1 905	1 083	76.6
西宁	7 447	18 486	3 375	2 636	9 600	11 364	58.7
银川	10 524	9 309	1 407	6 794	4 737	13 389	75.5
石嘴山	3 356	4 703	424	3 393	5 492	856	72.8
乌鲁木齐	19 287	16 215	4 940	5 909	12 780	28 713	64.3
克拉玛依	3 658	5 397	578	205	129	6 425	86.6

4

各工业行业环境统计

GEGONGYE HANGYE HUANJING TONGJI

ANNUAL STATISTIC REPORT ON ENVIRONMENT IN CHINA

2010

按行业分重点调查工业废水排放及处理情况（一）

（2010）

行业名称	汇总工业企业数（个）	工业废水排放量（万吨）	直接排入海的	工业废水排放达标量（万吨）	废水治理设施数（套）	废水治理设施处理能力（万吨/日）	本年运行费用（万元）
行业总计	**112 798**	**2 118 585.3**	**118 000.5**	**2 030 906.5**	**80 331**	**24 762.2**	**5 453 456.4**
煤炭开采和洗选业	4 623	104 765.3	36.1	97 541.5	3 605	1 395.2	139 888.1
石油和天然气开采业	228	11 554.6	401.6	11 530.3	826	390.4	174 405.4
黑色金属矿采选业	1 232	15 352.5	96.2	14 001.6	1 116	899.2	65 609.8
有色金属矿采选业	1 679	38 851.6	541.8	36 000.5	1 782	427.9	86 047.6
非金属矿采选业	763	7 683.1	53.0	7 312.6	506	85.7	12 228.8
其他采矿业	78	375.3	…	356.1	40	4.1	991.0
农副食品加工业	8 132	143 099.6	2 290.9	131 537.6	4 859	888.5	140 389.8
食品制造业	3 683	54 549.0	343.3	50 303.9	2 486	275.2	185 701.2
饮料制造业	2 968	75 519.2	979.7	70 858.7	1 931	369.4	102 276.5
烟草制品业	158	2 672.7	4.1	2 630.8	126	14.4	4 090.0
纺织业	7 891	245 469.5	5 941.4	239 135.1	7 458	1 073.8	401 227.7
纺织服装、鞋、帽制造业	1 063	12 038.6	128.8	11 613.0	633	66.6	19 463.5
皮革毛皮羽毛（绒）及其制品业	1 537	28 173.3	149.7	26 527.2	1 169	128.2	55 946.5
木材加工及木竹藤棕草制品业	1 797	5 036.0	7.6	4 607.7	577	19.3	6 306.4
家具制造业	487	2 146.0	12.0	2 114.4	145	5.3	2 655.0
造纸及纸制品业	5 570	393 699.1	8 040.7	378 018.5	5 551	2 548.2	649 143.8
印刷业和记录媒介的复制	436	1 578.2	6.4	1 524.3	170	5.0	4 590.8
文教体育用品制造业	238	1 070.8	38.2	1 033.2	140	3.5	1 911.1
石油加工、炼焦及核燃料加工业	1 139	70 024.3	25 175.5	68 699.9	1 796	355.8	368 473.9
化学原料及化学制品制造业	10 157	309 006.2	5 520.9	296 943.4	9 874	3 115.6	651 898.3
医药制造业	2 920	52 605.5	119.6	51 246.5	2 565	158.9	129 512.4
化学纤维制造业	370	42 370.5	2 744.6	40 130.8	389	177.2	84 688.3
橡胶制品业	825	7 042.4	139.4	6 930.7	359	61.7	10 840.1
塑料制品业	1 354	4 962.2	53.6	4 743.0	561	14.8	9 393.0
非金属矿物制品业	23 328	32 313.1	94.1	30 313.8	4 782	424.8	96 789.8
黑色金属冶炼及压延加工业	3 513	116 947.7	2 100.7	114 992.3	4 294	9 298.3	959 773.9
有色金属冶炼及压延加工业	2 722	31 118.2	85.5	29 906.0	2 228	428.3	196 851.1
金属制品业	6 221	30 151.5	556.8	28 962.4	5 086	180.8	143 827.7
通用设备制造业	3 660	13 055.1	116.7	12 638.2	1 418	49.5	28 403.2
专用设备制造业	1 251	9 714.4	38.5	9 440.4	782	189.2	15 197.6
交通运输设备制造业	2 388	26 219.4	1 351.7	25 346.8	2 040	84.8	122 093.1
电气机械及器材制造业	1 725	11 652.4	130.9	10 983.6	1 318	44.4	30 385.5
通信计算机及其他电子设备制造业	1 862	35 965.2	597.9	35 427.2	2 065	202.5	154 387.4
仪器仪表及文化办公用机械制造业	400	4 965.3	46.0	4 907.5	347	19.6	10 634.6
工艺品及其他制造业	680	2 559.4	134.0	2 389.5	452	10.9	7 245.0
废弃资源和废旧材料回收加工业	221	1 147.3	14.6	1 030.4	107	4.9	4 804.4
电力、热力的生产和供应业	4 132	129 623.5	59 682.7	127 132.4	6 120	1 201.9	269 954.8
燃气生产和供应业	92	1 931.4	46.0	1 781.6	58	5.2	5 935.1
水的生产和供应业	205	31 189.1	160.5	30 907.7	109	113.8	52 872.6
其他行业	1 070	10 386.5	18.9	9 405.8	461	19.5	46 621.6

按行业分重点调查工业废水排放及处理情况（二）

（2010）　　　　单位：吨

行业名称	工业废水中污染物排放量					
	汞	镉	六价铬	铅	砷	挥发酚
行业总计	**1.047**	**30.129**	**54.804**	**140.805**	**118.092**	**966.037**
煤炭开采和洗选业	…	0.001	0.053	0.014	0.153	260.134
石油和天然气开采业	0.003	…	…	…	…	7.120
黑色金属矿采选业	…	0.161	0.097	4.519	0.639	0.011
有色金属矿采选业	0.185	11.891	1.563	86.741	42.904	0.954
非金属矿采选业	0.010	0.033	0.541	0.504	0.490	2.010
其他采矿业	…	…	…	…	…	0
农副食品加工业	0.016	0.015	0.176	0.070	0.026	2.035
食品制造业	…	…	0.003	…	…	1.843
饮料制造业	…	…	…	…	…	0.612
烟草制品业	…	…	…	…	…	0.007
纺织业	0.001	0.033	0.814	0.102	0.003	3.993
纺织服装、鞋、帽制造业	…	…	0.039	…	…	0.833
皮革毛皮羽毛（绒）及其制品业	…	0.800	7.237	0.001	…	0.308
木材加工及木竹藤棕草制品业	…	…	…	0.069	…	0.634
家具制造业	…	0.001	0.003	…	…	0.001
造纸及纸制品业	…	…	0.062	0.006	0.006	42.947
印刷业和记录媒介的复制	…	0.002	0.034	…	…	0.787
文教体育用品制造业	…	…	0.122	0.003	…	0.004
石油加工、炼焦及核燃料加工业	0.008	0.108	0.287	0.597	0.425	447.434
化学原料及化学制品制造业	0.167	3.776	1.911	3.751	48.367	86.34
医药制造业	…	…	0.007	0.002	0.075	23.296
化学纤维制造业	…	…	0.002	…	…	1.827
橡胶制品业	…	…	0.006	…	…	0.005
塑料制品业	…	0.002	0.043	0.013	…	0.007
非金属矿物制品业	…	0.025	1.175	0.478	0.006	7.003
黑色金属冶炼及压延加工业	0.299	0.458	4.152	8.386	0.967	45.758
有色金属冶炼及压延加工业	0.331	12.259	2.724	28.546	23.508	5.178
金属制品业	…	0.239	21.761	0.919	0.318	0.692
通用设备制造业	…	0.038	3.069	0.113	0.093	0.439
专用设备制造业	0.018	0.012	2.805	0.350	0.006	0.586
交通运输设备制造业	0.004	0.098	3.020	0.171	0.002	0.994
电气机械及器材制造业	0.001	0.056	0.991	3.324	0.003	0.665
通信计算机及其他电子设备制造业	0.001	0.089	1.520	1.965	0.043	1.056
仪器仪表及文化办公用机械制造业	…	0.011	0.274	0.072	0.010	0.023
工艺品及其他制造业	…	0.008	0.132	0.042	…	3.876
废弃资源和废旧材料回收加工业	…	0.014	0.001	0.014	…	0
电力、热力的生产和供应业	0.004	…	0.027	0.033	0.015	2.242
燃气生产和供应业	…	…	…	…	…	1.219
水的生产和供应业	…	…	0.114	…	0.033	4.809
其他行业	…	…	0.040	…	…	8.360

按行业分重点调查工业废水排放及处理情况（三）

（2010）

单位：吨

行业名称	工业废水中污染物排放量			
	氰化物	化学需氧量	石油类	氨氮
行业总计	**241.8**	**3 656 347.0**	**10 139.3**	**245 372.3**
煤炭开采和洗选业	1.8	100 174.0	858.2	5 669.4
石油和天然气开采业	0.1	32 593.1	408.7	1 001.6
黑色金属矿采选业	…	10 031.3	24.7	1 797.9
有色金属矿采选业	2.4	32 176.9	24.4	1 553.3
非金属矿采选业	…	8 360.7	9.1	363.9
其他采矿业	…	341.7	…	12.1
农副食品加工业	8.4	495 714.9	114.1	21 090.8
食品制造业	…	124 749.8	195.4	8 375.7
饮料制造业	0.6	219 009.5	50.7	8 407.4
烟草制品业	…	3 484.5	6.0	185.8
纺织业	…	300 607.6	142.5	17 409.7
纺织服装、鞋、帽制造业	…	13 188.8	8.2	754.7
皮革毛皮羽毛（绒）及其制品业	…	64 325.3	95.9	5 065.6
木材加工及木竹藤棕草制品业	…	10 820.3	8.4	454.2
家具制造业	…	4 437.6	5.5	110.2
造纸及纸制品业	0.1	952 199.5	187.3	24 999.7
印刷业和记录媒介的复制	…	1 599.8	23.8	121.6
文教体育用品制造业	0.2	917.7	7.0	82.1
石油加工、炼焦及核燃料加工业	22.9	81 718.9	1 388.0	7 706.3
化学原料及化学制品制造业	88.5	446 906.5	1 705.7	76 031.7
医药制造业	0.1	108 352.8	182.0	6 807.0
化学纤维制造业	2.3	125 048.3	176.8	3 882.2
橡胶制品业	…	8 042.7	30.3	530.1
塑料制品业	…	7 276.2	11.1	541.2
非金属矿物制品业	1.5	31 161.5	167.8	2 028.4
黑色金属冶炼及压延加工业	35.8	106 491.9	1 774.0	8 575.4
有色金属冶炼及压延加工业	0.6	29 628.0	290.7	6 762.3
金属制品业	39.1	25 307.4	293.2	1 889.2
通用设备制造业	0.5	14 924.5	356.9	996.0
专用设备制造业	0.1	10 225.4	157.5	925.2
交通运输设备制造业	30.4	53 822.8	721.0	2 605.8
电气机械及器材制造业	0.6	12 769.5	124.5	727.9
通信计算机及其他电子设备制造业	3.2	28 571.7	119.3	2 738.1
仪器仪表及文化办公用机械制造业	0.1	4 020.3	32.1	383.2
工艺品及其他制造业	0.2	3 311.0	6.8	171.2
废弃资源和废旧材料回收加工业	…	1 553.2	0.8	86.9
电力、热力的生产和供应业	0.1	54 843.8	304.4	4 677.9
燃气生产和供应业	1.4	2 721.3	7.7	954.6
水的生产和供应业	0.7	25 441.1	106.4	969.5
其他行业	…	99 475.1	12.5	17 926.4

按行业分重点调查工业废水排放及处理情况（四）

（2010）

单位：吨

行业名称	工业废水中污染物去除量				
	挥发酚	氰化物	化学需氧量	石油类	氨氮
行业总计	**73 271.7**	**7 206.4**	**14 153 851.3**	**354 246.6**	**826 486.7**
煤炭开采和洗选业	16.5	2.1	136 901.5	839.6	2 304.3
石油和天然气开采业	117.9	0.1	123 759.2	172 504.0	5 869.7
黑色金属矿采选业	13 434.3	…	62 730.2	292.8	46.9
有色金属矿采选业	1.3	278.6	105 589.8	14.6	1 910.4
非金属矿采选业	…	…	16 862.0	49.1	416.0
其他采矿业	…	…	925.0	…	8.7
农副食品加工业	4.7	6.7	1 633 059.0	185.8	34 943.6
食品制造业	0.1	0.6	611 459.2	128.5	47 592.7
饮料制造业	0.6	…	1 094 975.9	54.9	14 013.4
烟草制品业	…	…	11 947.1	4.7	82.1
纺织业	37.0	7.0	1 439 892.8	85.5	31 919.4
纺织服装、鞋、帽制造业	0.1	…	48 921.1	2.8	1 265.9
皮革毛皮羽毛（绒）及其制品业	2.3	…	334 118.7	273.7	12 133.9
木材加工及木竹藤棕草制品业	3.4	…	19 057.7	6.1	430.6
家具制造业	…	…	10 695.4	22.1	79.9
造纸及纸制品业	33.9	0.2	4 459 122.2	165.6	24 989.0
印刷业和记录媒介的复制	0.1	…	2 873.4	59.3	119.2
文教体育用品制造业	0.4	6.1	1 283.5	21.4	84.1
石油加工、炼焦及核燃料加工业	20 398.2	1 461.5	486 362.1	98 029.9	150 595.6
化学原料及化学制品制造业	4 821.7	1 368.0	1 418 349.2	28 846.5	382 911.3
医药制造业	180.0	7.4	684 600.7	4 257.1	41 612.8
化学纤维制造业	0.4	12.8	306 474.1	470.4	4 238.3
橡胶制品业	…	…	13 212.2	46.3	270.9
塑料制品业	…	0.4	14 772.0	10.3	674.5
非金属矿物制品业	17.6	3.9	51 724.6	660.8	1 466.3
黑色金属冶炼及压延加工业	32 813.6	1 490.8	490 313.0	38 917.9	22 145.0
有色金属冶炼及压延加工业	3.9	125.4	40 519.0	632.8	17 383.4
金属制品业	1.6	2 112.9	58 969.1	1 222.5	1 997.4
通用设备制造业	240.2	48.8	18 854.8	437.4	1 886.2
专用设备制造业	3.0	2.5	16 844.3	258.0	592.9
交通运输设备制造业	3.3	29.5	63 655.8	3 071.7	4 491.6
电气机械及器材制造业	2.6	16.4	26 522.8	422.5	588.2
通信计算机及其他电子设备制造业	1.9	95.9	81 993.6	413.5	4 901.2
仪器仪表及文化办公用机械制造业	0.3	2.7	9 075.2	48.5	401.4
工艺品及其他制造业	0.5	17.8	9 710.8	31.3	186.8
废弃资源和废旧材料回收加工业	…	0	3 797.4	7.4	134.3
电力、热力的生产和供应业	18.8	0.2	86 776.2	420.3	3 811.2
燃气生产和供应业	756.9	27.4	10 518.8	49.1	540.3
水的生产和供应业	354.7	80.8	108 336.2	1 240.5	2 678.2
其他行业	…	0.2	38 296.1	41.7	4 769.0

按行业分重点调查工业废气排放及处理情况（一）

（2010）

行业名称	工业废气排放总量（标态）（亿米³）	燃料燃烧废气排放量	生产工艺废气排放量	燃料煤消费量（万吨）	原料煤消费量（万吨）	燃料油消费量（万吨）
行业总计	**519 166.6**	**303 896.0**	**215 270.6**	**224 584.1**	**90 739.3**	**2 087.4**
煤炭开采和洗选业	2 323.8	2 061.2	262.6	1 729.2	13 096.7	20.3
石油和天然气开采业	1 026.3	959.2	67.2	208.9	…	102.0
黑色金属矿采选业	2 471.5	734.2	1 737.3	163.5	93.8	4.8
有色金属矿采选业	468.6	182.2	286.4	126.0	39.4	4.3
非金属矿采选业	861.3	436.2	425.1	374.0	53.8	0.4
其他采矿业	29.7	18.6	11.1	20.8	2.5	…
农副食品加工业	4 154.3	3 775.6	378.7	1 637.1	0.8	41.2
食品制造业	3 270.2	3 184.6	85.6	1 101.9	41.3	38.0
饮料制造业	3 097.1	3 082.0	15.1	1 126.7	4.7	16.1
烟草制品业	505.4	197.6	307.9	78.0	…	5.0
纺织业	3 258.0	3 031.1	227.0	2 443.9	1.9	23.4
纺织服装、鞋、帽制造业	175.5	156.0	19.5	114.2	…	3.6
皮革毛皮羽毛（绒）及其制品业	176.4	146.1	30.3	117.5	…	5.1
木材加工及木竹藤棕草制品业	1 567.4	646.4	921.0	222.9	2.7	1.0
家具制造业	134.2	30.3	104.0	18.4	0.1	1.0
造纸及纸制品业	7 697.0	7 363.9	333.2	4 920.3	14.0	21.9
印刷业和记录媒介的复制	110.1	35.1	75.1	20.4	…	5.4
文教体育用品制造业	53.4	25.1	28.3	7.6	…	2.1
石油加工炼焦及核燃料加工业	18 711.8	7 623.6	11 088.1	3 885.6	29 295.4	775.8
化学原料及化学制品制造业	25 741.2	12 885.4	12 855.8	9 289.5	7 832.3	155.6
医药制造业	1 603.0	1 316.5	286.5	823.1	3.7	32.1
化学纤维制造业	2 766.9	1 712.5	1 054.3	1 190.7	34.3	23.1
橡胶制品业	1 020.8	514.9	506.0	433.1	2.6	5.3
塑料制品业	766.8	314.6	452.1	268.2	0.1	6.1
非金属矿物制品业	87 263.1	29 282.7	57 980.4	9 183.4	14 562.6	355.6
黑色金属冶炼及压延加工业	122 928.3	29 857.7	93 070.7	5 964.2	23 060.5	21.0
有色金属冶炼及压延加工业	24 299.2	6 114.2	18 185.0	3 781.2	1 108.5	106.3
金属制品业	2 077.1	561.4	1 515.7	314.7	2.3	18.8
通用设备制造业	2 429.9	1 040.8	1 389.0	358.0	225.5	14.8
专用设备制造业	1 938.8	790.5	1 148.3	503.0	381.1	7.0
交通运输设备制造业	4 183.5	643.7	3 539.9	459.1	7.8	24.0
电气机械及器材制造业	1 062.8	177.6	885.2	103.6	0.5	14.1
通信计算机及其他电子设备制造业	6 368.7	1 812.9	4 555.7	62.5	0	16.4
仪器仪表及文化办公用机械制造业	554.8	30.4	524.4	15.4	0.1	1.7
工艺品及其他制造业	154.1	64.7	89.4	55.1	35.6	1.4
废弃资源和废旧材料回收加工业	85.3	60.7	24.6	20.4	1.2	0.5
电力、热力的生产和供应业	182 550.4	182 477.2	73.2	172 991.9	52.9	208.5
燃气生产和供应业	834.3	218.3	616.0	167.3	770.8	0.1
水的生产和供应业	10.8	10.7	0.1	6.3	…	2.8
其他行业	434.9	320.0	114.9	276.5	9.8	0.8

按行业分重点调查工业废气排放及处理情况（二）

（2010）

单位：吨

行业名称	二氧化硫去除量	燃料燃烧过程中去除的	生产工艺过程中去除的	二氧化硫排放量	燃料燃烧过程中排放的	生产工艺过程中排放的
行业总计	**33 039 968.4**	**22 312 865.4**	**10 727 102.9**	**17 054 509.9**	**14 416 983.3**	**2 637 811.7**
煤炭开采和洗选业	49 633.1	47 889.7	1 743.4	160 254.5	140 170.6	20 083.9
石油和天然气开采业	60 876.4	9 566.1	51 310.3	35 589.4	24 861.8	10 727.6
黑色金属矿采选业	20 655.7	13 051.8	7 603.9	52 769.2	28 356.5	24 412.7
有色金属矿采选业	560 344.3	4 393.3	555 951.0	111 246.5	12 869.0	98 377.4
非金属矿采选业	21 514.0	21 391.3	122.7	41 345.0	36 864.0	4 481.0
其他采矿业	110.7	110.7	…	1 309.4	1 053.1	256.3
农副食品加工业	68 836.0	66 710.7	2 125.3	169 325.3	168 722.7	602.5
食品制造业	47 507.9	47 122.2	385.7	115 645.9	115 562.0	83.9
饮料制造业	54 865.7	54 005.6	860.0	111 783.4	111 757.1	26.3
烟草制品业	15 185.3	14 893.9	291.3	10 013.1	9 979.9	33.2
纺织业	99 646.3	99 086.7	559.5	247 218.3	246 899.4	318.9
纺织服装、鞋、帽制造业	2 846.4	2 844.9	1.5	11 193.2	11 190.5	2.7
皮革毛皮羽毛（绒）及其制品业	3 293.0	3 293.0	…	14 016.3	14 011.1	5.2
木材加工及木竹藤棕草制品业	6 558.6	6 539.6	19.0	32 566.4	32 021.6	544.8
家具制造业	563.8	563.8	…	2 211.5	2 199.8	11.7
造纸及纸制品业	231 805.1	230 259.9	1 545.2	508 205.9	501 327.0	6 878.8
印刷业和记录媒介的复制	472.6	472.6	…	2 995.2	2 979.3	15.9
文教体育用品制造业	247.2	243.1	4.1	1 094.7	1 091.2	3.5
石油加工炼焦及核燃料加工业	2 388 861.2	272 813.3	2 116 047.9	635 334.4	310 221.9	325 112.6
化学原料及化学制品制造业	1 232 995.5	614 652.0	618 343.5	1 040 039.7	853 242.5	186 868.2
医药制造业	42 501.6	42 200.8	300.8	79 394.7	79 130.0	264.7
化学纤维制造业	172 634.7	162 690.9	9 943.8	106 883.5	103 696.1	3 187.4
橡胶制品业	36 871.7	36 870.1	1.6	39 475.3	39 102.5	372.8
塑料制品业	13 678.7	13 666.2	12.5	29 451.6	29 423.3	28.3
非金属矿物制品业	404 773.3	157 795.7	246 977.6	1 686 183.0	1 097 843.8	588 483.8
黑色金属冶炼及压延加工业	826 440.2	359 231.6	467 208.6	1 766 510.8	806 034.9	960 541.7
有色金属冶炼及压延加工业	7 125 182.6	494 113.7	6 631 068.9	803 326.3	422 199.0	381 127.3
金属制品业	10 732.6	10 303.0	429.6	35 033.2	34 195.4	837.8
通用设备制造业	23 263.3	16 483.0	6 780.2	50 474.6	39 691.5	10 786.6
专用设备制造业	24 638.4	24 613.9	24.6	39 112.4	38 832.0	280.3
交通运输设备制造业	16 667.4	15 846.7	820.7	33 874.0	33 534.6	339.4
电气机械及器材制造业	6 505.5	6 476.3	29.1	13 488.3	12 410.7	1 077.6
通信计算机及其他电子设备制造业	3 814.7	3 781.1	33.7	6 523.1	6 473.0	50.1
仪器仪表及文化办公用机械制造业	1 115.5	1 115.4	0.1	1 400.8	1 389.0	11.9
工艺品及其他制造业	403.9	403.9	…	8 827.6	8 533.8	293.8
废弃资源和废旧材料回收加工业	4 924.0	2 256.3	2 667.7	2 221.4	2 018.2	203.2
电力、热力的生产和供应业	19 417 438.2	19 415 231.4	2 206.8	8 997 910.7	8 996 565.6	1 345.1
燃气生产和供应业	15 777.9	14 095.5	1 682.4	20 091.5	13 073.6	7 017.9
水的生产和供应业	361.4	361.4	…	2 067.5	518.7	1 548.8
其他行业	25 424.2	25 424.1	0.1	28 102.4	26 936.3	1 166.1

按行业分重点调查工业废气排放及处理情况（三）

（2010）

单位：吨

行业名称	氮氧化物去除量	氮氧化物排放量	烟尘去除量	烟尘排放量	粉尘去除量	粉尘排放量
行业总计	**686 963.8**	**13 749 015.4**	**389 413 777.2**	**5 492 438.9**	**95 017 184.4**	**4 089 407.3**
煤炭开采和洗选业	661.8	73 945.1	1 186 506.4	116 162.1	161 705.7	149 118.8
石油和天然气开采业	103.3	27 192.6	83 519.9	12 836.7	52.9	12.3
黑色金属矿采选业	9.4	15 451.8	137 939.0	18 868.6	160 786.1	53 334.5
有色金属矿采选业	296.7	6 730.2	165 635.1	12 830.6	52 814.8	8 585.0
非金属矿采选业	365.4	18 424.3	262 617.8	24 712.6	42 848.9	28 589.3
其他采矿业	56.9	662.8	9 353.6	1 928.7	9 768.3	1 264.4
农副食品加工业	6 999.1	111 312.4	1 238 499.5	119 662.9	102 259.4	4 570.9
食品制造业	1 786.7	65 176.4	902 607.9	58 518.5	70 887.6	782.1
饮料制造业	1 655.3	65 323.5	746 876.1	71 876.9	6 632.8	766.1
烟草制品业	360.9	4 559.1	22 395.7	3 959.7	19 721.9	827.4
纺织业	3 118.6	129 287.5	1 246 885.5	119 850.5	3 269.9	1 188.5
纺织服装、鞋、帽制造业	121.4	6 770.9	75 129.2	7 835.1	363.3	202.5
皮革毛皮羽毛（绒）及其制品业	134.1	7 533.8	38 927.3	8 427.9	47.2	48.8
木材加工及木竹藤棕草制品业	121.6	15 759.8	145 975.9	43 125.9	540 601.0	24 960.8
家具制造业	2.9	1 370.5	11 843.7	1 659.2	3 449.0	1 237.4
造纸及纸制品业	10 056.8	236 312.8	3 751 233.9	195 708.2	46 337.1	5 679.0
印刷业和记录媒介的复制	15.4	1 568.5	6 492.4	1 703.2	15.2	7.0
文教体育用品制造业	0.5	504.1	1 773.9	794.2	1 797.4	231.7
石油加工炼焦及核燃料加工业	73 991.3	347 195.2	3 237 247.0	230 269.6	859 311.1	192 105.8
化学原料及化学制品制造业	54 583.0	563 414.0	7 852 093.0	435 553.3	1 172 198.8	141 193.7
医药制造业	1 218.6	47 292.6	494 844.3	44 044.2	673.8	658.7
化学纤维制造业	1 783.9	77 972.8	700 703.3	27 625.2	38 012.6	666.2
橡胶制品业	220.0	25 092.7	316 112.0	16 691.2	12 242.2	737.6
塑料制品业	78.4	14 939.1	75 889.3	10 711.1	1 376.0	223.2
非金属矿物制品业	47 825.8	1 593 760.5	20 521 182.6	1 069 290.9	56 431 463.8	2 324 691.5
黑色金属冶炼及压延加工业	25 305.6	931 009.6	21 995 428.8	562 751.3	28 634 591.3	934 691.0
有色金属冶炼及压延加工业	15 867.2	226 600.2	6 747 473.7	149 764.3	5 428 639.9	103 847.9
金属制品业	1 076.0	18 186.4	83 816.4	19 069.8	64 436.9	7 168.1
通用设备制造业	297.2	25 599.5	132 946.4	37 802.2	450 471.3	34 634.5
专用设备制造业	97.8	43 330.9	212 846.7	11 666.3	103 074.8	17 738.7
交通运输设备制造业	749.3	37 033.6	368 310.9	26 986.1	158 391.8	32 071.3
电气机械及器材制造业	1 098.3	7 797.3	55 816.3	6 469.8	2 465.2	855.7
通信计算机及其他电子设备制造业	440.3	8 515.0	17 750.4	3 397.2	4 258.1	640.9
仪器仪表及文化办公用机械制造业	8.4	1 000.9	3 791.9	522.9	2 085.4	330.3
工艺品及其他制造业	2.5	2 216.9	13 976.4	3 075.6	315.8	235.9
废弃资源和废旧材料回收加工业	77.3	1 442.3	6 293.1	749.4	2 086.1	2 814.3
电力、热力的生产和供应业	433 230.3	8 957 688.9	316 321 472.0	1 989 462.6	4 696.2	6 216.5
燃气生产和供应业	386.9	10 811.6	166 683.8	7 702.2	390.9	3 216.5
水的生产和供应业	…	652.8	1 235.2	527.5	…	…
其他行业	2 758.9	19 576.4	53 651.2	17 844.8	422 644.2	3 262.8

按行业分重点调查工业废气排放及处理情况（四）

（2010）

行业名称	废气治理设施数（套）	脱硫设施数	废气治理设施处理能力（标态）（万米³/时）	脱硫设施脱硫能力（吨/时）	废气治理设施运行费用（万元）	脱硫设施运行费用
行业总计	**187 400**	**26 611**	**965 616.8**	**31 358.3**	**10 545 218.8**	**4 349 929.9**
煤炭开采和洗选业	4 449	1 547	5 472.3	1 892.3	49 007.6	23 386.8
石油和天然气开采业	345	105	1 121.9	16.3	11 250.7	3 851.0
黑色金属矿采选业	685	121	3 550.5	4.5	23 382.9	4 691.3
有色金属矿采选业	637	150	946.7	174.5	34 184.0	24 639.8
非金属矿采选业	508	111	1 577.3	25.4	10 715.1	3 636.8
其他采矿业	21	4	61.1	0.1	160.4	18.8
农副食品加工业	5 205	875	12 350.0	413.2	79 657.0	22 405.2
食品制造业	2 993	615	6 798.7	380.1	27 423.7	8 906.6
饮料制造业	2 398	736	5 416.9	268.2	30 828.8	10 852.7
烟草制品业	811	89	1 031.2	99.9	13 080.8	963.3
纺织业	8 233	1 898	10 845.9	671.7	413 472.5	66 264.5
纺织服装、鞋、帽制造业	770	127	802.5	64.5	4 815.4	1 173.8
皮革毛皮羽毛（绒）及其制品业	1 222	212	639.0	15.3	5 702.8	1 419.6
木材加工及木竹藤棕草制品业	2 090	196	6 158.4	22.0	14 307.2	1 186.6
家具制造业	824	25	1 454.5	0.3	4 396.6	208.0
造纸及纸制品业	5 281	1 112	14 433.0	560.4	141 794.3	51 135.6
印刷业和记录媒介的复制	191	57	205.5	55.6	44 211.5	648.3
文教体育用品制造业	226	23	168.5	12.1	1 155.0	137.6
石油加工炼焦及核燃料加工业	2 761	744	18 338.2	2 274.3	498 975.3	268 086.0
化学原料及化学制品制造业	17 241	3 345	57 436.4	2 706.1	523 021.1	143 333.6
医药制造业	3 179	648	5 795.6	395.6	40 594.7	10 772.0
化学纤维制造业	880	260	3 841.1	136.1	50 326.7	16 366.6
橡胶制品业	8 507	275	1 851.9	71.1	22 897.1	5 823.9
塑料制品业	1 552	234	3 997.6	1 500.7	17 970.8	2 481.4
非金属矿物制品业	52 931	1 983	155 710.5	1 247.6	1 012 009.9	47 433.7
黑色金属冶炼及压延加工业	15 105	1 020	218 010.9	891.0	2 076 732.4	266 324.4
有色金属冶炼及压延加工业	7 034	1 281	43 157.6	1 568.4	567 605.2	241 824.6
金属制品业	4 500	382	6 696.1	58.5	30 702.8	3 380.7
通用设备制造业	3 954	655	8 262.6	104.9	39 997.4	6 642.5
专用设备制造业	1 965	258	3 878.7	553.3	17 414.3	3 717.9
交通运输设备制造业	6 692	538	10 340.2	109.8	56 535.1	5 763.6
电气机械及器材制造业	2 978	159	4 610.8	6.0	18 549.7	1 274.6
通信计算机及其他电子设备制造业	3 951	132	9 276.3	432.3	149 325.6	1 698.4
仪器仪表及文化办公用机械制造业	537	32	911.6	11.1	7 119.3	296.7
工艺品及其他制造业	323	47	294.7	2.3	1 856.0	161.0
废弃资源和废旧材料回收加工业	167	17	207.3	9.7	2 856.7	544.2
电力、热力的生产和供应业	15 173	6 198	338 051.1	14 497.9	4 286 210.3	3 090 608.4
燃气生产和供应业	392	125	928.9	86.8	5 897.6	2 479.8
水的生产和供应业	27	2	20.6	0.1	122.2	46.2
其他行业	662	273	963.7	18.2	208 952.3	5 343.4

按行业分重点调查工业固体废物产生及处置利用情况（一）

（2010）

单位：万吨

行业名称	工业固体废物产生量	危险废物	冶炼废渣	粉煤灰	炉渣	煤矸石	尾矿	放射性废物	脱硫石膏	其他废物
行业总计	**225 094**	**1 587**	**32 215**	**42 907**	**21 899**	**25 600**	**65 548**	**28**	**5 941**	**29 385**
煤炭开采和洗选业	27 316	0	9	154	356	24 003	1 537	0	5	1 252
石油和天然气开采业	207	18	0	13	31	0	0	0	0	145
黑色金属矿采选业	31 969	0	50	18	40	5	28 040	0	1	3 813
有色金属矿采选业	29 338	164	117	11	71	11	25 066	20	1	3 877
非金属矿采选业	1 780	93	7	28	62	1	1 326	0	1	262
其他采矿业	73	0	1	4	1	1	54	0	0	11
农副食品加工业	2 120	0	0	143	328	0	1	0	12	1 635
食品制造业	668	1	0	183	203	1	0	0	9	272
饮料制造业	929	0	0	78	238	0	0	0	7	605
烟草制品业	41	0	0	3	20	0	0	0	0	17
纺织业	754	17	0	109	494	3	0	0	4	127
纺织服装、鞋、帽制造业	56	0	0	9	21	0	3	0	0	22
皮革毛皮羽毛（绒）及其制品业	75	3	0	3	29	0	0	0	0	40
木材加工及木竹藤棕草制品业	210	0	0	14	60	0	0	0	0	136
家具制造业	18	0	0	1	5	0	0	0	0	12
造纸及纸制品业	2 321	124	9	538	762	0	1	0	19	868
印刷业和记录媒介的复制	13	1	0	1	4	0	0	0	0	7
文教体育用品制造业	4	0	0	0	2	0	0	0	0	2
石油加工炼焦及核燃料加工业	3 513	177	62	464	286	1 125	118	1	15	1 265
化学原料及化学制品制造业	14 359	390	702	1 203	2 646	93	1 508	0	276	7 541
医药制造业	406	34	2	76	179	2	2	0	2	110
化学纤维制造业	461	49	0	156	156	0	6	0	12	82
橡胶制品业	141	0	0	28	93	0	0	0	3	17
塑料制品业	75	2	0	6	55	0	0	0	0	11
非金属矿物制品业	5 161	12	307	809	1 351	187	361	0	80	2 053
黑色金属冶炼及压延加工业	38 008	103	28 132	800	2 297	54	3 156	0	67	3 398
有色金属冶炼及压延加工业	8 791	182	2 336	715	448	13	4 334	7	94	662
金属制品业	364	37	157	4	92	0	2	0	1	71
通用设备制造业	573	8	241	14	95	1	0	0	2	212
专用设备制造业	211	4	32	34	59	2	0	0	0	80
交通运输设备制造业	562	20	7	30	116	0	0	0	0	388
电气机械及器材制造业	81	13	0	4	25	0	2	0	0	36
通信计算机及其他电子设备制造	166	96	0	1	14	0	0	0	0	55
仪器仪表及文化办公用机械制造	31	15	0	0	4	0	0	0	0	11
工艺品及其他制造业	34	0	0	5	17	0	0	0	0	12
废弃资源和废旧材料回收加工业	66	1	41	0	6	0	2	0	0	17
电力、热力的生产和供应业	53 823	17	0	37 203	11 125	97	1	0	5 328	66
燃气生产和供应业	73	1	0	35	35	0	0	0	1	2
水的生产和供应业	22	1	0	0	2	0	0	0	0	19
其他行业	283	0	1	7	71	0	29	0	0	174

按行业分重点调查工业固体废物产生及处置利用情况（二）

（2010）

单位：万吨

行业名称	工业固体废物综合利用量	危险废物	冶炼废渣	粉煤灰	炉渣	煤矸石	尾矿	脱硫石膏	其他废物
行业总计	**150 899**	**977**	**29 474**	**30 857**	**20 001**	**19 962**	**17 161**	**5 176**	**22 578**
煤炭开采和洗选业	20 906	0	9	117	359	18 460	971	1	988
石油和天然气开采业	78	9	0	5	25	0	0	0	39
黑色金属矿采选业	6 658	0	48	16	36	3	3 159	1	3 394
有色金属矿采选业	11 582	81	43	10	28	25	9 206	1	2 189
非金属矿采选业	933	20	7	22	64	3	641	1	175
其他采矿业	52	0	1	4	1	1	38	0	7
农副食品加工业	2 073	0	0	141	315	0	1	12	1 604
食品制造业	653	0	0	181	201	1	0	9	261
饮料制造业	910	0	0	79	235	0	0	7	588
烟草制品业	34	0	0	3	17	0	0	0	13
纺织业	706	9	1	108	481	2	0	4	100
纺织服装、鞋、帽制造业	53	0	0	9	20	0	3	0	21
皮革毛皮羽毛（绒）及其制品业	60	0	0	3	28	0	0	0	29
木材加工及木竹藤棕草制品业	207	0	0	13	59	0	0	0	135
家具制造业	18	0	0	1	5	0	0	0	12
造纸及纸制品业	2 107	122	7	533	745	0	0	15	685
印刷业和记录媒介的复制	11	0	0	1	4	0	0	0	6
文教体育用品制造业	4	0	0	0	2	0	0	0	2
石油加工炼焦及核燃料加工业	2 761	124	59	389	240	1 012	114	11	812
化学原料及化学制品制造业	10 179	236	651	1 155	2 522	95	685	164	4 671
医药制造业	371	25	2	71	175	1	2	2	94
化学纤维制造业	436	47	0	148	155	0	6	9	71
橡胶制品业	137	0	0	28	89	2	0	3	15
塑料制品业	70	0	0	6	55	0	0	0	8
非金属矿物制品业	5 235	9	351	872	1 352	188	375	80	2 007
黑色金属冶炼及压延加工业	33 319	80	26 398	690	2 252	56	531	55	3 255
有色金属冶炼及压延加工业	4 353	105	1 425	540	403	12	1 421	58	387
金属制品业	322	15	157	4	91	0	2	1	54
通用设备制造业	536	4	238	14	91	1	0	1	187
专用设备制造业	183	2	26	29	58	2	0	0	66
交通运输设备制造业	496	4	6	30	113	0	0	0	343
电气机械及器材制造业	70	7	0	4	24	0	2	0	33
通信计算机及其他电子设备制造	123	64	0	1	14	0	0	0	45
仪器仪表及文化办公用机械制造	15	2	0	0	4	0	0	0	9
工艺品及其他制造业	33	0	0	5	17	0	0	0	11
废弃资源和废旧材料回收加工业	64	0	41	0	6	0	2	0	15
电力、热力的生产和供应业	44 820	8	3	25 583	9 614	96	0	4 740	62
燃气生产和供应业	66	1	0	35	28	0	0	0	2
水的生产和供应业	13	0	0	0	2	0	0	0	11
其他行业	255	0	1	7	71	0	2	0	174

按行业分重点调查工业固体废物产生及处置利用情况（三）

（2010）　　　　单位：万吨

行业名称	工业固体废物贮存量	危险废物贮存量	工业固体废物处置量	危险废物处置量	工业固体废物排放量	危险废物排放量
行业总计	**22 117**	**166**	**53 652**	**513**	**438**	**0.00 015**
煤炭开采和洗选业	1 627	0	5 327	0	188	0
石油和天然气开采业	7	0	123	9	0	0
黑色金属矿采选业	5 463	0	19 841	5	22	0
有色金属矿采选业	4 881	18	12 908	72	78	0
非金属矿采选业	411	73	432	0	6	0
其他采矿业	5	0	15	0	0	0
农副食品加工业	8	0	35	0	5	0
食品制造业	1	0	13	1	2	0
饮料制造业	1	0	16	0	2	0
烟草制品业	1	0	6	0	1	0
纺织业	0	0	47	8	0	0
纺织服装、鞋、帽制造业	0	0	2	0	0	0
皮革毛皮羽毛（绒）及其制品业	0	0	14	3	0	0
木材加工及木竹藤棕草制品业	0	0	2	0	0	0
家具制造业	0	0	0	0	0	0
造纸及纸制品业	57	0	161	4	3	0
印刷业和记录媒介的复制	0	0	2	1	0	0
文教体育用品制造业	0	0	0	0	0	0
石油加工炼焦及核燃料加工业	56	0	711	54	2	0
化学原料及化学制品制造业	1 476	32	2 788	133	12	0
医药制造业	2	0	32	9	1	0
化学纤维制造业	8	0	15	2	1	0
橡胶制品业	0	0	3	0	0	0
塑料制品业	0	0	5	2	0	0
非金属矿物制品业	37	0	235	3	27	0
黑色金属冶炼及压延加工业	1 359	1	3 371	27	11	0
有色金属冶炼及压延加工业	1 182	38	3 261	72	27	0
金属制品业	1	0	41	23	0	0
通用设备制造业	2	0	35	4	1	0
专用设备制造业	9	3	17	2	1	0
交通运输设备制造业	1	0	64	16	1	0
电气机械及器材制造业	0	0	11	7	0	0
通信计算机及其他电子设备制造业	0	0	43	33	0	0
仪器仪表及文化办公用机械制造业	0	0	16	13	0	0
工艺品及其他制造业	0	0	1	0	0	0
废弃资源和废旧材料回收加工业	1	0	2	0	0	0
电力、热力的生产和供应业	5 494	0	4 040	8	45	0.00 015
燃气生产和供应业	0	0	7	0	0	0
水的生产和供应业	0	0	9	1	0	0
其他行业	27	0	1	0	0	0

按行业分重点调查工业汇总情况（一）

（2010）

行业名称	工业用水总量（万吨）			工业总产值（现价）（万元）	专职环保人员数（人）
		新鲜水量	重复用水量		
行业总计	**38 863 997.5**	**5 439 513.8**	**33 424 483.7**	**2 814 519 985.8**	**316 358**
煤炭开采和洗选业	266 118.8	96 990.7	169 128.1	103 548 151.8	10 805
石油和天然气开采业	152 332.5	30 496.4	121 836.1	60 239 721.4	2 623
黑色金属矿采选业	242 488.0	35 606.1	206 881.9	11 854 778.2	2 571
有色金属矿采选业	165 014.8	52 572.0	112 442.8	15 322 722.0	4 392
非金属矿采选业	61 663.4	10 235.3	51 428.0	2 759 827.8	1 268
其他采矿业	1 103.8	625.9	477.9	245 994.4	89
农副食品加工业	405 930.5	171 941.3	233 989.2	99 213 915.1	18 007
食品制造业	143 409.6	68 257.9	75 151.6	43 773 344.4	7 438
饮料制造业	272 145.2	101 301.4	170 843.9	55 421 731.1	7 423
烟草制品业	12 095.7	4 007.5	8 088.2	50 450 360.5	320
纺织业	375 532.7	289 712.1	85 820.6	107 520 409.5	21 056
纺织服装、鞋、帽制造业	20 258.4	14 810.9	5 447.5	8 153 752.9	2 023
皮革毛皮羽毛（绒）及其制品业	38 602.0	32 643.6	5 958.5	12 546 284.9	3 652
木材加工及木竹藤棕草制品业	11 638.9	7 116.1	4 522.8	7 957 468.9	2 521
家具制造业	4 059.9	2 569.5	1 490.4	4 508 846.5	526
造纸及纸制品业	1 233 873.6	461 484.6	772 389.0	51 494 130.7	18 304
印刷业和记录媒介的复制	4 053.0	1 998.9	2 054.1	5 403 396.6	648
文教体育用品制造业	1 947.5	1 507.6	439.8	2 784 546.7	398
石油加工、炼焦及核燃料加工业	2 978 306.9	153 308.7	2 824 998.2	257 760 449.8	14 613
化学原料及化学制品制造业	5 438 564.8	475 452.1	4 963 112.7	206 533 854.6	28 016
医药制造业	352 904.7	66 472.0	286 432.7	52 019 896.3	8 815
化学纤维制造业	550 510.0	57 657.2	492 852.8	32 806 954.7	2 653
橡胶制品业	68 596.2	10 633.4	57 962.9	28 321 554.2	1 341
塑料制品业	16 458.9	7 468.0	8 990.9	16 119 680.5	2 298
非金属矿物制品业	379 480.8	79 293.5	300 187.4	95 820 489.8	56 858
黑色金属冶炼及压延加工业	7 681 928.7	374 005.4	7 307 923.3	382 754 911.6	18 496
有色金属冶炼及压延加工业	647 417.8	67 811.5	579 606.3	127 225 654.0	8 262
金属制品业	171 522.9	40 917.5	130 605.3	57 916 683.7	14 337
通用设备制造业	106 158.2	20 031.1	86 127.0	78 436 105.0	9 692
专用设备制造业	69 294.5	13 389.8	55 904.6	63 956 576.6	2 640
交通运输设备制造业	151 151.7	36 210.0	114 941.7	264 774 719.0	4 983
电气机械及器材制造业	46 446.8	16 049.1	30 397.7	92 694 307.0	3 600
通信计算机及其他电子设备制造业	195 271.6	45 678.2	149 593.4	200 492 205.3	10 985
仪器仪表及文化办公用机械制造业	29 622.3	6 380.3	23 242.0	23 083 569.8	1 119
工艺品及其他制造业	4 048.8	3 190.9	857.9	8 485 869.8	997
废弃资源和废旧材料回收加工业	2 925.4	1 931.7	993.7	5 115 743.0	443
电力、热力的生产和供应业	16 418 804.6	2 499 402.4	13 919 402.2	149 244 645.4	10 024
燃气生产和供应业	58 345.3	3 998.3	54 346.9	1 612 993.9	232
水的生产和供应业	67 560.7	62 044.9	5 515.8	19 564 476.8	1 229
其他行业	16 407.9	14 310.0	2 098.0	6 579 261.6	10 661

按行业分重点调查工业汇总情况（二）

（2010）

行业名称	三废综合利用产品产值（万元）	工业锅炉		锅炉烟尘排放达标的		工业炉窑数（座）	
		台数（台）	蒸吨数（蒸吨）	台数（台）	蒸吨数（蒸吨）		烟尘排放达标的
行业总计	**17 785 033.7**	**85 484**	**2 292 179**	**79 297**	**2 230 530**	**83 295**	**70 878**
煤炭开采和洗选业	299 842.0	5 623	27 933	4 634	23 623	280	191
石油和天然气开采业	235 806.7	2 463	23 265	2 379	23 193	3 585	3 443
黑色金属矿采选业	36 844.3	342	2 049	300	1 907	214	199
有色金属矿采选业	325 170.6	467	2 213	388	1 548	181	147
非金属矿采选业	30 890.9	190	2 563	177	2 544	443	298
其他采矿业	1 457.1	18	71	14	67	43	7
农副食品加工业	513 158.9	6 660	46 457	5 618	44 522	745	297
食品制造业	198 922.0	3 776	22 852	3 443	21 847	177	138
饮料制造业	284 846.1	3 106	22 609	2 783	21 568	569	519
烟草制品业	8 031.2	292	3 330	290	3 314	11	9
纺织业	278 412.9	8 678	51 492	8 324	49 759	328	319
纺织服装、鞋、帽制造业	2 965.2	975	3 418	944	3 360	14	13
皮革毛皮羽毛（绒）及其制品业	19 013.8	1 488	3 357	1 419	3 225	20	17
木材加工及木竹藤棕草制品业	109 438.7	2 007	8 691	1 729	7 881	71	48
家具制造业	6 853.3	211	648	184	607	63	39
造纸及纸制品业	2 298 091.3	5 553	54 370	5 228	52 696	253	228
印刷业和记录媒介的复制	6 487.0	281	1 013	268	999	3	1
文教体育用品制造业	829.8	127	335	117	301	14	11
石油加工炼焦及核燃料加工业	1 118 173.9	1 414	49 866	1 356	48 654	3 915	3 365
化学原料及化学制品制造业	1 357 444.8	9 193	117 244	8 791	112 437	6 324	5 692
医药制造业	144 234.1	2 823	15 525	2 634	14 755	101	83
化学纤维制造业	111 178.8	656	16 334	652	15 533	225	221
橡胶制品业	90 802.0	832	7 268	786	7 134	69	61
塑料制品业	65 850.0	1 134	5 239	1 081	5 075	100	92
非金属矿物制品业	4 303 749.6	2 710	12 860	2 489	12 284	30 988	23 809
黑色金属冶炼及压延加工业	2 782 605.1	1 739	59 164	1 665	58 339	11 751	10 849
有色金属冶炼及压延加工业	1 238 398.6	1 192	31 619	1 130	29 180	12 426	11 511
金属制品业	150 870.7	1 649	3 514	1 546	3 343	1 734	1 563
通用设备制造业	123 277.9	1 205	6 733	1 139	6 609	3 895	3 481
专用设备制造业	72 677.0	864	6 317	840	6 270	1 668	1 373
交通运输设备制造业	234 705.1	1 887	13 979	1 851	13 914	1 595	1 537
电气机械及器材制造业	77 390.2	724	2 685	695	2 620	472	460
通信计算机及其他电子设备制造业	301 913.3	779	2 512	754	2 445	373	359
仪器仪表及文化办公用机械制造业	18 899.6	202	925	195	908	61	40
工艺品及其他制造业	6 514.0	230	558	211	516	120	101
废弃资源和废旧材料回收加工业	140 314.8	39	228	37	220	118	69
电力、热力的生产和供应业	742 238.1	12 912	1 653 066	12 255	1 617 703	127	117
燃气生产和供应业	2 342.1	255	5 160	248	5 005	79	64
水的生产和供应业	537.5	46	178	37	164	8	8
其他行业	43 854.7	742	4 539	666	4 461	132	99

按行业分重点调查工业“三废”治理效率

（2010）

单位：%

行业名称	工业废水排放达标率	工业用水重复用水率	二氧化硫排放达标率	氮氧化物排放达标率	烟尘排放达标率	工业粉尘排放达标率	工业固体废物	
							综合利用率	处置率
行业总计	**95.9**	**86.0**	**93.3**	**88.2**	**91.3**	**92.2**	**66.6**	**23.8**
煤炭开采和洗选业	93.1	63.6	86.9	79.5	85.6	74.9	75.6	19.2
石油和天然气开采业	99.8	80.0	95.9	92.4	99.7	100.0	37.7	59.3
黑色金属矿采选业	91.2	85.3	97.2	90.7	81.2	68.6	20.8	62.1
有色金属矿采选业	92.7	68.1	65.2	56.2	80.1	77.4	39.3	44.0
非金属矿采选业	95.2	83.4	91.8	77.0	86.8	58.6	52.4	24.3
其他采矿业	94.9	43.3	90.6	79.1	87.0	81.5	72.0	20.9
农副食品加工业	91.9	57.6	86.6	77.1	87.0	95.4	97.8	1.6
食品制造业	92.2	52.4	85.6	72.8	90.9	91.4	97.7	2.0
饮料制造业	93.8	62.8	87.3	84.5	89.4	95.2	97.9	1.7
烟草制品业	98.4	66.9	92.8	86.1	94.4	99.0	82.7	14.3
纺织业	97.4	22.9	94.4	89.2	96.0	92.7	93.7	6.2
纺织服装、鞋、帽制造业	96.5	26.9	94.8	90.6	98.1	98.7	94.9	4.3
皮革毛皮羽毛（绒）及其制品业	94.2	15.4	90.1	78.9	96.1	63.1	80.3	19.2
木材加工及木竹藤棕草制品业	91.5	38.9	88.2	83.1	91.0	73.1	98.9	1.0
家具制造业	98.5	36.7	91.1	80.4	83.5	98.0	97.9	1.6
造纸及纸制品业	96.0	62.6	94.2	86.1	93.0	85.0	90.5	6.9
印刷业和记录媒介的复制	96.6	50.7	90.5	91.9	93.2	96.4	83.1	16.7
文教体育用品制造业	96.5	22.6	84.6	85.6	90.2	98.0	88.0	11.0
石油加工、炼焦及核燃料加工业	98.1	94.9	89.6	89.6	92.6	91.2	78.2	20.2
化学原料及化学制品制造业	96.1	91.3	91.8	80.1	90.4	86.8	70.4	19.4
医药制造业	97.4	81.2	92.7	85.9	90.1	93.5	91.4	7.8
化学纤维制造业	94.7	89.5	98.4	88.5	98.5	100.0	94.5	3.4
橡胶制品业	98.4	84.5	96.9	93.9	98.1	97.5	97.3	2.5
塑料制品业	95.6	54.6	97.0	93.0	95.8	98.4	93.6	6.4
非金属矿物制品业	93.8	79.1	88.3	82.9	86.9	94.7	96.1	4.5
黑色金属冶炼及压延加工业	98.3	95.1	95.8	85.8	95.3	92.8	87.5	8.9
有色金属冶炼及压延加工业	96.1	89.5	91.7	73.4	95.7	87.3	49.3	37.1
金属制品业	96.1	76.1	94.7	80.9	93.2	96.6	88.6	11.3
通用设备制造业	96.8	81.1	93.3	74.0	90.3	92.0	93.4	6.0
专用设备制造业	97.2	80.7	97.6	31.6	96.8	98.8	86.8	8.2
交通运输设备制造业	96.7	76.0	96.3	88.2	95.1	97.3	88.2	11.4
电气机械及器材制造业	94.3	65.4	90.6	89.6	94.5	94.3	86.7	13.2
通信计算机及其他电子设备制造业	98.5	76.6	97.7	96.5	95.7	98.9	74.2	25.6
仪器仪表及文化办公用机械制造业	98.8	78.5	96.7	90.3	98.4	100.0	48.8	51.3
工艺品及其他制造业	93.4	21.2	92.1	82.4	87.5	70.8	97.2	2.5
废弃资源和废旧材料回收加工业	89.8	34.0	93.7	91.3	92.0	99.9	95.8	2.6
电力、热力的生产和供应业	98.1	84.8	95.0	91.0	92.4	95.3	82.4	7.5
燃气生产和供应业	92.2	93.1	87.6	81.9	95.2	98.5	89.8	10.2
水的生产和供应业	99.1	8.2	20.4	93.7	57.2		59.8	39.7

各地区火电行业工业废气排放及处理情况（一）

（2010）

地 区 名 称	汇总工业企业数（个）	燃料煤消费量（万吨）	废气治理设施数（套）	脱硫设施数	废气治理设施运行费用（万元）	脱硫设施运行费用
全 国	**1 642**	**166 288**	**6 444**	**3 266**	**4 108 345.6**	**3 004 520.3**
北 京	7	956	66	17	43 831.0	31 901.0
天 津	14	2 543	78	30	58 194.7	41 482.7
河 北	88	9 997	370	196	265 584.3	224 802.4
山 西	78	9 933	284	151	207 140.3	166 077.6
内蒙古	87	15 201	352	162	228 445.4	140 487.5
辽 宁	65	7 333	290	146	103 566.6	80 200.1
吉 林	33	4 143	151	43	34 179.1	21 412.8
黑龙江	76	5 084	295	46	40 183.4	20 698.5
上 海	21	3 352	80	32	90 724.0	64 871.4
江 苏	200	14 743	790	489	469 141.9	297 854.5
浙 江	128	9 193	533	396	337 986.8	253 510.8
安 徽	50	6 374	168	90	117 060.2	96 292.9
福 建	27	3 794	100	56	107 176.0	80 719.3
江 西	15	2 538	59	27	73 023.7	58 570.3
山 东	255	14 494	1 065	589	390 742.2	287 978.5
河 南	90	10 496	332	169	266 619.8	215 801.0
湖 北	30	3 627	119	58	177 557.7	99 269.6
湖 南	23	3 170	123	47	94 015.6	70 963.9
广 东	95	9 791	280	133	341 622.8	250 386.3
广 西	13	2 157	53	25	30 651.0	24 762.1
海 南	5	427	13	6	12 817.0	11 649.0
重 庆	31	1 811	94	53	81 431.2	73 036.5
四 川	44	3 229	119	46	98 635.2	74 330.5
贵 州	20	4 996	102	52	128 547.8	111 236.1
云 南	12	3 131	59	29	69 444.7	50 199.8
陕 西	49	5 346	159	67	93 378.4	69 214.1
甘 肃	18	2 458	71	32	42 358.5	31 256.5
青 海	4	543	11	4	9 735.0	5 015.0
宁 夏	16	2 899	59	26	73 801.7	36 075.6
新 疆	48	2 531	169	49	20 749.6	14 464.0

各地区火电行业工业废气排放及处理情况（二）

（2010）

单位：吨

地区名称	工业二氧化硫排放量	工业二氧化硫去除量	工业氮氧化物排放量	工业氮氧化物去除量	工业烟尘排放量	工业烟尘去除量
全　国	**8 347 206**	**18 999 558**	**8 526 602**	**424 526**	**1 668 883**	**309 788 641**
北　京	9 351	79 533	58 464	0	2 042	1 585 176
天　津	64 729	209 766	89 474	13 056	9 078	4 073 250
河　北	364 646	1 334 797	450 785	52 750	57 541	23 390 942
山　西	692 500	1 266 366	492 399	61 227	99 651	21 134 617
内蒙古	650 302	1 302 032	818 990	9 556	160 304	21 875 651
辽　宁	332 066	481 001	354 105	29 032	90 632	17 578 943
吉　林	163 768	124 711	322 894	0	89 810	7 258 153
黑龙江	255 683	70 456	244 040	5 226	162 792	11 163 207
上　海	140 525	277 350	140 436	4 736	17 959	3 567 897
江　苏	501 393	1 490 978	624 503	30 015	107 000	18 574 065
浙　江	345 014	762 463	444 107	22 288	66 933	10 714 782
安　徽	143 777	428 569	293 052	48 591	54 630	15 653 285
福　建	123 540	323 214	182 794	2 655	12 237	4 952 702
江　西	171 072	307 473	96 042	0	20 676	6 503 178
山　东	702 315	2 025 037	667 028	1 129	102 675	28 563 761
河　南	566 865	803 512	638 655	2 568	184 565	24 861 232
湖　北	225 183	572 852	213 757	18	31 494	6 642 283
湖　南	157 135	473 384	209 692	3 671	62 365	6 613 363
广　东	332 316	878 787	440 276	93 700	32 757	12 494 883
广　西	227 430	420 912	63 204	6 087	8 514	3 428 906
海　南	14 621	80 429	15 215	133	1 158	661 978
重　庆	220 604	713 551	121 192	0	29 939	2 764 458
四　川	364 989	590 471	326 248	8 739	63 929	8 538 261
贵　州	468 388	1 867 028	135 376	9 809	21 422	12 047 206
云　南	231 479	451 633	174 286	0	11 647	9 294 492
陕　西	446 310	839 987	309 878	7 960	29 812	6 790 961
甘　肃	134 021	246 862	114 980	11 580	22 073	3 003 179
青　海	43 046	25 458	40 173	0	5 993	903 722
宁　夏	98 893	445 236	236 664	0	32 585	12 295 223
新　疆	155 245	105 710	207 895	0	76 671	2 858 884

5

流域及入海陆源废水排放统计

LIUYU JI RUHAI LUYUAN FEISHUI PAIFANG TONGJI

ANNUAL STATISTIC REPORT ON ENVIRONMENT IN CHINA

2010

流域接纳工业废水及处理情况（一）

（2010）

流域	地区名称	汇总工业企业数（个）	工业废水排放量（万吨）	工业废水排放达标量（万吨）	工业废水处理量（万吨）	废水治理设施数（套）	本年运行费用（万元）
总计		**88 608**	**1 903 155**	**1 829 396**	**5 106 167**	**65 093**	**4 461 664.3**
辽河	内蒙古	318	5 846	5 510	6 756	159	7 311.2
	辽宁	3 681	35 872	32 992	319 532	2 220	250 640.6
	吉林	275	2 757	1 913	7 428	184	4 797.2
	合计	**4 274**	**44 474**	**40 415**	**333 716**	**2 563**	**262 749.0**
海河	北京	718	8 198	8 096	31 304	481	68 058.5
	天津	1 954	19 680	19 671	87 629	912	69 321.1
	河北	5 279	103 377	101 840	806 972	3 584	345 584.8
	山西	1 424	17 588	16 454	42 112	932	46 120.7
	内蒙古	79	669	496	896	46	2 045.6
	山东	1 082	50 455	48 599	46 744	676	83 875.2
	河南	1 635	55 526	54 177	75 931	897	92 789.6
	合计	**12 171**	**255 492**	**249 333**	**1 091 588**	**7 528**	**707 795.5**
淮河	江苏	2 539	60 265	57 762	114 038	1 753	101 424.3
	安徽	1 877	29 821	29 386	95 035	812	69 358.1
	山东	1 727	59 386	58 808	75 564	1 378	102 926.9
	河南	1 529	52 775	51 364	71 861	1 122	64 574.0
	合计	**7 672**	**202 246**	**197 320**	**356 499**	**5 065**	**338 283.3**
松花江	内蒙古	175	4 296	3 476	3 146	52	3 417.1
	吉林	658	27 133	24 979	36 677	404	38 921.3
	黑龙江	1 234	33 716	32 067	124 488	1 009	135 152.1
	合计	**2 067**	**65 144**	**60 522**	**164 311**	**1 465**	**177 490.5**
珠江	江西	87	894	858	791	49	840.4
	湖南	157	1 946	1 342	3 379	179	1 781.6
	广东	8 693	161 210	154 425	294 832	7 852	464 390.0
	广西	3 500	151 923	147 369	263 694	1 945	113 493.6
	贵州	1 495	6 123	4 194	12 093	785	16 571.8
	云南	604	6 219	5 935	79 144	667	43 918.3
	合计	**14 536**	**328 315**	**314 122**	**653 933**	**11 477**	**640 995.7**

流域接纳工业废水及处理情况（一）（续表）

（2010）

流域	地区名称	汇总工业企业数（个）	工业废水排放量（万吨）	工业废水排放达标量（万吨）	工业废水处理量（万吨）	废水治理设施数（套）	本年运行费用（万元）
长江	上海	1 871	33 993	33 279	147 245	1 749	157 318.3
	江苏	5 853	202 965	200 212	288 478	5 143	370 469.2
	浙江	2 232	55 696	54 242	91 608	1 886	122 003.6
	安徽	2 476	42 574	41 437	174 860	1 337	119 077.8
	福建	186	2 525	2 392	2 111	106	2 257.9
	江西	3 301	72 284	68 067	140 508	1 992	120 707.7
	河南	369	18 176	17 842	17 480	449	18 557.0
	湖北	2 568	90 470	87 653	328 138	2 048	136 324.7
	湖南	3 929	95 605	89 583	305 098	3 155	129 125.3
	广东	60	883	748	652	32	1 170.0
	广西	175	1 097	1 063	5 569	144	1 962.0
	重庆	2 489	45 180	42 798	61 331	1 498	56 491.5
	四川	6 520	92 110	88 871	194 681	4 365	325 363.5
	贵州	2 749	11 428	9 438	53 899	1 419	124 581.5
	云南	878	6 744	6 507	43 065	958	38 429.0
	西藏	4	66		5	2	64.6
	陕西	866	4 978	4 790	22 619	424	13 813.2
	甘肃	319	1 291	1 134	1 560	130	1 782.7
	青海	15	122	67	192	2	503.0
	合计	**36 860**	**778 187**	**750 124**	**1 879 189**	**26 839**	**1 748 002.5**
黄河	山西	2 878	30 990	29 177	192 727	1 770	148 774.0
	内蒙古	951	19 036	18 183	92 343	475	47 124.1
	山东	1 811	67 208	66 081	176 142	1 595	182 342.8
	河南	1 528	34 850	33 436	55 568	1 101	50 956.2
	四川	8	310	280	60	3	55.3
	陕西	1 835	40 791	39 833	50 531	4 448	76 875.0
	甘肃	895	6 654	5 429	20 313	335	17 524.9
	青海	456	4 884	3 702	12 538	91	5 121.5
	宁夏	666	24 575	21 440	26 708	338	57 574.0
	合计	**11 028**	**229 297**	**217 560**	**626 931**	**10 156**	**586 347.8**

流域接纳工业废水及处理情况（二）

（2010）

单位：吨

流域	地区名称	工业废水中污染物排放量					
		汞	镉	六价铬	铅	砷	挥发酚
	总　计	**0.870**	**28.162**	**35.455**	**133.343**	**116.125**	**648.744**
辽河	内蒙古	…	0.015	0.021	0.030	0.030	0.139
	辽　宁	…	0.029	1.230	0.319	0.097	3.788
	吉　林	…	…	…	…	…	0.001
	合　计	…	**0.044**	**1.251**	**0.348**	**0.127**	**3.928**
海河	北　京			0.022	0.016	…	0.110
	天　津	…	…	0.078	0.002	0.022	1.519
	河　北	…	…	0.904	2.567	…	6.843
	山　西	0.010	0.001	0.078	0.045	0.009	3.996
	内蒙古	…	0.002		0.001		0.293
	山　东	…	…	0.020	0.002	0.007	2.302
	河　南	…	0.009	0.279	0.010	…	8.911
	合　计	**0.010**	**0.012**	**1.382**	**2.642**	**0.038**	**23.974**
淮河	江　苏	…	0.007	0.504	0.404	0.701	8.843
	安　徽	…	…	0.043	0.003	…	4.520
	山　东	…	…	0.210	0.011	0.094	1.064
	河　南	…	0.003	0.051	0.004	0.003	1.342
	合　计	…	**0.009**	**0.807**	**0.421**	**0.797**	**15.769**
松花江	内蒙古			…			2.754
	吉　林	…	0.004	0.627	0.048	…	46.254
	黑龙江	0.001	0.055	0.263	0.131	0.223	378.098
	合　计	**0.001**	**0.059**	**0.890**	**0.179**	**0.223**	**427.106**
珠江	江　西					0.085	
	湖　南	0	0.325	0.002	3.790	2.726	0.950
	广　东	0.024	0.836	5.250	6.333	1.834	6.053
	广　西	0.064	3.726	1.319	21.294	23.615	13.682
	贵　州	0.002	0.002	0.027	0.010	0.136	0.108
	云　南	0	0.030	0.014	0.253	0.067	2.200
	合　计	**0.089**	**4.919**	**6.610**	**31.681**	**28.462**	**22.994**

流域接纳工业废水及处理情况（二）（续表）

（2010）　　单位：吨

流域	地区名称	工业废水中污染物排放量					
		汞	镉	六价铬	铅	砷	挥发酚
长江	上　海	0.002	0.010	0.722	0.078	0.013	4.930
	江　苏	0.001	0.039	6.740	2.026	0.140	16.488
	浙　江	…	…	0.165	0.037	…	0.168
	安　徽	0.002	0.164	0.061	1.232	1.887	4.612
	福　建	…	0.015	…	0.051	0.067	0.111
	江　西	0.001	4.212	2.267	10.792	13.110	17.563
	河　南	…	0.003	0.042	0.003	0.017	1.952
	湖　北	0.008	0.347	3.230	1.906	2.906	23.424
	湖　南	0.523	13.042	3.880	32.243	58.891	38.036
	广　东	…	0.006	0.038	0.198	0.632	0.748
	广　西	…	0.010	…	0.243	0.037	0
	重　庆	…	0.069	3.582	0.094	…	1.350
	四　川	0.008	0.354	0.815	1.769	0.525	9.634
	贵　州	…	0.005	0.096	0.083	0.006	0.160
	云　南	0.022	0.150	0.024	1.843	0.723	0.585
	西　藏						
	陕　西	0.187	0.036	0.033	0.518	0.126	0.421
	甘　肃	0.001	0.063	0.021	3.688	0.076	0.125
	青　海						
	合　计	**0.754**	**18.523**	**21.714**	**56.803**	**79.164**	**120.304**
黄河	山　西	…	0.015	0.417	0.144	0.015	5.104
	内蒙古	…	…	0.334	0.475	0	1.934
	山　东	…	…	0.195	0.011	0.166	2.085
	河　南	0.003	0.153	0.113	2.308	0.232	1.860
	四　川		0.008	0.060	0.090		
	陕　西	0.010	0.793	1.225	8.409	1.690	2.062
	甘　肃	0.003	3.477	0.152	29.613	4.946	1.916
	青　海	…	0.119	0.020	0.202	0.131	0.394
	宁　夏		0.030	0.284	0.016	0.134	19.316
	合　计	**0.017**	**4.596**	**2.800**	**41.267**	**7.314**	**34.671**

流域接纳工业废水及处理情况（三）

（2010）

单位：吨

流域	地区名称	工业废水中污染物排放量			
		氰化物	化学需氧量	石油类	氨氮
总 计		**163.8**	**3 507 663.0**	**8 095.7**	**229 781.1**
辽河	内蒙古	…	18 886.9	1.4	961.9
	辽 宁	3.3	141 796.0	568.8	7 307.4
	吉 林	0	6 111.8	0.8	391.0
	合 计	**3.3**	**166 794.7**	**571.0**	**8 660.3**
海河	北 京	0.1	4 882.3	57.6	392.1
	天 津	0.1	22 217.7	97.2	3 196.8
	河 北	4.0	194 433.5	470.9	16 329.6
	山 西	9.9	49 139.9	124.4	5 586.3
	内蒙古		4 725.0	2.8	32.0
	山 东	0.6	95 090.2	22.9	3 911.9
	河 南	3.5	129 539.6	298.4	7 767.0
	合 计	**18.2**	**500 028.2**	**1 074.3**	**37 215.6**
淮河	江 苏	2.9	76 051.7	279.0	5 749.3
	安 徽	3.4	40 020.0	83.3	7 540.9
	山 东	0.7	57 908.0	56.1	3 303.6
	河 南	0.9	82 258.5	56.1	8 824.8
	合 计	**8.0**	**256 238.3**	**474.6**	**25 418.7**
松花江	内蒙古	…	13 813.1	71.4	408.1
	吉 林	4.7	84 103.5	521.3	2 022.6
	黑龙江	2.0	95 738.5	392.5	4 315.3
	合 计	**6.7**	**193 655.1**	**985.2**	**6 746.1**
珠江	江 西		2 750.4	0.3	1 107.3
	湖 南	5.5	2 564.0	4.8	313.8
	广 东	10.8	169 451.6	292.7	7 576.0
	广 西	17.1	404 357.3	253.5	13 045.6
	贵 州	0.4	7 068.9	32.4	251.9
	云 南	4.3	12 818.8	39.1	2 518.3
	合 计	**38.1**	**599 011.0**	**622.7**	**24 813.0**

流域接纳工业废水及处理情况（三）（续表）

（2010）　　　　单位：吨

流域	地区名称	工业废水中污染物排放量			
		氰化物	化学需氧量	石油类	氨氮
长江	上　海	4.7	21 583.5	415.5	3 003.7
	江　苏	7.6	178 537.0	593.2	8 685.1
	浙　江	0.1	53 526.7	19.0	3 909.3
	安　徽	3.2	78 816.6	108.4	4 267.5
	福　建	…	4 520.7	4.9	667.3
	江　西	6.9	117 576.8	428.0	8 648.4
	河　南	1.2	45 968.6	33.4	3 203.7
	湖　北	19.1	154 738.2	488.5	14 637.5
	湖　南	25.1	187 885.9	430.6	17 448.7
	广　东	…	1 144.6	0.1	59.7
	广　西	…	2 355.3	…	200.6
	重　庆	0.4	86 549.0	221.2	5 440.1
	四　川	0.8	246 403.0	371.4	16 422.9
	贵　州	0.3	12 703.3	92.0	629.7
	云　南	0.6	10 684.7	21.1	1 677.0
	西　藏		61.7		0.7
	陕　西	2.3	9 873.4	8.6	2 328.1
	甘　肃	…	2 413.2	1.7	13.1
	青　海		142.9	4.3	15.5
	合　计	**72.4**	**1 215 485.1**	**3 241.9**	**91 258.5**
黄河	山　西	5.5	88 295.4	225.6	5 805.9
	内蒙古	1.4	37 386.6	47.7	3 318.0
	山　东	0.5	110 614.1	265.1	5 001.9
	河　南	1.4	74 303.7	100.7	4 311.8
	四　川		140.1		3.4
	陕　西	6.8	112 636.8	335.4	5 007.3
	甘　肃	0.1	15 858.9	60.6	1 176.2
	青　海	…	31 103.6	29.8	1 416.2
	宁　夏	1.4	106 111.3	61.1	9 628.3
	合　计	**17.1**	**576 450.6**	**1 126.1**	**35 669.0**

流域接纳工业废水及处理情况（四）

（2010） 单位：吨

流域	地区名称	工业废水中污染物去除量				
		氰化物	化学需氧量	石油类	氨氮	挥发酚
总计		**5 312.9**	**11 080 958.6**	**309 904.7**	**634 864.3**	**67 090.7**
辽河	内蒙古		6 583.9	13.8	73.4	…
	辽宁	46.9	195 771.5	4 692.1	3 458.6	14 590.2
	吉林	…	30 573.5	22.2	167.0	…
	合计	**46.9**	**232 928.8**	**4 728.2**	**3 699.0**	**14 590.2**
海河	北京	47.0	37 491.1	1 336.1	1 252.6	651.3
	天津	0.2	52 377.8	464.7	1 057.5	37.5
	河北	534.3	692 229.1	11 142.5	23 367.2	3 775.6
	山西	9.4	32 128.7	286.1	1 086.7	16.7
	内蒙古		1 368.0	6.3	4.9	…
	山东	3.7	701 168.4	428.0	95 297.7	33.4
	河南	44.0	330 382.1	944.4	6 799.9	9 692.9
	合计	**638.5**	**1 847 145.2**	**14 608.1**	**128 866.6**	**14 207.4**
淮河	江苏	257.1	315 281.5	1 274.5	8 163.4	1 093.3
	安徽	10.8	242 016.1	490.8	61 900.6	102.8
	山东	7.0	701 993.5	4 610.4	50 477.1	1 928.4
	河南	23.7	323 366.1	303.8	11 271.4	851.6
	合计	**298.6**	**1 582 657.2**	**6 679.5**	**131 812.6**	**3 976.1**
松花江	内蒙古		249 625.6	1.2	808.8	
	吉林	11.7	185 024.2	1 324.9	1 970.9	532.9
	黑龙江	113.0	334 510.2	154 126.7	43 293.0	5 796.7
	合计	**124.7**	**769 160.0**	**155 452.8**	**46 072.7**	**6 329.6**
珠江	江西		3 487.4		3 982.7	
	湖南	0.1	8 488.6	1.7	472.5	
	广东	993.2	635 797.6	2 052.8	26 929.7	143.3
	广西	237.5	831 120.6	935.5	7 398.0	548.9
	贵州	0.1	8 947.8	268.9	104.8	1.6
	云南	1.1	60 022.2	21 391.8	714.5	66.0
	合计	**1 232.0**	**1 547 864.3**	**24 650.8**	**39 602.1**	**759.8**

流域接纳工业废水及处理情况（四）（续表）

（2010）

单位：吨

流域	地区名称	工业废水中污染物去除量				
		氰化物	化学需氧量	石油类	氨氮	挥发酚
长江	上海	44.4	260 218.8	14 537.9	7 318.8	777.7
	江苏	142.6	900 725.8	25 265.6	38 182.1	3 050.6
	浙江	2.2	310 900.9	187.7	6 669.2	223.0
	安徽	410.7	183 728.0	6 214.9	13 042.7	2 761.2
	福建	…	37 362.1	134.7	10 144.5	
	江西	315.4	140 387.0	9 719.2	11 543.6	2 217.0
	河南	0.2	288 891.8	87.6	4 430.9	7.3
	湖北	119.4	367 603.6	2 396.5	12 424.3	1 343.0
	湖南	174.8	459 032.5	3 047.0	14 957.0	1 552.0
	广东		2 789.8		23.9	2.7
	广西	…	2 875.8	…	145.6	…
	重庆	17.8	126 030.3	885.9	6 289.4	1 034.8
	四川	14.4	317 938.9	972.6	7 194.3	698.9
	贵州	329.8	57 435.5	280.6	1 029.0	133.4
	云南	19.4	40 994.3	455.1	15 883.2	481.6
	西藏		3.7		1.2	
	陕西	0.1	18 280.8	4.0	2 387.1	…
	甘肃	0.3	11 068.5	0.5	9.8	…
	青海		1 470.9		19.8	
	合计	**1 591.5**	**3 527 739.0**	**64 189.7**	**151 696.5**	**14 283.2**
黄河	山西	1 068.4	105 299.2	1 893.1	15 266.0	5 341.3
	内蒙古	192.7	93 516.2	849.6	9 466.4	2 939.1
	山东	17.3	707 364.3	19 760.5	35 531.8	3 087.2
	河南	73.8	127 670.5	10 601.8	5 781.0	149.3
	四川		20.0			
	陕西	14.0	226 393.1	4 189.4	6 754.0	114.2
	甘肃	0.4	16 693.2	878.4	571.1	43.9
	青海	…	3 295.3	1.7	39.1	…
	宁夏	14.3	293 212.3	1 420.8	59 705.5	1 269.4
	合计	**1 380.7**	**1 573 464.1**	**39 595.5**	**133 114.9**	**12 944.4**

流域工业污染治理情况

（2010）

流域	地区名称	工业废水治理施工项目数（个）	工业废水治理竣工项目数（个）	工业废水治理项目完成投资（万元）	废水治理竣工项目新增处理能力（万吨/日）
	总计	**1 935**	**1 666**	**996 148.9**	**726.0**
辽河	内蒙古	1	1	10.0	
	辽宁	16	12	36 057.7	10.2
	吉林	9	8	2 809.5	0.6
	合计	**26**	**21**	**38 877.2**	**10.8**
海河	北京	5	5	1 761.6	0.2
	天津	45	34	47 138.8	6.9
	河北	34	32	24 357.9	7.9
	山西	63	54	26 168.6	11.7
	内蒙古				
	山东	9	7	5 988.0	1.3
	河南	6	5	1 265.0	0.8
	合计	**162**	**137**	**106 679.9**	**28.7**
淮河	江苏	39	34	20 212.0	25.7
	安徽	14	12	10 922.2	2.5
	山东	71	65	42 780.6	102.9
	河南	45	37	36 992.1	9.3
	合计	**169**	**148**	**110 906.9**	**140.4**
松花江	内蒙古	0	0	0	0
	吉林	34	31	19 538.3	17.2
	黑龙江	18	14	25 338.7	6.3
	合计	**52**	**45**	**44 877.0**	**23.5**
珠江	江西	0	0	0	0
	湖南	3	1	1 703.0	6.6
	广东	255	233	139 651.1	169.5
	广西	88	83	44 430.3	27.3
	贵州	25	21	3 005.6	1.2
	云南	30	29	6 255.3	1.1
	合计	**401**	**367**	**195 045.3**	**205.7**

流域工业污染治理情况（续表）

（2010）

流域	地区名称	工业废水治理施工项目数（个）	工业废水治理竣工项目数（个）	工业废水治理项目完成投资（万元）	废水治理竣工项目新增处理能力（万吨/日）
长江	上海	65	56	6 308.1	1.6
	江苏	143	137	49 826.6	40.0
	浙江	48	39	9 591.1	2.3
	安徽	21	17	3 327.6	1.6
	福建	6	6	802.7	0.2
	江西	78	53	19 864.9	25.4
	河南	12	11	4 953.0	0.2
	湖北	94	80	40 469.9	14.7
	湖南	67	58	59 247.5	45.3
	广东	1	1	105.7	0
	广西	1	1	32.0	0
	重庆	63	46	37 760.6	33.1
	四川	84	78	29 833.9	15.3
	贵州	42	36	18 178.9	0.9
	云南	78	73	10 691.2	4.0
	西藏				
	陕西	8	3	4 629.0	0.5
	甘肃	30	27	6 748.2	0.2
	青海				
	合计	**841**	**722**	**302 370.9**	**185.2**
黄河	山西	79	59	27 401.0	7.2
	内蒙古	49	43	30 785.9	2.3
	山东	56	50	32 764.1	94.8
	河南	33	29	8 406.5	3.9
	四川	2	1	715.0	0.1
	陕西	47	29	68 363.5	8.1
	甘肃	8	8	10 088.7	2.5
	青海	3	3	1 018.6	0.4
	宁夏	7	4	23 478.0	12.3
	合计	**284**	**226**	**203 021.3**	**131.7**

流域生活及其他污染情况

（2010）

流域	地区名称	污水处理厂数（座）	污水处理厂处理能力（万吨/日）	城镇生活污水排放量（万吨）	处理生活污水量（万吨）	城镇生活污水中化学需氧量排放量（吨）	城镇生活污水中氨氮排放量（吨）
总计		**2 543**	**10 633**	**3 242 628**	**2 428 405**	**6 745 320**	**775 692**
辽河	内蒙古	18	70	12 246	9 845	24 746	6 812
	辽宁	49	341	110 288	82 715	271 652	36 855
	吉林	8	50	14 608	11 501	48 506	6 392
	合计	**75**	**461**	**137 142**	**104 061**	**344 904**	**50 059**
海河	北京	68	370	128 408	111 229	87 115	11 733
	天津	28	202	48 516	39 785	109 751	16 577
	河北	168	661	137 004	115 194	301 220	33 860
	山西	42	94	24 642	20 963	66 247	9 087
	内蒙古			904	0	8 158	856
	山东	32	121	27 868	21 232	50 720	6 037
	河南	27	151	38 635	29 930	53 844	8 809
	合计	**365**	**1 597**	**405 976**	**338 333**	**677 056**	**86 960**
淮河	江苏	84	250	104 962	63 518	248 046	27 924
	安徽	39	172	52 012	43 086	98 149	13 432
	山东	55	241	64 037	55 293	105 212	16 542
	河南	75	282	93 613	69 539	156 167	23 328
	合计	**253**	**944**	**314 624**	**231 437**	**607 575**	**81 227**
松花江	内蒙古	6	19	4 669	2 107	19 339	5 285
	吉林	35	203	55 233	31 368	140 878	17 558
	黑龙江	33	200	65 941	38 509	258 277	29 548
	合计	**74**	**422**	**125 842**	**71 984**	**418 494**	**52 391**
珠江	江西			1 566		8 810	685
	湖南	4	12	4 693	2 145	18 235	1 310
	广东	264	1 474	445 624	365 193	414 403	59 251
	广西	80	252	113 329	43 652	335 111	24 890
	贵州	35	35	13 416	6 929	58 687	4 195
	云南	16	40	12 336	9 365	51 182	4 484
	合计	**399**	**1 813**	**590 965**	**427 284**	**886 428**	**94 815**

流域生活及其他污染情况（续表）

（2010）

流域	地区名称	污水处理厂数（座）	污水处理厂处理能力（万吨/日）	城镇生活污水排放量（万吨）	处理生活污水量（万吨）	城镇生活污水中化学需氧量排放量（吨）	城镇生活污水中氨氮排放量（吨）
长江	上海	52	685	211 554	178 024	201 262	24 275
	江苏	288	849	187 913	161 758	285 775	21 240
	浙江	45	238	63 499	59 356	48 245	8 087
	安徽	41	230	64 667	49 012	200 483	19 304
	福建	4	6	2 985	1 072	13 504	1 029
	江西	46	214	87 765	61 026	311 487	25 731
	河南	20	69	22 242	13 388	42 169	6 192
	湖北	102	510	173 104	129 361	400 474	45 615
	湖南	118	477	172 505	108 703	610 247	57 832
	广东	3	4	1 252	801	2 326	424
	广西	7	8	3 398	572	12 842	974
	重庆	79	231	82 933	66 926	147 983	19 723
	四川	165	406	161 491	100 770	485 392	43 831
	贵州	84	152	42 817	28 963	173 726	14 451
	云南	25	117	36 966	32 489	61 321	6 482
	西藏	1	1				
	陕西	4	16	10 737	4 791	33 390	3 746
	甘肃	3	9	4 178	1 512	14 226	2 494
	青海	1	5	1 649	1 057	1 780	447
	合计	**1 088**	**4 228**	**1 331 654**	**999 582**	**3 046 631**	**301 876**
黄河	山西	67	164	46 533	33 901	138 187	21 373
	内蒙古	25	84	23 905	18 879	84 741	13 638
	山东	65	296	73 555	65 927	117 272	15 573
	河南	43	246	78 012	65 957	109 055	15 446
	四川	1	1	430	0	2 912	228
	陕西	52	211	59 805	38 458	160 272	21 888
	甘肃	12	68	24 366	14 652	89 621	10 882
	青海	9	28	10 998	4 523	33 206	5 142
	宁夏	15	69	18 822	13 431	28 966	4 195
	合计	**289**	**1 167**	**336 425**	**255 726**	**764 231**	**108 365**

湖泊接纳工业废水及处理情况（一）

（2010）

流域	地区名称	汇总工业企业数（个）	工业废水排放量（万吨）	工业废水排放达标量（万吨）	工业废水处理（万吨）	废水治理设施数（套）	本年运行费用（万元）
总计		**7 131**	**217 802**	**214 471**	**265 833**	**5 735**	**375 070.7**
滇池	云南	237	1 285	1 285	2 397	253	8 150.6
巢湖	安徽	714	6 096	5 603	22 912	265	12 608.6
洞庭湖	湖南	405	15 726	15 108	13 855	187	14 689.0
鄱阳湖	江西	161	2 561	2 279	2 469	56	2 898.9
太湖	上海	163	1 145	1 072	924	128	4 218.6
	江苏	3 409	136 950	136 368	138 615	3 140	215 995.2
	浙江	2 042	54 039	52 756	84 660	1 706	116 509.8
	合计	**5 614**	**192 134**	**190 196**	**224 199**	**4 974**	**336 723.6**

湖泊接纳工业废水及处理情况（二）

（2010）　　单位：吨

流域	地区名称	工业废水中污染物排放量					
		汞	镉	六价铬	铅	砷	挥发酚
总计		**0.0 001**	**0.034**	**4.974**	**1.414**	**0.395**	**7.806**
滇池	云南	0.0 001	0.030	0.013	0.098	0.143	0.016
巢湖	安徽	…	…	0.004	…	0.208	0.057
洞庭湖	湖南	…	…	0.007	0.023	…	1.553
鄱阳湖	江西	…	0.005	0.002	0.095	0.003	0.853
太湖	上海	…	…	0.003	…	…	0
	江苏	…	…	4.784	1.161	0.040	5.159
	浙江	…	…	0.161	0.037	…	0.168
	合计	…	…	**4.948**	**1.198**	**0.040**	**5.327**

湖泊接纳工业废水及处理情况（三）

（2010）　　单位：吨

流域	地区名称	工业废水中污染物排放量			
		氰化物	化学需氧量	石油类	氨氮
总计		**4.6**	**231 532.0**	**316.5**	**11 881.2**
滇池	云南	…	1 064.8	1.9	73.7
巢湖	安徽	…	8 760.2	12.2	474.1
洞庭湖	湖南	1.3	53 042.4	56.5	2 287.0
鄱阳湖	江西	…	2 750.7	3.9	103.9
太湖	上海	…	592.6	3.8	77.3
	江苏	3.1	113 803.6	220.6	5 240.0
	浙江	0.1	51 517.7	17.5	3 625.2
	合计	**3.2**	**165 913.9**	**241.9**	**8 942.5**

湖泊接纳工业废水及处理情况（四）

（2010）　　单位：吨

流域	地区名称	工业废水中污染物去除量				
		氰化物	化学需氧量	石油类	氨氮	挥发酚
总　计		**52.1**	**1 077 288.0**	**7 926.2**	**32 150.0**	**815.3**
滇　池	云　南	2.9	4 115.4	25.2	28.8	…
巢　湖	安　徽	3.8	20 076.1	977.6	464.9	102.0
洞庭湖	湖　南	2.0	45 758.1	2 055.5	1 466.3	258.9
鄱阳湖	江　西	…	6 040.7	213.7	37.8	0.2
太　湖	上　海	…	3 544.4	9.7	205.4	0.1
	江　苏	41.2	692 794.2	4 485.3	23 551.5	231.2
	浙　江	2.2	304 959.0	159.1	6 395.4	223.0
	合　计	**43.3**	**1 001 297.7**	**4 654.1**	**30 152.2**	**454.2**

湖泊工业污染治理投资情况

（2010）

流域	地区名称	工业废水治理施工项目数（个）	工业废水治理竣工项目数（个）	工业废水治理项目完成投资（万元）	废水治理竣工项目新增处理能力（万吨/日）
总　计		**157**	**140**	**61 365.5**	**24.4**
滇　池	云　南	27	25	2 711.1	0.6
巢　湖	安　徽	2	1	165.6	0
洞庭湖	湖　南	3	1	11 962.2	6.7
鄱阳湖	江　西	6	6	726.7	0.2
太　湖	上　海	6	5	471.6	0.2
	江　苏	75	72	36 411.7	14.5
	浙　江	38	30	8 916.6	2.3
	合　计	**119**	**107**	**45 799.9**	**17.0**

湖泊生活及其他污染情况

（2010）

流域	地区名称	污水处理厂数（座）	污水处理厂处理能力（万吨/日）	城镇生活污水排放量（万吨）	处理生活污水量（万吨）	城镇生活污水中化学需氧量排放量（吨）	城镇生活污水中氨氮排放量（吨）
总　计		**281**	**1 074**	**239 652**	**208 794**	**280 059**	**31 547**
滇　池	云　南	8	84	25 253	24 345	7 166	1 700
巢　湖	安　徽	11	95	23 095	2 214	46 553	4 990
洞庭湖	湖　南	7	28	11 432	7 227	58 561	4 549
鄱阳湖	江　西	5	13	4 143	2 440	13 992	1 163
太　湖	上　海	9	23	7 767	5 736	12 588	926
	江　苏	199	601	106 157	89 460	97 130	10 425
	浙　江	42	231	61 805	57 446	44 067	7 795
	合　计	**250**	**855**	**175 728**	**152 641**	**153 786**	**19 146**

三峡地区接纳工业废水及处理情况（一）

（2010）

地区		汇总企业数（个）	工业废水排放量（万吨）	工业废水处理量（万吨）	工业废水排放达标量（万吨）	工业废水排放达标率（%）
总计		**11 764**	**161 020**	**332 639**	**153 666**	**87.2**
库区	湖北	29	1 585	1 615	1 513	83.2
	重庆	1 438	28 351	40 336	27 254	91.3
	合计	**1 467**	**29 936**	**41 951**	**28 766**	**87.2**
影响区	湖北	143	7 648	9 049	7 583	98.6
	重庆	965	16 075	20 153	14 923	91.7
	四川	1 353	21 860	35 984	21 096	93.2
	贵州	213	1 664	1 406	1 615	88.1
	合计	**2 674**	**47 246**	**66 592**	**45 217**	**92.9**
上游区	重庆	58	357	377	229	75.5
	四川	5 248	71 531	159 338	69 045	89.3
	贵州	1 589	6 786	27 102	5 453	78.5
	云南	728	5 164	37 279	4 955	82.1
	合计	**7 623**	**83 837**	**224 096**	**79 683**	**81.4**

三峡地区接纳工业废水及处理情况（二）

（2010）

地区		工业废水中污染物排放量					
		汞	镉	六价铬	铅	砷	挥发酚
总计		**0.030**	**0.576**	**4.580**	**3.785**	**1.217**	**11.430**
库区	湖北	…	…	…	…	…	…
	重庆	…	0.002	0.152	0.061	…	1.230
	合计	…	**0.002**	**0.152**	**0.061**	…	**1.230**
影响区	湖北	…	…	0.005	0	…	0.013
	重庆	…	0.067	3.402	0.033	…	0.120
	四川	0.005	0	0.294	0.001	…	0.133
	贵州						
	合计	**0.005**	**0.067**	**3.701**	**0.034**	…	**0.266**
上游区	重庆	…	…	0.027	0	…	…
	四川	0.003	0.354	0.583	1.769	0.525	9.502
	贵州	…	0.004	0.094	0.080	0.001	0.109
	云南	0.022	0.150	0.024	1.842	0.691	0.324
	合计	**0.025**	**0.508**	**0.727**	**3.690**	**1.217**	**9.934**

三峡地区接纳工业废水及处理情况（三）

（2010）

地　区		工业废水中污染物排放量			
		氰化物	化学需氧量	石油类	氨氮
总　计		**1.7**	**361 913.1**	**699.4**	**23 572.2**
库　区	湖　北	…	928.7	…	20.1
	重　庆	0.2	57 090.3	71.9	3 991.4
	合　计	**0.2**	**58 019.0**	**71.9**	**4 011.5**
影响区	湖　北	…	7 583.6	16.9	464.5
	重　庆	…	28 812.1	148.7	1 409.9
	四　川	0.2	68 346.0	62.2	1 973.7
	贵　州		3 007.4		92.6
	合　计	**0.2**	**107 749.1**	**227.8**	**3 940.7**
上游区	重　庆	0.3	350.9	0.1	16.3
	四　川	0.7	182 460.1	310.3	14 465.7
	贵　州	0.1	6 011.6	71.1	422.2
	云　南	0.3	7 322.4	18.2	715.9
	合　计	**1.3**	**196 145.0**	**399.7**	**15 620.0**

三峡地区工业废水治理投资情况

（2010）

地　区		工业废水治理施工项目数（个）	工业废水治理竣工项目数（个）	工业废水治理项目完成投资（万元）	废水治理竣工项目新增处理能力（万吨/日）
总　计		**264**	**231**	**111 786.7**	**52.8**
库　区	湖　北	1	0	2 220.0	
	重　庆	34	25	25 002.2	27.0
	合　计	**35**	**25**	**27 222.2**	**27.0**
影响区	湖　北	11	10	7 220.2	1.4
	重　庆	29	21	12 758.4	6.1
	四　川	21	21	10 874.4	0.5
	贵　州	4	4	157.0	0.1
	合　计	**65**	**56**	**31 010.0**	**8.1**
上游区	重　庆				
	四　川	67	61	26 874.5	14.8
	贵　州	28	24	17 352.7	0.5
	云　南	69	65	9 327.3	2.5
	合　计	**164**	**150**	**53 554.5**	**17.8**

三峡地区接纳生活及其他污染情况

（2010）

地　区		污　水处理厂数（座）	污水处理厂处理能力（万吨/日）	城镇生活污水排放量（万吨）	处理生活污水量（万吨）	城镇生活污水中化学需氧量排放量（吨）	城镇生活污水中氨氮排放量（吨）
总　计		**338**	**897**	**320 507**	**227 661**	**827 159**	**82 484**
库　区	湖　北	11	7	1 740	1 511	1 677	292
	重　庆	57	188	58 628	53 718	89 447	12 762
	合　计	**68**	**194**	**60 368**	**55 229**	**91 124**	**13 053**
影响区	湖　北	10	34	10 158	8 456	9 474	1 474
	重　庆	22	44	20 761	11 572	51 743	6 280
	四　川	19	44	22 828	12 008	91 871	8 227
	贵　州	4	6	1 766	1 190	6 139	1 861
	合　计	**55**	**127**	**55 513**	**33 226**	**159 227**	**17 843**
上游区	重　庆			1 545	224	5 443	576
	四　川	146	362	139 558	88 762	398 099	35 952
	贵　州	50	104	29 238	19 845	122 820	9 432
	云　南	19	110	34 285	30 375	50 447	5 628
	合　计	**215**	**576**	**204 626**	**139 206**	**576 808**	**51 588**

“南水北调”东线工程沿线地区工业废水及处理情况（一）

（2010）

地　区名　称	汇总企业数（个）	工业废水排放量（万吨）	工业废水处理量（万吨）	工业废水排放达标量（万吨）	工业废水排放达标率（%）
总　计	**8 570**	**215 315**	**457 979**	**211 409**	**97.4**
天　津	1 885	19 022	83 429	19 017	100.0
河　北	1 113	10 813	6 610	10 668	96.1
江　苏	1 172	30 787	78 116	29 540	95.2
安　徽	379	15 367	29 753	15 202	98.7
山　东	2 786	89 138	190 890	87 973	97.7
河　南	1 235	50 188	69 180	49 009	96.7

“南水北调”东线工程沿线地区工业废水及处理情况（二）

（2010）　　单位：吨

地区名称	工业废水中污染物排放量					
	汞	镉	六价铬	铅	砷	挥发酚
总　计	**0.001**	**0.012**	**1.631**	**1.256**	**0.313**	**24.984**
天　津	…	…	0.078	0.002	0.022	1.519
河　北	…	…	0.550	0.850	…	0.648
江　苏	…	…	0.303	0.391	0.195	6.929
安　徽			0.020	…		3.449
山　东	0.001	…	0.399	0.011	0.094	5.198
河　南	…	0.012	0.281	0.003	0.002	7.242

“南水北调”东线工程沿线地区工业废水及处理情况（三）

（2010）　　单位：吨

地区名称	工业废水中污染物排放量			
	氰化物	化学需氧量	石油类	氨氮
总　计	**7.1**	**351 478.5**	**1 019.3**	**27 016.6**
天　津	0.1	21 377.7	96.8	3 143.2
河　北	…	29 240.8	137.5	5 153.0
江　苏	0.7	40 833.7	194.3	2 468.5
安　徽	2.5	20 542.9	70.7	3 396.4
山　东	0.9	105 220.5	203.4	5 232.6
河　南	3.0	134 262.8	316.4	7 622.9

“南水北调”东线工程沿线地区工业废水治理投资情况

（2010）

地 区 名 称	工业废水治理施工 项目数（个）	工业废水治理竣工 项目数（个）	工业废水治理项目 完成投资（万元）	废水治理竣工项目新增 处理能力（万吨/日）
总 计	**191**	**166**	**103 721.6**	**135.1**
天 津	44	33	23 921.8	3.9
河 北	6	6	2 287.0	0.2
江 苏	20	16	7 344.0	23.4
安 徽	9	9	4 456.8	0.9
山 东	104	95	61 089.0	106.5
河 南	8	7	4 623.0	0.2

“南水北调”东线工程沿线地区生活及其他污染情况

（2010）

地 区 名 称	污 水 处理厂数 （座）	污水处理厂 处理能力 （万吨/日）	城镇生活 污水排放量 （万吨）	处理生活 污 水 量 （万吨）	城镇生活污 水中化学需 氧量排放量 （吨）	城镇生活 污水中氨氮 排 放 （吨）
总 计	**222**	**1 095**	**302 681**	**246 700**	**550 624**	**77 809**
天 津	26	196	47 181	38 696	104 812	15 914
河 北	23	67	14 362	11 976	37 429	4 557
江 苏	35	144	55 974	40 157	125 069	14 156
安 徽	16	93	26 575	22 348	36 210	5 964
山 东	90	432	112 614	100 888	175 835	25 428
河 南	32	163	45 975	32 634	71 269	11 789

入海陆源工业废水排放及处理情况（一）

（2010）

海域	地区名称	汇总工业企业数（个）	工业废水排放量（万吨）		工业废水排放达标量（万吨）	工业废水处理量（万吨）	废水治理设施数（套）	本年运行费用（万元）
				直接排入海的				
总计		**21 539**	**435 967.3**	**113 366.7**	**419 459.5**	**728 958.2**	**15 727**	**985 706.9**
渤海	天津	330	8 338.6	460.8	8 338.6	7 002.7	100	22 680.3
	河北	322	13 060.6	176.6	12 727.5	97 404.9	318	31 010.6
	辽宁	529	8 828.5	1 825.8	8 042.2	38 080.3	238	31 169.6
	山东	665	29 539.3	2 048.9	29 231.8	71 635.8	723	92 811.9
	合计	**1 846**	**59 767.0**	**4 512.0**	**58 340.0**	**214 123.7**	**1 379**	**177 672.4**
黄海	辽宁	365	27 842.9	22 795.4	26 360.1	5 948.0	242	4 693.4
	江苏	892	17 050.3	350.9	16 355.2	22 793.3	776	37 367.1
	山东	1 072	21 775.2	7 287.5	21 565.9	32 704.0	828	49 205.6
	合计	**2 329**	**66 668.4**	**30 433.8**	**64 281.1**	**61 445.4**	**1 846**	**91 266.1**
东海	上海	872	21 268.7	2 007.1	20 717.0	136 382.5	943	107 127.4
	浙江	5 554	92 584.0	11 835.5	88 800.7	93 866.4	3 990	244 612.3
	福建	3 103	85 994.5	59 112.2	85 602.9	97 472.3	1 364	63 520.2
	合计	**9 529**	**199 847.3**	**72 954.7**	**195 120.6**	**327 721.1**	**6 297**	**415 259.9**
南海	广东	7 210	98 601.2	4 452.5	91 098.7	111 523.2	5 788	286 006.1
	广西	306	7 271.3	755.7	6 932.6	8 415.8	151	6 468.1
	海南	319	3 812.2	257.9	3 686.4	5 729.0	266	9 034.3
	合计	**7 835**	**109 684.7**	**5 466.1**	**101 717.7**	**125 668.0**	**6 205**	**301 508.5**

入海陆源工业废水排放及处理情况（二）

（2010）

单位：吨

海域	地区名称	工业废水中污染物排放量									
		汞	镉	六价铬	铅	砷	挥发酚	氰化物	化学需氧量	石油类	氨氮
总计		**0.013**	**0.874**	**16.342**	**2.816**	**0.195**	**31.3**	**68.6**	**555 864.2**	**1 653.9**	**29 857.5**
渤海	天津	…	…	…	…	…	…	…	7 823.3	13.6	1 962.1
	河北	…	…	0.014	…	…	0.1	0.1	23 023.1	24.2	765.6
	辽宁	…	…	…	…	…	0.6	0.9	37 354.2	111.2	642.3
	山东	…	…	0.004	1.350	0.072	0.7	…	51 938.6	180.3	2 857.9
	合计	…	…	**0.019**	**1.350**	**0.072**	**1.4**	**1.0**	**120 139.1**	**329.4**	**6 228.0**
黄海	辽宁	…	…	0.603	…	…	9.8	0.4	42 493.5	242.0	1 048.6
	江苏	…	0.007	0.477	…	…	2.8	0.9	20 818.7	20.0	1 815.6
	山东	…	0.001	0.113	0.033	0.018	0.3	…	31 759.5	46.2	1 339.4
	合计	…	**0.008**	**1.194**	**0.033**	**0.018**	**12.8**	**1.3**	**95 071.8**	**308.3**	**4 203.6**
东海	上海	…	0.001	0.368	0.074	0.012	2.6	3.4	15 313.0	348.3	2 081.5
	浙江	0.008	…	5.014	0.217	0.007	0.5	5.6	118 868.3	77.5	7 444.8
	福建	…	0.002	7.486	0.303	0.041	8.0	51.9	38 429.4	333.7	2 781.7
	合计	**0.008**	**0.003**	**12.868**	**0.594**	**0.060**	**11.1**	**60.9**	**172 610.7**	**759.4**	**12 308.0**
南海	广东	0.004	0.863	2.246	0.791	0.021	2.5	5.3	118 452.1	243.4	5 692.8
	广西			0.016	0.023	0.024	3.4	0.2	43 524.6	12.7	1 117.3
	海南	…	…	…	0.025	…	…	…	6 065.9	0.8	307.7
	合计	**0.004**	**0.863**	**2.262**	**0.839**	**0.045**	**6.0**	**5.4**	**168 042.6**	**256.8**	**7 117.9**

入海陆源工业废水排放及处理情况（三）

（2010）

单位：吨

海域	地区名称	工业废水中污染物去除量				
		挥发酚	氰化物	化学需氧量	石油类	氨氮
总计		**2 929.3**	**1 559.1**	**2 964 649.8**	**71 676.2**	**165 594.3**
渤海	天津	12.8	…	17 616.0	389.9	238.0
	河北	558.5	19.2	91 030.7	3 820.4	984.4
	辽宁	347.5	12.8	57 130.5	3 880.2	4 428.6
	山东	156.0	242.6	352 804.9	13 290.2	7 245.7
	合计	**1 074.8**	**274.6**	**518 582.2**	**21 380.8**	**12 896.7**
黄海	辽宁	2.4	…	127 526.4	411.5	22 837.4
	江苏	129.8	1.5	85 212.1	41.1	1 700.3
	山东	127.4	12.0	199 616.1	1 054.6	4 784.7
	合计	**259.6**	**13.4**	**412 354.6**	**1 507.2**	**29 322.4**
东海	上海	557.0	42.1	221 452.2	14 200.9	3 893.6
	浙江	295.4	751.4	856 070.6	25 367.7	65 317.7
	福建	344.5	115.9	437 207.3	5 224.7	18 614.6
	合计	**1 196.8**	**909.4**	**1 514 730.1**	**44 793.4**	**87 825.9**
南海	广东	263.7	361.2	454 784.9	3 568.5	33 968.7
	广西	134.4	0.5	23 662.8	411.2	616.2
	海南	…	…	40 535.3	15.1	964.4
	合计	**398.1**	**361.7**	**518 982.9**	**3 994.8**	**35 549.3**

入海陆源环境污染治理投资情况

（2010）

流域	地区名称	工业废水治理施工项目数（个）	工业废水治理竣工项目数（个）	工业废水治理项目完成投资（万元）	废水治理竣工项目新增处理能力（万吨/日）
	总　计	**678**	**613**	**227 152.7**	**130.9**
渤海	天　津	6	5	12 641.6	0
	河　北	1	1	392.0	0
	辽　宁	6	7	19 477.0	3.0
	山　东	30	23	55 001.0	14.9
	合　计	**43**	**36**	**87 511.6**	**18.0**
黄海	辽　宁	10	5	3 086.3	0.1
	江　苏	24	24	12 734.7	1.3
	山　东	28	24	22 180.7	18.4
	合　计	**62**	**53**	**38 001.7**	**19.8**
东海	上　海	50	44	4 443.4	0.9
	浙　江	88	78	18 752.5	8.5
	福　建	269	245	50 653.3	42.7
	合　计	**407**	**367**	**73 849.2**	**52.0**
南海	广　东	147	141	22 966.9	39.7
	广　西	6	6	900.0	0.7
	海　南	13	10	3 923.3	0.7
	合　计	**166**	**157**	**27 790.2**	**41.1**

入海陆源生活及其他污染情况

（2010）

流域	地区名称	污水处理厂数（座）	污水处理厂处理能力（万吨/日）	城镇生活污水排放量（万吨）	处理生活污水量（万吨）	城镇生活污水中化学需氧量排放量（吨）	城镇生活污水中氨氮排放量（吨）
	总　计	**424**	**2 833.3**	**780 554**	**594 373**	**1 174 101**	**137 862**
渤海	天　津	6	11.9	7 621	3 394	27 266	2 744
	河　北	13	59.5	12 780	11 008	20 438	2 494
	辽　宁	10	49.5	17 821	11 720	50 875	7 150
	山　东	29	89.9	19 249	16 381	24 336	3 843
	合　计	**58**	**210.8**	**57 471**	**42 502**	**122 915**	**16 231**
黄海	辽　宁	15	75.7	20 379	18 985	26 188	4 015
	江　苏	41	67.2	30 014	16 908	78 972	8 792
	山　东	34	211.0	45 318	41 367	59 387	10 322
	合　计	**90**	**353.8**	**95 711**	**77 260**	**164 548**	**23 129**
东海	上　海	24	553.3	84 189	64 505	91 662	8 690
	浙　江	43	268.1	77 315	61 909	130 369	1 037
	福　建	38	189.0	59 887	43 321	143 037	12 564
	合　计	**105**	**1 010.4**	**221 391**	**169 735**	**365 068**	**31 634**
南海	广　东	142	1 142.2	359 252	284 724	392 544	56 617
	广　西	5	30.0	15 822	4 700	45 917	3 048
	海　南	24	86.1	30 907	15 451	83 108	7 203
	合　计	**171**	**1 258.3**	**405 981**	**304 875**	**521 570**	**66 867**

6

医院环境统计

YIYUAN HUANJING TONGJI

ANNUAL STATISTIC REPORT ON ENVIRONMENT IN CHINA

2010

医院废水排放及处理情况（一）

（2010）

地区名称	汇总医院数（家）	总床位数（张）	废水处理设施数（套）	废水处理设施能力（万吨/日）	废水处理设施运行费用（万元）
合　计	**10 332**	**2 547 283**	**9 703**	**411.0**	**1 011 512.1**
北　京	331	76 171	350	10.2	3 135.8
天　津	176	98 574	189	3.2	2 446.5
河　北	418	94 486	363	5.7	58 593.5
山　西	343	60 291	290	4.2	3 055.6
内蒙古	267	51 781	211	2.5	941.2
辽　宁	451	89 218	362	6.3	140 982.2
吉　林	263	59 379	212	4.3	1 323.3
黑龙江	327	162 947	242	3.2	11 300.8
上　海	355	93 136	352	149.2	3 284.6
江　苏	425	130 918	451	15.3	4 432.6
浙　江	358	108 424	369	11.1	29 475.3
安　徽	258	68 050	222	7.6	2 036.9
福　建	301	107 005	324	7.3	35 247.1
江　西	338	63 016	278	5.5	24 341.0
山　东	628	156 355	601	12.6	4 387.8
河　南	492	146 173	741	11.9	84 915.4
湖　北	333	74 460	292	7.6	2 866.4
湖　南	396	97 857	377	11.4	2 933.3
广　东	646	167 042	687	18.1	200 455.1
广　西	324	66 910	327	7.8	16 814.9
海　南	59	12 309	51	0.7	313.2
重　庆	283	89 798	283	4.1	2 511.6
四　川	813	161 626	814	16.9	28 457.1
贵　州	259	38 071	190	3.8	56 479.4
云　南	419	73 973	309	4.2	2 027.6
陕　西	344	77 141	269	65.5	92 685.5
甘　肃	204	38 548	157	2.1	119 389.1
青　海	82	11 537	65	0.6	8 928.5
宁　夏	103	16 454	85	2.4	13 488.9
新　疆	336	55 633	240	5.7	54 261.9

医院废水排放及处理情况（二）

（2010）

地区名称	废水排放量（万吨）	达标排放量	化学需氧量排放量（吨）	氨氮排放量（吨）	废水处理量（万吨）
合计	**45 202.5**	**40 245.2**	**61 708.8**	**7 877.5**	**43 361.1**
北京	1 804.7	1 759.0	2 341.9	298.8	1 793.2
天津	768.2	768.2	1 286.9	97.3	767.0
河北	1 251.5	1 124.0	1 642.3	535.9	1 212.4
山西	841.4	723.6	1 252.3	347.4	801.4
内蒙古	658.6	529.2	919.8	124.5	641.2
辽宁	1 338.3	1 167.6	1 076.0	135.6	1 299.7
吉林	637.2	523.1	922.8	87.5	642.1
黑龙江	573.0	508.0	942.5	42.1	546.8
上海	2 739.8	2 594.2	1 960.5	625.8	2 610.6
江苏	3 566.1	3 485.1	6 988.2	355.2	3 465.0
浙江	2 691.0	2 628.4	1 681.5	450.2	2 684.7
安徽	1 771.5	1 496.4	2 905.9	258.2	1 731.6
福建	1 409.1	1 247.2	1 553.9	189.6	1 392.2
江西	1 499.4	1 319.0	1 646.2	182.3	1 304.9
山东	2 432.6	2 367.9	1 857.8	214.8	2 429.6
河南	2 483.4	2 370.0	2 577.0	288.0	2 362.1
湖北	1 873.9	1 627.1	3 418.6	335.3	1 733.9
湖南	2 662.1	2 257.1	3 524.6	356.3	2 331.7
广东	4 408.8	3 645.8	2 964.1	495.6	4 307.5
广西	1 542.4	1 378.3	1 606.7	338.0	1 482.4
海南	223.5	193.2	103.2	15.8	192.6
重庆	979.1	956.3	7 093.5	218.0	975.9
四川	2 417.7	2 254.8	2 800.8	495.6	2 383.1
贵州	547.4	437.2	764.0	150.1	512.2
云南	942.4	706.8	1 763.9	221.4	905.3
陕西	1 318.5	1 086.0	1 697.3	184.6	1 260.5
甘肃	479.4	292.4	753.8	88.7	475.7
青海	258.7	118.6	1 150.7	156.5	155.7
宁夏	246.2	204.6	432.9	36.0	235.4
新疆	836.6	476.1	2 079.3	552.4	724.8

医院医疗废物排放及处理情况

（2010）

地区名称	医疗废物产生量（吨）	医疗废物处置量（吨）	送往集中处置厂处置量	医疗废物处理设施数（套）	医疗废物处理设施运行费用（万元）
合计	**336 209.3**	**332 216.6**	**247 466.3**	**4 945**	**935 763.7**
北京	14 853.4	14 853.4	14 845.6	72	478.8
天津	3 845.9	3 845.9	3 790.9	3	186.6
河北	6 354.3	6 291.5	4 070.3	265	16 943.4
山西	7 521.8	7 442.6	5 246.4	154	6 741.5
内蒙古	6 360.8	6 357.1	4 814.1	138	833.3
辽宁	8 098.1	7 806.5	6 030.8	180	33 157.8
吉林	3 676.5	3 666.0	3 106.5	161	378.9
黑龙江	7 168.3	6 941.5	5 390.3	160	959.5
上海	15 511.4	15 504.0	15 414.9	212	769.4
江苏	20 514.6	20 512.4	16 997.6	111	1 941.0
浙江	21 124.7	21 038.1	20 064.8	135	51 098.8
安徽	7 205.0	7 185.6	5 029.5	100	3 150.1
福建	8 023.7	8 023.4	7 951.2	65	1 168.5
江西	45 909.7	45 868.3	6 131.2	533	55 286.1
山东	18 863.7	18 842.6	17 949.0	134	83 657.4
河南	13 889.8	13 799.2	10 505.5	193	130 710.0
湖北	6 478.1	5 849.8	4 424.4	142	844.7
湖南	10 984.6	10 796.8	6 349.2	148	49 485.5
广东	33 200.7	33 182.8	32 563.1	439	74 786.8
广西	12 231.2	12 203.5	7 518.4	177	26 560.5
海南	1 838.4	1 838.4	1 833.9	71	208.8
重庆	8 222.1	8 222.1	6 523.8	56	1 294.5
四川	12 316.8	11 676.9	9 601.5	224	7 698.5
贵州	6 142.7	5 554.4	3 931.5	60	12 541.7
云南	8 337.8	8 044.8	5 943.2	182	1 273.0
陕西	9 757.2	9 729.5	8 930.4	176	145 693.0
甘肃	4 289.7	4 111.6	2 012.8	128	292.8
青海	1 993.5	1 917.0	1 488.0	9	39.6
宁夏	1 594.3	1 594.3	787.8	70	67 690.5
新疆	9 900.5	9 516.5	8 219.8	447	159 892.7

医院医疗放射源管理情况

（2010）

地区名称	放射源数（枚）	集中管理的	退役放射源数（枚）
合　计	**10 505**	**9 642**	**653**
北　京	340	340	1
天　津	73	72	2
河　北	605	578	15
山　西	344	268	36
内蒙古	181	172	5
辽　宁	384	329	3
吉　林	532	527	17
黑龙江	364	301	5
上　海	77	76	5
江　苏	471	455	11
浙　江	344	281	7
安　徽	286	257	18
福　建	236	199	27
江　西	290	254	24
山　东	714	702	17
河　南	403	344	79
湖　北	212	174	17
湖　南	666	627	30
广　东	955	869	102
广　西	244	214	26
海　南	66	64	66
重　庆	549	537	7
四　川	669	658	10
贵　州	199	174	16
云　南	101	92	8
陕　西	362	355	8
甘　肃	334	302	41
青　海	64	57	9
宁　夏	54	42	8
新　疆	386	322	33

7

环境管理统计

HUANJING GUANLI TONGJI

ANNUAL STATISTIC REPORT ON ENVIRONMENT IN CHINA

2010

各地区环境信访工作情况（一）

（2010）

地　区 名　称	来信总数 （封）	环境污染与生态破坏类（件）					
		水污染	大　气 污　染	固体废物 污　　染	噪　声 污　染	危险化学品	放射性污染
总　计	**701 073**	**91 967**	**262 953**	**12 908**	**262 389**	**702**	**284**
国家级	25 603	1 340	1 833	229	634	46	21
北　京	15 492	799	9 238	66	6 033	8	2
天　津	13 837	901	4 688	16	6 024	0	0
河　北	25 649	7 693	10 258	768	3 846	0	0
山　西	7 725	411	4 705	65	2 130	1	2
内蒙古	10 054	501	4 216	143	4 380	9	5
辽　宁	28 648	1 962	10 278	291	10 827	39	28
吉　林	5 556	390	2 529	84	1 909	7	0
黑龙江	10 346	3 081	3 751	0	3 514	0	0
上　海	25 709	1 900	8 150	208	5 707	26	4
江　苏	63 009	11 704	25 050	1 476	23 508	180	12
浙　江	51 454	10 912	24 120	630	11 602	24	40
安　徽	20 853	1 819	6 177	144	11 159	4	3
福　建	34 518	4 762	13 187	837	13 491	48	32
江　西	27 648	8 799	9 017	712	5 091	13	4
山　东	24 788	6 434	13 051	311	4 858	6	17
河　南	17 826	2 512	7 612	254	5 441	16	6
湖　北	26 170	2 838	9 662	410	12 070	91	13
湖　南	14 802	3 563	4 757	2 254	3 438	72	33
广　东	115 392	10 242	59 875	756	44 290	39	17
广　西	26 385	2 540	6 604	1 828	13 432	0	6
海　南	11 372	123	543	4	10 605	0	3
重　庆	37 812	1 696	7 332	95	25 341	0	0
四　川	11 607	2 291	3 692	381	2 546	18	6
贵　州	141	42	62	0	23	0	0
云　南	8 998	1 036	2 951	357	3 862	9	9
西　藏	467	40	106	16	268	0	0
陕　西	24 203	680	4 618	246	18 400	39	2
甘　肃	4 508	516	1 556	187	1 744	0	0
青　海	2 469	50	550	5	1 839	0	0
宁　夏	4 300	202	1 278	44	2 726	0	0
新　疆	3 732	188	1 507	91	1 651	7	19

各地区环境信访工作情况（二）

（2010）　　单位：件

地区名称	环境污染与生态破坏类				
	电磁辐射	三产污染	生态破坏	畜养、农药化肥污染	其他
总计	**4 173**	**27 613**	**1 546**	**2 521**	**9 655**
国家级	48	47	121	46	10
北京	55	0	1	5	143
天津	23	0	0	0	35
河北	1 280	0	512	3	4
山西	2	64	4	3	272
内蒙古	29	351	49	9	220
辽宁	160	2 518	24	124	497
吉林	2	237	0	24	85
黑龙江	0	0	0	0	0
上海	110	741	1	137	384
江苏	210	1 507	10	52	1 684
浙江	357	1 438	29	121	1 630
安徽	172	1 174	54	17	531
福建	110	634	150	502	634
江西	78	3 199	110	571	30
山东	160	37	30	61	138
河南	56	187	20	64	227
湖北	94	728	30	96	443
湖南	78	131	75	51	372
广东	508	8 399	74	349	1 493
广西	186	1 446	21	62	194
海南	0	0	1	4	71
重庆	270	1 978	4	0	61
四川	72	2 354	14	105	19
贵州	0	0	0	0	0
云南	31	234	55	55	87
西藏	0	0	11	0	26
陕西	6	13	29	6	115
甘肃	44	135	106	35	58
青海	1	0	7	1	18
宁夏	11	5	0	0	31
新疆	20	56	4	18	143

各地区环境信访工作情况（三）

（2010）

单位：件

地区名称	建设项目类	行业作风类	发明建议类	环境监测类	咨询类	其他	已办结数量
总计	**4 053**	**366**	**1 015**	**541**	**6 036**	**33 343**	**654 833**
国家级	269	93	341	5	6	21 320	3 089
北京	103	63	391	375	146	291	12 501
天津	27	0	3	0	336	1 834	13 797
河北	1 025	4	0	0	256	0	25 649
山西	0	0	0	2	35	29	7 725
内蒙古	29	5	3	4	73	28	9 728
辽宁	171	8	6	8	174	533	28 557
吉林	6	1	0	0	33	30	5 548
黑龙江	0	0	0	0	0	0	9 934
上海	125	24	72	5	2 416	5 699	23 323
江苏	627	11	39	1	185	460	61 755
浙江	388	7	19	4	44	511	50 205
安徽	63	3	1	3	50	14	20 425
福建	325	9	8	3	25	53	34 170
江西	19	1	0	0	12	2	27 597
山东	69	12	10	17	40	164	24 770
河南	63	6	11	0	1 523	304	17 766
湖北	44	7	4	0	30	515	24 142
湖南	93	13	2	26	37	58	14 712
广东	269	54	29	11	299	472	104 969
广西	28	10	11	5	19	22	25 876
海南	9	0	0	0	2	7	10 943
重庆	106	2	18	0	35	874	37 813
四川	68	5	1	2	3	30	11 607
贵州	0	0	0	0	0	14	126
云南	51	5	25	60	224	32	8 602
西藏	0	0	0	0	0	0	457
陕西	12	6	0	0	13	18	24 132
甘肃	57	15	14	5	18	18	4 491
青海	3	0	0	0	0	0	2 444
宁夏	0	1	1	1	0	0	4 294
新疆	4	1	6	4	2	11	3 686

各地区环境信访工作情况（四）

（2010）

地区名称	来访批次（批）	来访人次（人）	环境污染与生态破坏类（件）					
			水污染	大气污染	固体废物污染	噪声污染	危险化学品	放射性污染
总　计	**34 683**	**65 948**	**7 919**	**13 669**	**944**	**8 291**	**70**	**45**
国家级	489	2 103	178	221	59	107	2	8
北　京	133	331	18	52	1	51	1	0
天　津	351	566	43	151	0	150	0	0
河　北	617	1 055	185	246	18	92	0	0
山　西	320	473	27	208	5	71	1	0
内蒙古	1 052	1 190	133	607	19	189	0	0
辽　宁	2 722	6 541	614	893	95	583	5	7
吉　林	914	1 869	118	460	13	332	0	4
黑龙江	2 423	2 817	691	957	0	775	0	0
上　海	669	1 539	63	250	5	189	1	1
江　苏	1 123	2 354	253	458	27	243	0	0
浙　江	2 132	4 279	483	887	44	451	8	3
安　徽	921	1 653	249	360	23	189	2	1
福　建	926	1 932	203	353	28	250	0	1
江　西	3 257	7 928	916	1 042	56	564	11	0
山　东	877	1 843	347	426	11	113	4	0
河　南	1 078	1 766	270	453	17	255	0	1
湖　北	2 364	4 276	488	948	64	751	5	0
湖　南	3 450	6 432	902	1 186	175	672	21	9
广　东	2 541	5 688	554	1 398	53	629	1	1
广　西	1 692	2 477	418	537	23	377	2	0
海　南	42	71	2	10	0	12	0	0
重　庆	87	216	15	26	3	14	0	0
四　川	1 331	1 913	240	343	76	310	2	8
贵　州	27	64	10	7	1	4	0	0
云　南	1 651	2 635	291	645	82	401	0	0
西　藏	26	32	5	1	0	10	0	0
陕　西	329	437	52	94	5	142	0	0
甘　肃	397	534	57	162	18	114	0	1
青　海	89	117	10	47	2	18	0	0
宁　夏	211	354	25	56	7	104	0	0
新　疆	442	463	59	185	14	129	4	0

各地区环境信访工作情况（五）

（2010）

单位：件

地区名称	环境污染与生态破坏类				
	电磁辐射	三产污染	生态破坏	畜养、农药化肥污染	其　他
总　计	**361**	**2 323**	**124**	**524**	**517**
国家级	9	0	16	2	0
北　京	0	0	0	0	2
天　津	2	1	0	0	2
河　北	30	0	0	0	6
山　西	0	5	0	3	0
内蒙古	0	53	6	3	7
辽　宁	13	297	4	32	65
吉　林	1	12	1	5	6
黑龙江	0	0	0	0	0
上　海	11	46	0	13	42
江　苏	17	103	1	2	37
浙　江	37	106	2	18	47
安　徽	10	68	0	10	11
福　建	8	37	9	20	23
江　西	45	315	12	296	0
山　东	6	3	1	3	7
河　南	11	18	0	10	14
湖　北	14	128	6	19	17
湖　南	56	267	31	31	108
广　东	23	159	9	9	62
广　西	8	302	7	8	10
海　南	0	0	1	1	2
重　庆	6	9	1	0	4
四　川	35	285	1	15	6
贵　州	0	0	0	0	0
云　南	11	47	8	8	11
西　藏	0	0	1	1	2
陕　西	1	0	0	3	5
甘　肃	2	22	3	1	6
青　海	0	3	0	0	9
宁　夏	2	12	0	1	4
新　疆	3	25	4	10	2

各地区环境信访工作情况（六）

（2010）　　单位：件

地区名称	建设项目类	行业作风类	发明建议类	环境监测类	咨询类	其他	已办结数量
总计	**408**	**15**	**26**	**96**	**313**	**304**	**32 887**
国家级	11	2	6	0	3	57	232
北京	13	0	3	0	2	0	104
天津	2	0	1	1	1	1	350
河北	24	0	0	0	4	12	617
山西	0	0	0	0	0	0	318
内蒙古	8	0	0	0	26	1	1 017
辽宁	67	0	0	0	29	18	2 701
吉林	6	0	0	0	0	0	917
黑龙江	0	0	0	0	0	0	2 326
上海	25	0	2	0	9	17	519
江苏	45	0	4	0	9	10	1 085
浙江	45	0	0	0	0	59	2 058
安徽	10	0	0	0	3	0	897
福建	10	0	1	1	3	9	894
江西	0	0	0	0	0	0	3 246
山东	4	0	0	0	2	3	874
河南	7	0	2	0	0	32	1 009
湖北	8	0	0	0	2	14	2 194
湖南	49	6	1	2	117	18	3 423
广东	20	1	3	1	5	13	2 282
广西	20	3	2	0	1	4	1 510
海南	0	0	0	0	0	11	36
重庆	6	1	0	0	1	1	87
四川	9	0	0	0	1	0	1 331
贵州	0	0	0	0	1	3	21
云南	11	0	1	91	68	8	1 383
西藏	2	0	0	0	0	0	26
陕西	0	0	0	0	21	6	320
甘肃	2	2	0	0	4	3	393
青海	2	0	0	0	0	0	82
宁夏	0	0	0	0	0	0	211
新疆	2	0	0	0	1	4	424

各地区人大建议、政协提案情况

（2010）

单位：件

地区名称	人大建议			政协提案		
	人大建议数	已办理数	已办理完毕数	政协提案数	已办理数	已办理完毕数
总计	**4 971**	**4 827**	**4 787**	**6 918**	**6 773**	**6 760**
国家级	257	257	257	166	166	166
北京	86	86	86	68	68	68
天津	22	22	22	13	13	13
河北	19	19	19	25	25	25
山西	22	22	22	13	13	13
内蒙古	118	118	118	169	169	169
辽宁	186	186	186	197	197	197
吉林	16	16	12	29	29	29
黑龙江	126	126	126	119	119	119
上海	96	86	86	95	87	87
江苏	395	395	395	523	523	523
浙江	471	446	446	509	509	509
安徽	189	189	189	297	297	297
福建	276	276	276	387	387	387
江西	218	218	217	401	401	399
山东	115	115	115	251	251	251
河南	186	186	186	419	419	419
湖北	108	108	108	161	161	161
湖南	340	340	340	408	408	408
广东	333	322	302	611	592	592
广西	90	84	81	141	136	136
海南	27	27	27	42	42	42
重庆	256	256	256	330	330	330
四川	332	293	287	531	489	488
贵州	63	42	41	91	48	44
云南	216	216	216	334	334	334
西藏	2	2	2	8	8	8
陕西	93	67	67	122	108	108
甘肃	103	97	95	197	183	181
青海	26	26	24	30	30	30
宁夏	14	14	14	40	40	40
新疆	170	170	169	191	191	187

各地区环境保护档案工作情况（一）

（2010）

地区名称	档案机构（个）	现存档案资料		档案库房实用面积（米²）	现有专职档案人员（人）	兼职档案人员（人）
		全宗（个）	案卷（卷）			
总　计	**2 455**	**13 850**	**2 441 523**	**74 268**	**1 710**	**4 656**
国家级	11	46	84 718	660	12	106
北　京	27	29	163 498	949	21	174
天　津	15	22	46 401	768	12	63
河　北	155	1 605	129 085	3 873	157	236
山　西	142	137	98 420	2 661	147	217
内蒙古	85	106	86 619	2 298	92	160
辽　宁	69	95	92 060	3 199	76	196
吉　林	85	94	99 589	1 740	58	74
黑龙江	139	171	72 461	2 572	4	161
上　海	21	22	172 701	836	14	57
江　苏	149	480	158 207	5 901	124	349
浙　江	104	140	217 266	4 172	65	278
安　徽	73	88	64 423	2 118	42	116
福　建	58	99	69 987	1 633	22	85
江　西	36	533	15 831	1 513	20	100
山　东	207	209	189 597	6 565	204	601
河　南	195	505	43 707	4 011	144	157
湖　北	83	6 114	62 634	2 836	54	146
湖　南	99	125	38 046	3 134	30	107
广　东	122	116	254 881	4 493	93	223
广　西	60	235	24 196	1 697	13	114
海　南	15	21	8 015	954	25	23
重　庆	26	50	74 266	1 699	17	93
四　川	133	1 136	33 329	4 041	58	208
贵　州	12	152	16 534	574	12	53
云　南	91	883	28 855	2 448	62	172
西　藏						
陕　西	87	259	28 699	2 604	21	144
甘　肃	83	193	29 383	1 665	60	83
青　海	14	35	4 088	493	6	30
宁　夏	26	29	10 928	613	6	24
新　疆	33	121	23 099	1 548	39	106

各地区环境保护档案工作情况（二）

（2010）

地 区 名 称	本年接收 档案（卷）	本年借阅档案 人次（人次）	本年借阅 档案（卷）	本年编研 资料（种）	档案管理达标升级情况（个）			
					国家级	省一级	省二级	省三级
总 计	**214 531**	**193 392**	**286 475**	**3 767**	**54**	**490**	**407**	**263**
国家级	8 545	1 544	6 930	13	5	6	0	0
北 京	14 260	2 576	7 194	48	1	6	13	0
天 津	1 240	1 494	2 310	114	0	5	4	2
河 北	5 454	10 176	14 672	272	2	32	18	39
山 西	9 636	4 280	7 099	76	0	7	18	14
内蒙古	8 674	6 476	897	11	7	32	12	2
辽 宁	6 323	6 309	7 725	72	0	15	17	7
吉 林	12 490	2 977	6 030	86	0	12	28	16
黑龙江	3 320	3 950	5 536	53	5	44	2	5
上 海	6 177	2 004	6 622	29	1	4	6	0
江 苏	9 848	20 016	26 992	616	3	62	31	15
浙 江	23 574	19 667	28 128	186	1	23	22	3
安 徽	8 395	4 091	5 873	68	11	5	3	0
福 建	14 850	4 551	6 691	120	1	12	4	0
江 西	807	4 817	4 895	27	0	0	3	6
山 东	13 007	12 978	26 217	20	3	67	47	27
河 南	3 653	9 033	11 144	143	0	30	9	12
湖 北	7 731	7 297	11 418	206	2	34	47	1
湖 南	1 468	4 440	6 450	165	0	3	50	0
广 东	26 093	20 248	27 674	165	9	36	10	0
广 西	2 289	6 831	8 818	49	0	1	0	0
海 南	1 067	3 195	9 864	106	0	1	1	1
重 庆	4 445	3 029	4 204	44	0	4	6	6
四 川	3 538	10 546	15 493	364	0	10	29	68
贵 州	532	748	719	63	0	3	2	5
云 南	5 012	7 535	10 307	236	3	8	10	27
西 藏								
陕 西	3 700	2 285	5 024	26	0	2	3	6
甘 肃	3 235	5 548	6 229	115	0	13	5	0
青 海	755	700	714	12	0	6	3	0
宁 夏	1 303	1 102	1 576	4	0	2	0	0
新 疆	3 110	2 949	3 030	258	0	5	4	1

各地区环境法制工作情况（一）

（2010）

地区名称	颁布地方性法规 法规（件）	颁布地方性法规 行政规章（件）	当年受理环境行政处罚案件数（起）	当年受理环境行政复议案件数（起）	当年受理环境行政诉讼案件数（起）	当年受理环境犯罪案件数（起）
总　计	**22**	**20**	**116 820**	**694**	**228**	**11**
国家级		0	48	97	0	
北　京	1	0	1 774	1	3	1
天　津		0	68	1		
河　北		0	4 110	2	2	
山　西	1	0	2 742			
内蒙古	0	1	1 307	2	4	0
辽　宁	4	1	33 781	9	0	0
吉　林	0	4	1 236	1	0	0
黑龙江	2	2	929	4		
上　海	0	0	1 172	16	7	1
江　苏	0	0	9 704	56	42	2
浙　江	2	2	19 479	110	36	0
安　徽	1	0	757	11		
福　建	1	1	3 099	20	5	3
江　西		0	1 935	5		
山　东	1	0	6 195	10	5	1
河　南	0	2	4 726	19	13	0
湖　北		0	1 604	9	19	
湖　南		0	1 566	15	9	
广　东		1	11 977	194	62	2
广　西	0	0	705	6	1	0
海　南						
重　庆	0	1	924	54	3	0
四　川	1	0	1 133	31	10	1
贵　州	3	0	688	1		
云　南	2	0	943	4	2	0
西　藏	1	1	7			
陕　西		0	1 540	6	3	
甘　肃	0	2	381	4		
青　海	0	1	47			
宁　夏	1	0	295	5	2	
新　疆	1	1	1 948	1		

各地区环境法制工作情况（二）

（2010）　　　　单位：起

地　区 名　称	当年做出环境行政 处罚决定的案件数	当年做出环境行政 复议决定的案件数	当年做出判决的环境 行政诉讼案件数	当年做出判决的 环境犯罪案件数
总　计	**112 025**	**598**	**168**	**8**
国家级	38	81	0	
北　京	1 774	1	3	1
天　津	68			
河　北	4 110	2		
山　西	2 742			
内蒙古	1 262	0	10	0
辽　宁	33 719	2	6	0
吉　林	1 224	1	0	0
黑龙江	656	4	0	
上　海	1 134	6	4	1
江　苏	9 592	50	28	2
浙　江	18 424	101	30	0
安　徽	653	6		
福　建	3 051	19	4	3
江　西	1 847	4		
山　东	6 144	10	4	
河　南	4 128	20	7	0
湖　北	1 057	9	13	
湖　南	1 426	13	8	
广　东	11 268	163	32	0
广　西	606	6	1	0
海　南				
重　庆	883	53	3	0
四　川	1 118	31	9	1
贵　州	645			
云　南	449	4	2	0
西　藏	5			
陕　西	1 349	3	3	
甘　肃	373	4		
青　海	47			
宁　夏	295	4	1	
新　疆	1 938	1		

各地区环境保护系统年末机构总数（一）

（2010）

单位：个

地区名称	总计	国家、省级合计	环保局	监察机构	监测站	科研所	宣教中心	信息中心	其他
总计	**12 849**	**414**	**32**	**48**	**37**	**33**	**32**	**27**	**205**
国家级	43	43	1	13	1	3	1	1	23
北京	101	13	1	1	1	1	1	1	7
天津	86	13	1	1	1	1	1	1	7
河北	821	12	1	2	1	1	1	1	5
山西	582	19	1	1	1	1	1	1	13
内蒙古	390	11	1	3	2	1	1	0	3
辽宁	474	18	1	1	1	1	1	1	12
吉林	273	12	1	1	1	1	1	1	6
黑龙江	417	16	1	1	1	1	1	1	10
上海	76	8	1	1	1	1	1	1	2
江苏	722	17	1	1	1	1	1	1	11
浙江	572	9	1	1	2	1	1	1	2
安徽	387	11	1	1	1	1	1	1	5
福建	352	13	1	1	1	1	1	1	7
江西	405	11	1	1	1	1	1	1	5
山东	876	13	1	1	1	1	1	1	7
河南	797	9	1	1	1	1	1	1	3
湖北	484	10	1	1	1	1	1	1	4
湖南	460	6	1	1	1	1	1	0	1
广东	845	12	1	1	1	1	1	1	6
广西	335	12	1	1	2	1	1	0	6
海南	141	7	1	1	1	1	1	1	1
重庆	331	12	1	1	1	0	1	1	7
四川	641	13	1	1	2	1	1	1	6
贵州	359	13	1	1	1	1	1	1	7
云南	467	11	1	1	1	1	1	1	5
西藏	114	8	1	1	1	0	1	0	4
陕西	370	15	1	1	1	1	1	1	9
甘肃	350	11	1	1	1	1	1	1	5
青海	131	9	1	1	1	1	1	1	3
宁夏	65	11	1	1	1	1	1	0	6
新疆	382	16	1	2	2	2	1	1	7

各地区环境保护系统年末机构总数（二）

（2010）　　　　单位：个

地　区 名　称	地市级 合　计	环保局	监察机构	监测站	科研所	宣教中心	信息中心	其　他
总　计	**1 937**	**333**	**354**	**339**	**204**	**117**	**123**	**467**
北　京	0	0	0	0	0	0	0	0
天　津	0	0	0	0	0	0	0	0
河　北	80	11	12	11	10	5	6	25
山　西	101	11	11	11	11	7	10	40
内蒙古	71	12	13	13	7	5	2	19
辽　宁	114	14	14	14	14	12	9	37
吉　林	65	9	12	9	7	9	3	16
黑龙江	73	13	14	14	10	6	3	13
上　海	0	0	0	0	0	0	0	0
江　苏	90	13	14	13	11	8	6	25
浙　江	62	11	11	11	6	4	3	16
安　徽	86	17	17	17	12	2	13	8
福　建	57	9	9	10	9	8	7	5
江　西	71	11	11	12	10	4	4	19
山　东	106	17	17	17	7	6	7	35
河　南	107	17	19	17	13	6	2	33
湖　北	84	13	20	12	11	4	5	19
湖　南	82	14	14	14	10	2	1	27
广　东	145	21	22	21	18	11	14	38
广　西	61	14	14	15	9	1	4	4
海　南	11	2	2	3	2	0	1	1
重　庆	0	0	0	0	0	0	0	0
四　川	83	21	21	24	1	2	5	9
贵　州	48	9	9	9	3	3	2	13
云　南	68	16	15	16	6	3	3	9
西　藏	32	7	7	7	0	1	0	10
陕　西	58	10	13	10	6	4	1	14
甘　肃	80	14	15	15	6	0	7	23
青　海	24	8	8	5	0	0	1	2
宁　夏	17	5	5	5	1	0	1	0
新　疆	61	14	15	14	4	4	3	7

各地区环境保护系统年末机构总数（三）

（2010）　　　　单位：个

地区名称	县级合计	环保局	监察机构	监测站	其他	全省乡镇环保机构
总计	**8 606**	**2 810**	**2 658**	**2 211**	**927**	**1 892**
北京	88	16	21	18	33	0
天津	73	16	21	20	16	0
河北	530	172	150	139	69	199
山西	413	119	117	113	64	49
内蒙古	292	101	92	90	9	16
辽宁	311	100	90	78	43	31
吉林	196	56	70	55	15	0
黑龙江	326	118	101	90	17	2
上海	68	18	18	19	13	0
江苏	386	105	137	95	49	229
浙江	291	90	78	74	49	210
安徽	269	104	81	65	19	21
福建	276	84	87	77	28	6
江西	312	100	90	94	28	11
山东	555	140	154	131	130	202
河南	537	158	189	123	67	144
湖北	342	99	98	86	59	48
湖南	351	121	106	106	18	21
广东	425	110	89	105	121	263
广西	253	105	83	64	1	9
海南	48	16	15	15	2	75
重庆	127	40	40	40	7	192
四川	512	181	181	131	19	33
贵州	261	88	87	67	19	37
云南	351	129	109	105	8	37
西藏	74	73	0	0	1	0
陕西	297	107	109	79	2	0
甘肃	220	86	86	45	3	39
青海	98	43	40	11	4	0
宁夏	37	17	11	9	0	0
新疆	287	98	108	67	14	18

各地区环境保护系统年末实有人数

（2010）

单位：人

地区名称	年末实有人数	高级职称	中级职称	初级职称	环保局	监察机构	监测站	乡镇环保机构
总　计	**193 911**	**10 936**	**27 487**	**33 640**	**45 938**	**62 468**	**54 698**	**7 154**
国家级	2 584	667	546	300	309	415	154	0
北　京	2 582	202	420	661	544	355	521	0
天　津	1 918	139	311	368	376	239	825	0
河　北	16 199	570	1 268	2 662	3 963	4 846	3 871	1 723
山　西	11 431	377	1 652	2 281	1 713	4 165	3 444	562
内蒙古	5 308	364	737	658	1 587	1 635	1 574	15
辽　宁	8 272	700	1 619	1 303	1 456	2 667	2 328	2
吉　林	5 218	414	1 075	1 257	767	2 396	1 542	0
黑龙江	5 020	461	1 019	807	1 225	1 751	1 489	20
上　海	2 306	173	538	465	454	498	941	0
江　苏	10 101	698	2 276	2 132	2 274	3 010	3 500	377
浙　江	6 565	633	1 264	1 407	1 384	1 458	2 100	989
安　徽	5 658	358	816	1 428	1 181	2 108	1 740	110
福　建	3 815	400	697	725	818	1 161	1 254	22
江　西	5 130	266	550	866	1 588	1 586	1 350	26
山　东	13 563	788	1 960	2 672	2 805	4 045	4 071	765
河　南	21 910	575	2 086	3 459	4 245	10 539	5 216	576
湖　北	7 623	288	1 375	1 469	1 976	2 575	1 959	244
湖　南	9 011	351	1 279	1 534	2 640	2 954	2 667	107
广　东	10 793	612	1 400	1 696	3 108	2 221	2 834	789
广　西	3 641	338	916	901	1 147	901	1 302	16
海　南	1 445	35	78	90	501	325	243	191
重　庆	2 656	177	288	313	623	723	874	285
四　川	7 567	343	817	1 042	2 105	2 363	2 589	34
贵　州	3 095	82	299	531	1 134	788	730	74
云　南	4 620	324	689	665	1 840	920	1 370	67
西　藏	587	9	15	22	417	36	78	0
陕　西	5 784	173	391	589	1 223	2 450	1 643	0
甘　肃	4 090	101	312	519	1 192	1 577	935	68
青　海	865	67	127	114	227	286	256	0
宁　夏	1 008	82	154	219	257	277	306	0
新　疆	3 546	169	513	485	859	1 198	992	92

各地区环境保护系统各级机构人员数（一）

（2010）　　单位：人

地区名称	年末实有人数合计	国家级、省级合计	环保局	监察机构	监测站	科研所	宣教中心	信息中心	其他
总　计	**193 911**	**15 011**	**2 770**	**1 493**	**3 144**	**3 053**	**498**	**307**	**3 746**
国家级	2 584	2 584	309	415	154	585	31	34	1 056
北　京	2 582	817	121	58	149	237	46	8	198
天　津	1 918	648	101	60	181	161	21	22	102
河　北	16 199	369	75	54	80	49	15	13	83
山　西	11 431	386	78	33	72	45	9	11	138
内蒙古	5 308	292	62	45	84	26	15	0	60
辽　宁	8 272	406	62	21	76	104	14	6	123
吉　林	5 218	357	72	20	86	72	18	14	75
黑龙江	5 020	405	75	26	81	70	13	7	133
上　海	2 306	668	98	59	181	220	28	14	68
江　苏	10 101	492	125	17	73	69	13	21	174
浙　江	6 565	394	73	26	187	74	7	6	21
安　徽	5 658	338	63	38	87	79	12	12	47
福　建	3 815	370	70	39	63	46	9	6	137
江　西	5 130	385	59	43	87	110	9	12	65
山　东	13 563	481	84	35	78	71	22	13	178
河　南	21 910	319	108	35	65	36	22	8	45
湖　北	7 623	299	74	31	61	66	19	5	43
湖　南	9 011	410	61	18	173	135	22	0	1
广　东	10 793	452	99	43	97	26	11	9	167
广　西	3 641	296	44	23	102	43	17	0	67
海　南	1 445	256	92	25	8	101	7	13	10
重　庆	2 656	471	127	55	183	0	16	10	80
四　川	7 567	473	100	30	146	129	13	9	46
贵　州	3 095	360	69	44	72	85	8	9	73
云　南	4 620	451	88	18	92	149	15	10	79
西　藏	587	114	49	7	34	0	0	0	24
陕　西	5 784	445	94	47	81	43	13	16	151
甘　肃	4 090	318	68	18	85	74	11	7	55
青　海	865	213	47	24	70	20	8	6	38
宁　夏	1 008	297	55	35	57	30	22	0	98
新　疆	3 546	445	68	51	99	98	12	6	111

各地区环境保护系统各级机构人员数（二）

（2010）

单位：人

地区名称	地市级合计	环保局	监察机构	监测站	科研所	宣教中心	信息中心	其他
总计	**42 462**	**8 633**	**9 180**	**15 697**	**3 445**	**788**	**748**	**3 971**
北京	0	0	0	0	0	0	0	0
天津	0	0	0	0	0	0	0	0
河北	2 500	498	450	816	216	76	54	390
山西	1 899	352	388	534	173	37	80	335
内蒙古	1 549	283	271	669	132	23	18	153
辽宁	3 338	496	624	966	526	117	73	536
吉林	1 025	201	323	288	78	36	14	85
黑龙江	1 428	349	322	527	118	48	10	54
上海	0	0	0	0	0	0	0	0
江苏	2 457	486	518	1 023	188	45	24	173
浙江	1 437	271	288	608	127	31	20	92
安徽	1 648	319	423	684	114	12	56	40
福建	1 068	206	259	368	146	36	32	21
江西	1 282	295	303	428	121	24	25	86
山东	2 691	537	545	976	141	62	80	350
河南	3 318	648	789	1 220	185	61	21	394
湖北	1 722	312	418	597	223	24	16	132
湖南	1 940	388	379	791	208	18	4	152
广东	3 470	675	836	1 047	350	62	69	431
广西	1 308	301	251	615	96	10	25	10
海南	209	36	56	84	21	0	10	2
重庆	0	0	0	0	0	0	0	0
四川	2 045	485	465	895	63	16	17	104
贵州	649	162	118	290	16	7	8	48
云南	1 181	339	202	485	64	12	20	59
西藏	203	99	29	44	0	1	0	30
陕西	1 336	278	242	583	63	19	8	143
甘肃	1 133	263	252	453	22	0	38	105
青海	203	65	55	69	0	0	8	6
宁夏	344	79	105	142	12	0	6	0
新疆	1 079	210	269	495	42	11	12	40

各地区环境保护系统各级机构人员数（三）

（2010）　　　　单位：人

地区名称	县级合计	环保局	监察机构	监测站	其他	全省乡镇环保机构
总　计	**129 284**	**34 535**	**51 795**	**35 857**	**7 097**	**7 154**
北　京	1 765	423	297	372	673	0
天　津	1 270	275	179	644	172	0
河　北	11 607	3 390	4 342	2 975	900	1 723
山　西	8 584	1 283	3 744	2 838	719	562
内蒙古	3 452	1 242	1 319	821	70	15
辽　宁	4 526	898	2 022	1 286	320	2
吉　林	3 836	494	2 053	1 168	121	0
黑龙江	3 167	801	1 403	881	82	20
上　海	1 638	356	439	760	83	0
江　苏	6 775	1 663	2 475	2 404	233	377
浙　江	3 745	1 040	1 144	1 305	256	989
安　徽	3 562	799	1 647	969	147	110
福　建	2 355	542	863	823	127	22
江　西	3 437	1 234	1 240	835	128	26
山　东	9 626	2 184	3 465	3 017	960	765
河　南	17 697	3 489	9 715	3 931	562	576
湖　北	5 358	1 590	2 126	1 301	341	244
湖　南	6 554	2 191	2 557	1 703	103	107
广　东	6 082	2 334	1 342	1 690	716	789
广　西	2 021	802	627	585	7	16
海　南	789	373	244	151	21	191
重　庆	1 900	496	668	691	45	285
四　川	5 015	1 520	1 868	1 548	79	34
贵　州	2 012	903	626	368	115	74
云　南	2 921	1 413	700	793	15	67
西　藏	270	269	0	0	1	0
陕　西	4 003	851	2 161	979	12	0
甘　肃	2 571	861	1 307	397	6	68
青　海	449	115	207	117	10	0
宁　夏	367	123	137	107	0	0
新　疆	1 930	581	878	398	73	92

各地区环境科技工作情况（一）

（2010）

地区名称	科研课题项目数（项）	科研课题经费数（万元）	授权专利数（项）	授权发明专利数	获科学技术奖励数（项）	国家级	省（部）级一等	省（部）级二等	省（部）级三等
总计	**1 133**	**44 719.8**	**101**	**27**	**56**	**1**	**4**	**14**	**37**
北京									
天津	47	2 721.2	3	1	6	0	1	4	1
河北									
山西									
内蒙古	10	368.0	0	0	0	0	0	0	0
辽宁	103	3 560.0	9	6	9	0	0	2	7
吉林	88	2 385.0	0	0	0				
黑龙江									
上海	79	5 122.4	0	0	3	0	1	1	1
江苏	310	14 134.2	52	8	12	0	0	4	8
浙江	169	3 685.1	13	7	5	1	0	1	3
安徽									
福建									
江西									
山东									
河南	38	770.0	20	2	5	0	1	1	3
湖北									
湖南									
广东									
广西									
海南	30	1 154.3	0	0	3	0	0	0	3
重庆	30	1 716.5	0	0	2	0	0	0	2
四川	101	3 638.8	2	1	5	0	1	0	4
贵州	40	762.4	0	0	0	0	0	0	0
云南	48	1 938.9	2	2	0	0	0	0	0
西藏									
陕西									
甘肃	19	149.0	0	0	3	0	0	1	2
青海	9	195.0	0	0	1	0	0	0	1
宁夏									
新疆	12	2 419.0			2				2

各地区环境科技工作情况（二）

（2010）

地区名称	颁布地方环境标准数（项）	累积颁布地方环境标准总数（项）	从事科技活动人员数（人）	科技管理人员	科技人员	科技辅助人员	外聘流动研究人员	科研业务费支出（万元）
总计	**103**	**121**	**10 223**	**1 502**	**7 228**	**1 210**	**273**	**40 493.2**
北京								
天津	0	0	337	39	182	106	0	3 436.2
河北								
山西								
内蒙古	0	0	165	32	116	17	0	29.0
辽宁	0	0	1 076	153	802	106	15	8 250.7
吉林			534	40	461	26	7	1 408.0
黑龙江								
上海	1	8	196	7	188	1	0	18.0
江苏	0	0	2 505	338	1 728	373	66	11 361.2
浙江	0	0	1 922	381	1 380	146	15	5 384.0
安徽								
福建								
江西								
山东								
河南	0	0	715	106	487	93	29	974.5
湖北								
湖南								
广东								
广西								
海南	0	0	231	55	127	24	25	363.8
重庆	0	0	349	46	258	40	5	237.0
四川	98	98	966	170	601	151	44	3 695.5
贵州	0	0	387	55	279	53	0	3 427.8
云南	0	0	549	51	423	40	35	1 287.1
西藏								
陕西								
甘肃	1	1	132	13	74	18	27	151.0
青海	0	0	89	9	61	14	5	195.0
宁夏								
新疆	3	14	70	7	61	2	0	274.4

各地区环境监测工作情况（一）

（2010）

地区名称	监测用房总面积（米²）	环境监测经费（万元）	环境监测仪器数量（台/套）	环境空气监测点位（个）		酸雨监测点位（个）	开展酸雨监测的监测站数（个）
					国控监测点位		
总计	**1 890 983**	**670 740.3**	**1 697 951**	**4 468**	**851**	**1 612**	**1 038**
国家级	14 954	23 389.0	560				
北京	22 209	10 832.3	5 733	27	12	3	3
天津	27 089	15 976.8	2 982	110	15	9	9
河北	104 222	27 347.2	6 790	145	27	25	22
山西	83 592	22 054.7	7 148	264	33	24	14
内蒙古	41 387	14 132.8	3 926	82	17	22	9
辽宁	62 945	21 491.8	6 727	187	87	126	100
吉林	53 768	16 087.8	3 032	160	26	30	20
黑龙江	45 784	8 308.8	2 539	51	19	44	13
上海	41 616	21 014.1	3 760	49	10	22	19
江苏	159 061	61 242.7	12 685	219	52	116	82
浙江	98 426	43 801.3	9 958	196	32	102	87
安徽	47 440	11 688.6	3 574	58	21	38	17
福建	55 916	22 153.5	6 841	174	30	70	41
江西	58 571	15 532.2	4 850	189	18	24	12
山东	140 422	47 307.1	12 303	270	74	125	82
河南	98 694	22 261.6	8 406	272	47	79	40
湖北	59 341	18 629.1	5 209	203	32	66	47
湖南	95 648	24 938.6	5 604	266	38	98	64
广东	121 254	75 630.4	14 065	329	66	99	82
广西	43 616	15 614.7	4 004	121	31	67	35
海南	12 236	3 099.5	1 596	56	8	21	13
重庆	63 554	18 369	7 729	77	13	54	39
四川	95 315	37 666.8	11 171	417	39	168	73
贵州	26 650	21 220.0	1 019	52	14	12	12
云南	56 918	10 730.1	7 422	162	11	56	41
西藏	13 420	761.9	722	15	4	1	1
陕西	55 201	13 298.7	3 208	139	30	39	25
甘肃	22 438	6 261.7	2 134	68	25	27	14
青海	11 458	4 551.3	1 063	31	4	8	3
宁夏	20 828	2 815.2	1 051	18	9	12	5
新疆	37 010	12 531.0	1 984	61	7	25	14

各地区环境监测工作情况（二）

（2010）

地区名称	沙尘暴监测点位数（个）	开展沙尘暴监测的监测站数（个）	地表水水质监测断面（个）	其中：国控断面	开展饮用水源地水质监测的城市数（个）	近岸海域监测点位（个）	开展近岸海域监测的监测站数（个）
总　计	**117**	**83**	**11 349**	**745**	**1 121**	**836**	**172**
国家级							
北　京	1	1	231	7	1	0	0
天　津	1	1	192	15	5	17	1
河　北	3	3	391	38	36	20	3
山　西	2	2	210	8	32	0	0
内蒙古	18	15	108	23	23	0	0
辽　宁	41	23	288	27	32	184	19
吉　林	5	4	268	27	38	0	0
黑龙江	0	0	119	26	13	0	0
上　海		0	371	3	1	14	3
江　苏	1	0	1 589	122	58	25	6
浙　江	0	0	937	36	86	111	27
安　徽	0	0	214	68	29	0	0
福　建	0	0	481	18	67	82	7
江　西	0	0	424	17	69	0	0
山　东	2	2	928	42	83	215	36
河　南	1	1	638	36	58	0	0
湖　北	0	0	464	19	18	0	0
湖　南	0	0	392	26	93	0	0
广　东	0	0	770	9	21	52	52
广　西	0	0	191	14	33	38	5
海　南	0	0	131	4	18	78	13
重　庆	0	0	262	8	39	0	0
四　川	0	0	489	25	101	0	0
贵　州			74	10	9	0	
云　南	0	0	538	43	48	0	0
西　藏			6	7	1		
陕　西	4	3	260	9	34	0	0
甘　肃	10	8	85	11	28	0	0
青　海	3	2	37	5	8	0	0
宁　夏	5	5	30	4	5	0	0
新　疆	20	13	231	38	34	0	0

各地区环境监测工作情况（三）

（2010）

地　区 名　称	开展环境噪声监测的监测站数（个）	环境噪声自动监测系统（套）	开展污染源监督性监测的重点企业数量（个）	开展生态监测的监测站数（个）	开展土壤监测的监测站数（个）
总　计	**1 720**	**344**	**48 024**	**164**	**379**
国家级					
北　京	19	108	359	1	1
天　津	21	0	737	1	3
河　北	60	3	1 403	1	29
山　西	83	22	1 538	1	9
内蒙古	48	1	718	6	7
辽　宁	78	0	1 603	27	26
吉　林	51	0	873	1	4
黑龙江	13	0	216	14	14
上　海	19	37	1 081	1	12
江　苏	90	20	3 310	21	38
浙　江	72	7	2 991	9	48
安　徽	170	28	578	2	12
福　建	70	22	1 942	1	10
江　西	51	10	714	0	7
山　东	143	1	4 631	9	20
河　南	91	3	1 791	7	16
湖　北	71	1	976	1	11
湖　南	98	0	1 161	15	19
广　东	96	28	6 470	2	10
广　西	47	0	1 023	1	5
海　南	18	0	95	1	4
重　庆	41	21	8 828	2	4
四　川	89	5	1 705	24	35
贵　州	19	2	293	1	
云　南	53	13	712	1	8
西　藏	2		11		
陕　西	58	0	782	2	4
甘　肃	14	0	362	0	8
青　海	2	1	254	1	2
宁　夏	5	0	128	1	5
新　疆	28	10	739	10	8

各地区环境污染控制与管理情况

（2010）

地区名称	已发放危险废物经营许可证数（个）	环保模范城市数（个）	参加“城考”城市数（个）
总计	**1 529**	**71**	**655**
北京	14	0	1
天津	18	1	1
河北	53	1	33
山西	14	0	22
内蒙古	11	0	20
辽宁	61	2	31
吉林	52	1	28
黑龙江	23	1	30
上海	50	0	1
江苏	367	19	40
浙江	81	7	33
安徽	29	1	22
福建	23	3	23
江西	36	0	21
山东	103	18	48
河南	56	0	38
湖北	51	1	36
湖南	121	0	29
广东	130	9	44
广西	14	1	21
海南	6	1	8
重庆	33	0	1
四川	50	2	32
贵州	22	0	13
云南	27	0	17
西藏	0	0	2
陕西	30	1	13
甘肃	14	0	16
青海	5	0	3
宁夏	9	0	7
新疆	26	2	21

各地区自然生态保护情况（一）

（2010）

地区名称	自然保护区数（个）	国家级	省级	地市级	县级	保护区面积（公顷）	国家级
总计	**2 588**	**319**	**859**	**418**	**992**	**149 440 589**	**92 675 600**
北京	20	2	12	6	0	133 966	26 403
天津	8	3	5	0	0	91 115	37 862
河北	35	11	20	2	2	587 268	217 177
山西	46	5	41	0	0	1 153 899	82 936
内蒙古	185	23	61	25	76	13 823 738	3 843 723
辽宁	98	12	27	35	24	2 661 940	936 405
吉林	38	13	16	3	6	2 303 900	921 188
黑龙江	202	23	78	32	69	6 408 475	2 337 122
上海	4	2	2	0	0	93 821	66 175
江苏	30	3	10	10	7	564 983	336 211
浙江	31	9	9	0	13	194 502	98 064
安徽	98	6	27	2	63	502 364	131 410
福建	92	12	26	10	44	445 478	205 521
江西	178	8	32	2	136	1 116 087	144 486
山东	86	7	33	24	22	1 134 854	256 766
河南	34	11	21	0	2	734 657	426 316
湖北	64	10	19	24	11	959 446	239 899
湖南	123	17	31	1	74	1 244 703	500 686
广东	367	11	63	115	178	3 591 949	225 534
广西	78	16	50	3	9	1 451 033	308 255
海南	50	9	24	12	5	2 736 508	106 526
重庆	48	3	19	1	25	827 598	195 512
四川	166	23	66	29	48	8 904 075	2 774 776
贵州	129	8	4	16	101	951 766	243 539
云南	167	16	44	59	48	2 988 168	1 427 075
西藏	47	9	13	3	22	41 502 766	37 153 065
陕西	54	12	35	4	3	1 161 099	350 605
甘肃	59	15	40	0	4	7 346 761	4 784 193
青海	11	5	6	0	0	21 822 201	20 252 490
宁夏	13	6	7	0	0	507 104	439 529
新疆	27	9	18	0	0	21 494 365	13 606 151

各地区自然生态保护情况（二）

（2010）

地区名称	保护区面积（公顷）			保护区面积占辖区面积比（%）
	省级	地市级	县级	
总计	**41 748 151**	**4 682 371**	**10 334 467**	**14.9**
北京	71 413	36 150	0	7.97
天津	53 253	0	0	8.06
河北	351 109	8 806	10 176	3.05
山西	1 070 963	0	0	7.4
内蒙古	7 089 975	315 783	2 574 257	11.69
辽宁	830 976	794 093	100 466	12.46
吉林	1 339 167	20 564	22 981	12.29
黑龙江	2 890 577	349 247	831 529	14.09
上海	27 646	0	0	5.22
江苏	85 448	122 827	20 497	4.08
浙江	61 231	0	35 207	1.5
安徽	289 796	490	80 668	3.6
福建	130 628	37 652	71 677	3.23
江西	389 132	3 609	578 860	6.69
山东	498 690	244 476	134 922	4.92
河南	306 941	0	1 400	4.4
湖北	395 859	185 024	138 664	5.16
湖南	438 480	464	305 073	5.88
广东	644 388	374 346	2 347 681	6.95
广西	903 794	118 947	120 037	5.97
海南	2 615 742	13 395	845	7.01
重庆	339 405	3 686	288 995	9.92
四川	2 952 277	1 327 778	1 849 244	18.42
贵州	56 965	215 109	436 153	5.39
云南	883 443	443 521	234 129	7.8
西藏	4 343 327	4 870	1 504	34.02
陕西	714 358	61 534	34 602	5.64
甘肃	2 447 668	0	114 900	16.17
青海	1 569 711	0	0	30.21
宁夏	67 575	0	0	9.79
新疆	7 888 214	0	0	12.95

各地区建设项目环境影响评价执行情况

（2010）

地区名称	申报项目投资总额（亿元）	申请项目环保投资总额（亿元）	审批环境影响评价的项目数（个）	编制报告书	填报报告表	填报登记	环评制度执行率（%）
总计	**257 744.3**	**7 779.4**	**390 190**	**23 952**	**169 793**	**196 445**	**99.9**
国家项目	19 469.4	823.3	378	313	50	15	
北京	8 986.2	219.6	13 199	501	4 034	8 664	100.0
天津	5 953.2	107.3	3 607	522	2 290	795	100.0
河北	9 311.9	227.8	16 189	904	8 200	7 085	99.9
山西	4 337.6	292.0	4 023	536	2 390	1 097	100.0
内蒙古	6 911.8	247.4	7 946	644	4 001	3 301	99.6
辽宁	6 988.7	150.1	16 332	1 087	6 892	8 353	99.0
吉林	4 321.8	72.3	7 664	748	3 677	3 239	100.0
黑龙江	7 513.4	120.9	4 748	516	2 920	1 312	100.0
上海	6 273.5	150.3	15 264	484	7 001	7 779	100.0
江苏	34 690.2	824.1	39 734	2 243	18 712	18 779	100.0
浙江	17 860.5	211.0	31 548	1 445	16 092	14 011	100.0
安徽	7 106.3	157.9	8 576	1 144	4 706	2 726	99.1
福建	7 805.1	154.5	17 492	824	9 011	7 657	100.0
江西	3 055.9	112.0	4 967	711	2 542	1 714	99.9
山东	17 385.2	610.3	22 514	1 518	11 294	9 702	100.0
河南	6 601.6	166.8	14 929	642	5 137	9 150	99.9
湖北	4 666.4	111.0	7 298	691	3 010	3 597	100.0
湖南	5 776.8	157.1	7 479	865	2 593	4 021	99.5
广东	31 029.6	1 269.4	68 417	1 640	31 698	35 079	100.0
广西	6 032.9	163.3	12 237	691	3 491	8 055	100.0
海南	2 036.6	49.0	2 682	186	1 271	1 225	98.9
四川	5 936.6	125.8	9 516	1 419	4 198	3 899	100.0
重庆	3 970.6	102.7	6 679	657	2 575	3 447	100.0
贵州	2 365.0	94.5	8 105	399	2 246	5 460	100.0
云南	4 218.8	244.0	12 953	670	2 841	9 442	100.0
西藏			2 488	119	266	2 103	
陕西	7 723.8	382.9	5 759	645	1 842	3 272	100.0
甘肃	1 586.5	38.6	3 497	212	1 347	1 938	100.0
青海	935.2	30.2	1 785	121	387	1 277	100.0
宁夏	2 296.8	74.4	2 369	210	1 050	1 109	100.0
新疆	4 596.4	288.9	9 816	645	2 029	7 142	99.0

各地区建设项目“三同时”执行情况

（2010）

地　区 名　称	当年完成 环保验收 项 目 数	按审批权限			实际执行“三同 时”项目投资 总额（亿元）	实际执行“三同 时”项目环保投资 （万元）	“三同时” 执行率 （%）
		省	市	县			
全　国	**106 765**	**3 321**	**25 375**	**77 795**	**49 853.7**	**20 330 039.8**	**98.0**
国家项目	274				14 103.4	3 860 489.9	
北　京	3 800	242		3 558	1 061.4	565 029.3	95.2
天　津	1 087	183		904	842.9	274 302.0	100.0
河　北	3 368	235	867	2 266	852.1	801 273.0	99.7
山　西	1 276	220	443	613	811.4	928 049.6	99.2
内蒙古	1 498	86	452	960	794.6	500 577.9	99.4
辽　宁	6 526	27	3 568	2 931	1 329.6	501 173.9	95.7
吉　林	1 747	147	246	1 354	546.5	470 592.1	94.8
黑龙江	1 643	123	432	1 088	581.1	535 251.0	95.9
上　海	4 375	146		4 229	1 299.8	374 486.0	100.0
江　苏	9 571	51	947	8 573	3 994.1	1 477 886.8	99.6
浙　江	11 360	72	1 340	9 948	2 993.1	2 265 102.5	99.5
安　徽	2 354	47	1 251	1 056	808.1	415 997.5	99.0
福　建	6 207	99	877	5 231	1 477.7	362 869.5	96.3
江　西	1 352	109	383	860	337.5	246 074.5	99.1
山　东	7 123	213	1 615	5 295	2 047.4	1 536 080.4	60.0
河　南	3 760	106	942	2 712	985.2	487 274.0	99.8
湖　北	3 625	81	274	3 270	624.9	291 550.4	99.5
湖　南	2 174	124	400	1 650	574.0	306 759.8	91.9
广　东	13 815	35	6 254	7 526	8 708.4	1 224 446.9	95.3
广　西	4 200	132	1 413	2 655	616.2	557 818.3	99.1
海　南	285	22	195	68	634.7	115 723.3	98.6
重　庆	2 356	129		2 227	956.2	438 826.1	100.0
四　川	2 203	210	632	1 361	679.9	375 268.4	100.0
贵　州	2 647	106	123	2 418	104.4	164 659.3	100.0
云　南	2 272	77	370	1 825	412.9	361 356.9	100.0
西　藏	609	17	462	130			
陕　西	924	68	403	453	817.1	364 527.6	98.9
甘　肃	547	28	514	5	79.6	71 314.3	100.0
青　海	440	35	112	293	116.3	97 073.6	90.5
宁　夏	622	31	577	14	90.9	103 461.2	79.6
新　疆	2 725	120	283	2 322	572.3	254 743.8	97.5

各地区排污费征收情况（一）

（2010）

地区名称	排污费征收		排污费解缴入库		排污费解缴入库明细	
					污水类解缴入库	
	开单户数（户）	金额（万元）	户数（户）	金额（万元）	户数（户）	金额（万元）
总　计	**413 575**	**1 881 899.9**	**401 172**	**1 779 326.3**	**182 048**	**223 822.6**
北　京	1 921	3 574.8	1 696	3 104.7	873	375.7
天　津	3 984	18 041.9	3 693	17 360.0	1 833	1 338.1
河　北	27 973	137 247.7	27 973	137 213.5	5 135	6 358.8
山　西	10 446	166 299.6	9 374	159 366.6	2 499	4 217.9
内蒙古	6 017	102 934.2	5 651	79 734.6	2 136	3 215.1
辽　宁	27 568	120 630.4	27 193	114 131.8	9 927	13 775.6
吉　林	24 187	41 876.5	24 014	40 529.8	6 842	3 911.2
黑龙江	10 984	43 882.7	10 800	43 459.3	3 898	2 675.3
上　海	5 455	24 827.9	5 372	24 738.7	3 123	2 743.7
江　苏	34 566	202 611.5	34 440	198 790.7	21 741	24 216.6
浙　江	30 536	101 472.6	28 327	96 046.8	13 285	13 192.6
安　徽	14 404	52 856.2	14 209	51 700.1	5 802	6 264.6
福　建	20 006	35 245.1	19 960	35 400.9	10 431	9 458.2
江　西	7 521	47 806.9	7 455	49 075.9	3 025	10 651.4
山　东	17 685	150 685.9	17 073	136 627.7	3 719	8 633.1
河　南	16 489	91 791.1	16 227	84 769.6	5 854	10 161.4
湖　北	9 969	37 542.0	9 678	37 002.4	4 851	8 631.1
湖　南	12 433	54 260.9	12 402	53 074.6	6 569	15 068.6
广　西	62 578	94 138.4	58 390	88 401.1	45 035	25 177.1
广　东	8 833	46 673.2	8 472	42 023.2	2 588	3 478.6
海　南	919	3 604.5	912	3 529.7	491	622.3
重　庆	6 928	37 726.1	6 819	35 924.3	3 541	9 664.6
四　川	7 602	58 566.5	7 457	55 117.2	3 142	7 690.7
贵　州	7 595	45 304.3	7 557	44 882.5	3 715	13 717.5
云　南	7 395	29 214.3	7 390	28 180.0	2 841	2 741.1
西　藏	4 688	881.9	4 647	826.4		
陕　西	5 177	47 893.4	5 130	46 420.8	1 576	7 312.4
甘　肃	8 489	21 306.7	8 443	21 172.6	3 318	1 915.3
青　海	1 241	6 624.3	1 210	6 274.9	407	1 442.9
宁　夏	2 897	14 690.2	2 722	12 205.0	613	690.2
新　疆	7 089	41 688.1	6 486	32 240.7	3 238	4 480.6

各地区排污费征收情况（二）

（2010）

地区名称	排污费解缴入库明细					
	废气类解缴入库		噪声类解缴入库		危险废物解缴入库	
	户数（户）	金额（万元）	户数（户）	金额（万元）	户数（户）	金额（万元）
总　计	**193 734**	**1 392 160.2**	**93 347**	**140 064.9**	**809**	**450.1**
北　京	1 453	3 181.2	54	14.0		
天　津	3 111	16 331.8	139	257.7	1	0.0
河　北	19 577	123 806.6	4 899	5 895.7	8	5.1
山　西	8 290	84 436.4	678	1 412.1	13	2.1
内蒙古	3 818	98 062.1	455	618.9	43	5.7
辽　宁	11 819	91 221.5	9 056	14 862.6	3	1.1
吉　林	12 299	30 988.9	6 134	5 013.5	54	9.1
黑龙江	5 372	39 939.6	2 155	854.6	15	3.6
上　海	2 931	20 146.2	813	883.0		
江　苏	18 252	162 775.5	3 434	10 268.2	18	16.5
浙　江	18 004	71 999.3	5 791	12 971.3	88	36.1
安　徽	5 774	41 013.3	4 421	5 148.6	52	32.9
福　建	7 695	17 295.5	6 856	7 171.5	7	2.2
江　西	2 988	27 211.6	2 392	9 308.0	57	17.1
山　东	11 225	132 867.6	6 040	9 356.2		
河　南	9 669	76 428.1	4 057	3 050.9	53	27.8
湖　北	3 294	22 554.1	3 780	4 937.6	72	61.4
湖　南	4 592	34 119.3	3 144	3 958.4	69	125.6
广　西	11 284	53 962.5	15 575	11 254.4		
广　东	3 897	39 995.3	3 250	2 774.6		
海　南						
重　庆	3 179	20 841.6	2 529	7 630.4	56	29.8
四　川	4 332	32 506.7	2 322	12 335.1	5	0.7
贵　州	3 433	30 024.5	1 044	1 422.2	116	32.5
云　南	2 953	19 472.1	1 756	2 915.8	9	0.9
西　藏						
陕　西	3 273	34 201.6	1 042	3 644.8	22	6.4
甘　肃	4 691	14 128.7	259	236.9	12	20.9
青　海	878	5 607.3	105	212.3	1	0.2
宁　夏	2 387	13 693.4	391	296.6		
新　疆	3 264	33 348.0	776	1 358.9	35	12.2

各地区污染源自动监控情况（一）

（2010）

地 区 名 称	已实施自动监控国家重点监控企业数（家）	已实施自动监控国家重点监控企业中		国控企业排放口自动监控已验收数（个）	国控企业自动监控已验收排放口中		国控企业排放口自动监控已审核数（个）	国控企业自动监控已审核排放口中	
		水排放口数	气排放口数		水排放口数	气排放口数		水排放口数	气排放口数
总 计	**7 988**	**5 890**	**5 816**	**10 113**	**5 189**	**4 924**	**8 932**	**4 645**	**4 287**
北 京	33	24	11	33	22	11	33	22	11
天 津	54	37	88	95	35	60	88	34	54
河 北	637	492	694	861	353	508	732	316	416
山 西	742	393	685	943	357	586	952	366	586
内蒙古	217	128	337	404	117	287	388	112	276
辽 宁	179	97	209	261	86	175	190	73	117
吉 林	133	90	124	137	59	78	87	36	51
黑龙江	189	131	149	233	105	128	202	97	105
上 海	76	63	43	99	63	36	94	63	31
江 苏	710	576	343	901	564	337	891	562	329
浙 江	342	248	129	371	245	126	299	205	94
安 徽	248	208	211	419	208	211	419	208	211
福 建	183	150	91	207	130	77	198	125	73
江 西	207	146	166	304	144	160	268	136	132
山 东	613	413	377	767	406	361	726	394	332
河 南	598	612	459	1 018	593	425	1 009	592	417
湖 北	239	192	115	224	139	85	158	101	57
湖 南	404	323	207	418	265	153	353	226	127
广 东	363	293	219	490	280	210	461	261	200
广 西	444	332	229	539	325	214	335	190	145
海 南	28	25	5	25	20	5	5	5	0
重 庆	88	50	72	122	50	72	122	50	72
四 川	374	282	164	349	223	126	284	198	86
贵 州	120	78	104	129	46	83	114	37	77
云 南	105	90	79	82	55	27	57	40	17
西 藏									
陕 西	280	179	178	157	78	79	66	32	34
甘 肃	97	87	108	181	84	97	159	71	88
青 海	50	21	53	64	18	46	61	18	43
宁 夏	79	41	57	97	41	56	94	40	54
新 疆	156	89	110	183	78	105	87	35	52

各地区污染源自动监控情况（二）

（2010） 单位：家

地区名称	重点监控企业自动监控已联网数	已联网的自动监控企业中			污染源监控中心已建成数	已建成的污染源监控中心中		污染源监控中心已联网数	已联网的污染源监控中心中	
		实时数据正常企业数	历史数据交换企业数	历史数据交换75%企业数		省级监控中心数	地市级监控中心数		省级监控中心数	地市级监控中心数
总　计	**9 129**	**4 172**	**5 959**	**5 133**	**343**	**31**	**312**	**314**	**31**	**284**
北　京	33	0	26	26	1	1	0	1	1	0
天　津	69	49	62	60	1	1	0	1	1	0
河　北	818	207	587	484	12	1	11	12	1	11
山　西	837	154	336	315	12	1	11	12	1	11
内蒙古	263	162	157	146	13	1	12	12	1	11
辽　宁	324	90	41	38	15	1	14	10	1	9
吉　林	168	42	124	101	10	1	9	10	1	9
黑龙江	184	88	124	89	15	1	14	15	1	14
上　海	76	13	73	34	1	1	0	1	1	0
江　苏	726	235	509	499	14	1	13	14	1	13
浙　江	368	288	283	281	12	1	11	12	1	11
安　徽	220	11	100	100	18	1	17	1	1	0
福　建	145	39	124	120	10	1	9	10	1	9
江　西	200	162	176	168	12	1	11	11	1	10
山　东	590	512	527	495	18	1	17	18	1	17
河　南	582	353	0	0	18	1	17	17	1	16
湖　北	274	158	181	120	14	1	13	13	1	12
湖　南	515	369	391	364	15	1	14	15	1	14
广　东	373	25	331	288	21	1	20	21	1	20
广　西	463	304	372	328	15	1	14	15	1	14
海　南	108	29	64	57	3	1	2	2	1	1
重　庆	78	1	71	71	8	1	7	8	1	7
四　川	602	348	486	457	21	1	20	21	1	20
贵　州	264	203	213	183	10	1	9	10	1	9
云　南	149	21	90	78	2	1	1	3	1	2
西　藏										
陕　西	264	191	218	3	11	1	10	11	1	10
甘　肃	127	77	99	86	14	1	13	14	1	13
青　海	60	26	13	11	5	1	4	5	1	4
宁　夏	82	4	47	20	6	1	5	5	1	4
新　疆	167	11	134	111	16	2	14	14	2	13

各地区排污申报核定情况

（2010）

地区名称	一般工业企业排污单位申报核定户数（户）	污水处理厂（场）申报核定户数（户）	固体废物专业处置单位申报核定户数（户）	COD排放量申报核定（吨）	SO_2排放量申报核定（吨）
总　计	**510 088**	**2 096**	**772**	**2 597 143.9**	**7 638 610.0**
北　京	12 239	56	27	22 208.3	12 231.9
天　津	10 602	29	12	28 409.3	85 235.0
河　北	28 256	114	8	100 057.9	616 473.9
山　西	16 835	47	5	28 902.3	352 898.3
内蒙古	6 760	24	3	30 566.2	372 336.9
辽　宁	32 983	29	11	63 509.6	467 711.1
吉　林	24 820	41	32	181 932.0	185 486.7
黑龙江	16 132	20	8	27 038.2	184 024.0
上　海	5 480	49	5	30 436.5	81 868.5
江　苏	49 594	345	66	308 097.3	498 803.1
浙　江	44 285	212	54	471 239.6	364 493.5
安　徽	15 755	58	18	43 515.9	173 620.0
福　建	22 308	78	28	75 102.1	137 293.9
江　西	6 242	28	19	95 640.2	269 205.7
山　东	25 118	136	18	134 152.0	563 653.2
河　南	25 509	141	39	170 809.2	535 932.4
湖　北	7 948	29	5	60 864.8	172 510.1
湖　南	22 072	114	24	174 453.9	271 880.2
广　东	65 501	212	63	190 572.1	241 889.6
广　西	8 096	26	8	34 420.3	232 772.0
海　南					
重　庆	9 029	37	13	35 580.9	168 865.4
四　川	8 842	85	28	66 504.1	260 156.9
贵　州	7 788	44	10	24 121.5	232 302.0
云　南	7 390	61	242	81 580.0	422 225.5
西　藏					
陕　西	5 392	15	3	37 643.0	211 847.4
甘　肃	11 041	26	7	21 416.7	147 627.8
青　海	983	3	1	4 821.0	30 364.7
宁　夏	2 380	10	1	9 941.2	73 284.4
新　疆	10 708	27	14	43 607.7	271 615.9

各地区突发环境事件情况（一）

（2010）

地区名称	事件次数（次）	直接经济损失（万元）	按事故程度分							
			特别重大环境事件		重大环境事件		较大环境事件		一般环境事件	
			次数	经济损失	次数	经济损失	次数	经济损失	次数	经济损失
总　计	**420**	**2 256.9**	**0**	**0**	**3**	**0**	**12**	**200.0**	**405**	**2 056.9**
北　京	30								30	
天　津	0									
河　北	7	600.0					1		6	600.0
山　西	9								9	
内蒙古	5								5	
辽　宁	10								10	
吉　林	3	616.0							3	616.0
黑龙江	0									
上　海	161								161	
江　苏	7		0		0		0		7	
浙　江	35	180.1	0	0	0	0	0	0	35	180.1
安　徽	30	231.6							30	231.6
福　建	4				2				2	
江　西	9	7.3							9	7.3
山　东	0									
河　南	18	124.5	0	0	1	0	0	0	17	124.5
湖　北	27						7		20	
湖　南	1								1	
广　东	2		0	0	0		0		2	
广　西	4	10.0					2	0	2	10.0
海　南	0									
重　庆	23								23	
四　川	1	200.0					1	200.0		
贵　州	5	10.5	0	0	0	0	0	0	5	10.5
云　南	0									
西　藏	0									
陕　西	9	5.2					1	0	8	5.2
甘　肃	10	8.7							10	8.7
青　海	1	240.0							1	240.0
宁　夏	3								3	
新　疆	6	23.0							6	23.0

各地区突发环境事件情况（二）

（2010） 单位：万元

地区名称	按事故类型分					
	水污染				大气污染	
	次数	经济损失	危险化学品污染		次数	经济损失
			次数	经济损失		
总 计	**135**	**932.7**	**42**	**781.6**	**157**	**863.5**
北 京	1		0		11	
天 津						
河 北	2		0		3	600.0
山 西	2				2	
内蒙古					4	
辽 宁	1	0.8	1	0.8	6	
吉 林	3	616	3	616.0		
黑龙江						
上 海	26		0		71	
江 苏	2		0		2	
浙 江	19	164.7	7	100.6	10	3.4
安 徽	10	66.7			8	17.6
福 建	2				1	
江 西	5	3.7	3	0.7	1	2.0
山 东						
河 南	8	53.5	7	53.5	8	35.7
湖 北	15		11		10	
湖 南						
广 东	2					
广 西	3	10.0			1	0
海 南						
重 庆	16		9		7	
四 川					1	200.0
贵 州	2	10.0	1	10.0	2	0.5
云 南						
西 藏						
陕 西	7	4.2	0		0	
甘 肃	3	3.1			6	4.3
青 海						
宁 夏					3	
新 疆	6					

各地区突发环境事件情况（三）

（2010）　　　　单位：万元

地区名称	按事故类型分							
	海洋污染				固体废物污染		噪声污染与振动危害	
	次数	经济损失	溢油及化学品污染		次数	经济损失	次数	经济损失
			次数	经济损失				
总　计	**3**	**0**	**3**	**0**	**35**	**127.3**	**1**	**0**
北　京					13			
天　津								
河　北	2		2					
山　西								
内蒙古					1			
辽　宁	1		1		1			
吉　林								
黑龙江								
上　海	0		0		9		0	
江　苏	0		0		0		0	
浙　江	0	0	0	0	1	0	0	0
安　徽					5	125.0		
福　建								
江　西	0	0	0	0	1	0	0	0
山　东								
河　南	0	0	0	0	0	0	1	0
湖　北					1			
湖　南								
广　东								
广　西								
海　南								
重　庆								
四　川								
贵　州	0	0	0	0	0	0	0	0
云　南								
西　藏								
陕　西	0		0		2	1.0	0	
甘　肃					1	1.3		
青　海								
宁　夏								
新　疆								

各地区突发环境事件情况（四）

（2010） 单位：万元

地区名称	按事故类型分					
	生态破坏		核与辐射		其他	
	次数	经济损失	次数	经济损失	次数	经济损失
总计	**0**	**0**	**5**	**0**	**84**	**311.2**
北京			1		4	
天津						
河北						
山西					5	
内蒙古						
辽宁			1			
吉林						
黑龙江						
上海	0		0		55	
江苏	0		0		3	
浙江	0	0	0	0	5	12.0
安徽					7	22.3
福建			1			
江西	0	0	0	0	2	1.6
山东						
河南	0	0	0	0	1	35.3
湖北			1			
湖南					1	
广东						
广西						
海南						
重庆						
四川						
贵州	0	0	1	0	0	0
云南						
西藏						
陕西	0		0		0	
甘肃						
青海					1	240.0
宁夏						
新疆						

各地区突发环境事件情况（五）

（2010）　　单位：万元

地区名称	按事件起因分					
	违法排污引发环境事件		生产事故引发环境事件		交通事故引发环境事件	
	次数	经济损失	次数	经济损失	次数	经济损失
总　计	**16**	**2.1**	**135**	**790.5**	**96**	**959.7**
北　京	0		11		11	
天　津						
河　北			6	600.0		
山　西			3		6	
内蒙古			2		3	
辽　宁	1		2		5	
吉　林					2	525.0
黑龙江						
上　海			31		9	
江　苏	0		6		1	
浙　江	3	0	17	107.5	11	64.6
安　徽	1	0	12	49.1	9	57.3
福　建					1	
江　西	3	1.6	0	0	3	3.5
山　东						
河　南	0	0	5	26.0	11	55.2
湖　北			15		6	
湖　南	1					
广　东	1					
广　西	3		1			
海　南						
重　庆	1		6		10	
四　川						
贵　州	1	0	1	0.5	2	10.0
云　南						
西　藏						
陕　西	1	0.5	4	3.7	2	0
甘　肃			5	3.7	3	4.1
青　海					1	240.0
宁　夏			3			
新　疆			5			

各地区突发环境事件情况（六）

（2010） 单位：万元

地区名称	按事件起因分			
	自然灾害引发环境事件		其他	
	次数	经济损失	次数	经济损失
总　计	**14**	**93.2**	**158**	**371.2**
北　京			8	
天　津				
河　北			1	
山　西				
内蒙古				
辽　宁	1	0.8	1	
吉　林	1	91.0		
黑龙江				
上　海			121	
江　苏	0		0	
浙　江	1	0	3	0
安　徽	1	0.2	7	125.0
福　建			3	
江　西	1	0.2	2	2.0
山　东				
河　南	0	0	2	43.3
湖　北			6	
湖　南				
广　东			1	
广　西				
海　南				
重　庆	6			
四　川			1	200.0
贵　州	0	0	0	
云　南				
西　藏				
陕　西	2	1.0	0	
甘　肃			2	0.9
青　海				
宁　夏				
新　疆	1			

各地区突发环境事件情况（七）

（2010）

地区名称	环境问题引发群体性事件（起）	伤亡人数（人）	死亡人数	中毒人数	突发环境事件罚款总数（万元）	污染损害赔款总数（万元）
总　计	**1**	**241**	**32**	**135**	**1 139.0**	**3 485.7**
北　京						
天　津						
河　北		72	19		45.0	
山　西						
内蒙古						
辽　宁		4	2	2		
吉　林		2	1		48.0	
黑龙江						
上　海	0	0	0	0	20.0	
江　苏		0	0	0		
浙　江	1	0	0	0	0	4.7
安　徽					23.0	133.0
福　建					956.0	3 187.0
江　西					20.0	1.0
山　东						
河　南	0	7	5	0	0	0
湖　北		110	0	110	0	0
湖　南						
广　东						
广　西						10.0
海　南						
重　庆			0			
四　川		21		21		150.0
贵　州						
云　南						
西　藏						
陕　西	0	0	0	0	27.0	0
甘　肃						
青　海		4	2	0		
宁　夏		21	3	2		
新　疆						

各地区环境宣教情况

（2010）

地 区 名 称	召开新闻发布会 （次）	发布新闻通稿 （篇）	组织宣传活动 （次）
总 计	**249**	**55 786**	**11 296**
国家级	3	28	12
北 京	4	1 011	248
天 津	1	470	657
河 北	3	19	15
山 西	22	1 144	162
内蒙古	5	2 577	200
辽 宁	9	2 670	567
吉 林	3	4 471	967
黑龙江	0	43	37
上 海	3	526	282
江 苏	34	5 815	1 221
浙 江	20	6 580	1 319
安 徽	5	2 690	200
福 建	10	560	507
江 西	2	3 454	254
山 东	3	1 581	276
河 南	5	21	266
湖 北	4	2 359	354
湖 南			
广 东	45	3 614	638
广 西	6	3 149	312
海 南	1	4	8
重 庆	28	1 341	475
四 川	4	156	736
贵 州	4	3 418	169
云 南			
西 藏	1	4	108
陕 西	16	1 695	442
甘 肃	2	828	296
青 海	2	323	82
宁 夏	0	161	138
新 疆	4	5 074	348

8

附　表

FUBIAO

ANNUAL STATISTIC REPORT ON ENVIRONMENT IN CHINA

2010

全国行政区划

（2010 年底）　　　　　　　　　单位：个

省级区划名称	地级区划数	地级市	县级区划数	县级市	市辖区	县	自治县
全　国	**333**	**283**	**2 856**	**370**	**853**	**1 461**	**117**
北京市			16		14	2	
天津市			16		13	3	
河北省	11	11	172	22	36	108	6
山西省	11	11	119	11	23	85	
内蒙古自治区	12	9	101	11	21	17	
辽宁省	14	14	100	17	56	19	8
吉林省	9	8	60	20	20	17	3
黑龙江省	13	12	128	18	64	45	1
上海市			18		17	1	
江苏省	13	13	105	26	55	24	
浙江省	11	11	90	22	32	35	1
安徽省	17	17	105	5	44	56	
福建省	9	9	85	14	26	45	
江西省	11	11	100	11	19	70	
山东省	17	17	140	31	49	60	
河南省	17	17	159	21	50	88	
湖北省	13	12	103	24	38	38	2
湖南省	14	13	122	16	34	65	7
广东省	21	21	121	23	54	41	3
广西壮族自治区	14	14	109	7	34	56	12
海南省	2	2	20	6	4	4	6
重庆市			40		19	17	4
四川省	21	18	181	14	43	120	4
贵州省	9	4	88	9	10	56	11
云南省	16	8	129	9	12	79	29
西藏自治区	7	1	73	1	1	71	
陕西省	10	10	107	3	24	80	
甘肃省	14	12	86	4	17	58	7
青海省	8	1	43	2	4	30	7
宁夏回族自治区	5	5	22	2	9	11	
新疆维吾尔自治区	14	2	98	19	11	62	6
香港特别行政区							
澳门特别行政区							
台湾省							

注：数据摘自《中国统计摘要》（中国统计出版社），以下同。

国民经济与社会发展总量指标摘要

指　标	单　位	2005年	2006年	2007年	2008年	2009年	2010年
人口							
年底总人口	万人	130 756	131 448	132 129	132 802	133 474	133 972
其中：城镇人口	万人	56 212	57 706	59 379	60 667	62 186	68 685
乡村人口	万人	74 544	73 742	72 750	72 135	71 288	65 287
国民经济核算							
国内生产总值	亿元	184 937.4	216 314.4	265 810.3	314 045.4	340 902.8	397 983.3
其中：第一产业	亿元	23 070.4	24 737.0	28 627.0	34 000.0	35 226.0	40 497.0
第二产业	亿元	87 364.6	103 162.0	124 799.0	146 183.4	157 638.8	186 480.9
第三产业	亿元	73 432.9	82 972.0	103 879.6	120 486.6	148 038.0	171 005.4
人均国内生产总值	元/人	14 185	16 500	20 169	23 708	25 605	29 762
支出法国内生产总值	亿元	188 692.1	221 170.5	263 093.8	306 859.8	345 024	394 307.6
其中：最终消费	亿元	97 822.7	110 413.2	128 793.8	149 112.6	165 527	186 905.3
资本形成总额	亿元	80 646.3	94 103.2	110 919.4	133 612.3	164 464	191 690.8
固定资产投资							
全社会固定资产投资总额	亿元	88 773.6	109 998.2	137 323.9	172 291.1	224 599	278 139.8
城镇	亿元	75 095.1	93 368.7	117 464.5	148 167.2	193 920	241 414.9
其中：房地产开发	亿元	15 909.2	19 422.9	25 288.8	30 579.8	36 242	48 267.1
农村	亿元	13 678.5	16 629.5	19 859.5	24 123.0	30 678	36 724.9
主要农业、工业产品产量							
粮食	万吨	48 402.2	49 747.9	50 160.3	52 870.9	53 028	54 647.7
棉花	万吨	571.4	674.6	762.4	749.2	638	596.1
油料	万吨	3 077.1	3 059.4	2 568.7	2 952.8	3 154	3 230.1
肉类	万吨	7 743.1	8 051.4	6 865.7	7 278.3	7 650	7 925.8
原煤	亿吨	22.05	23.73	25.26	27.93	29.73	32.40
原油	万吨	18 135.29	18 476.57	18 632	18 973	18 949	20 301
发电量	亿千瓦时	25 002.60	28 657.26	32 816	34 669	37 147	42 065
粗钢	万吨	35 324.0	41 914.9	48 929	50 092	57 218	62 696
水泥	万吨	106 884.8	123 676.5	136 117	140 000	164 398	188 000

国民经济与社会发展总量指标摘要（续表）

指　标	单　位	2005年	2006年	2007年	2008年	2009年	2010年
国内商业和对外贸易							
社会消费品零售总额	亿元	67 177	76 410	89 210	108 488	132 678	156 998
进出口总额	亿美元	14 219	17 604	21 737.3	25 616.3	22 075.4	29 727.6
其中：出口额	亿美元	7 620	9 689	12 177.8	14 285.5	12 016.1	15 779.3
进口额	亿美元	6 600	7 915	9 558.5	11 330.9	10 059.2	13 948.3
利用外资							
实际利用外资额	亿美元	638.1	735.2	783.4	952.6	918.0	1 088.2
其中：外商直接投资	亿美元	603.3	694.7	747.7	924.0	900.3	1 057.3
外商其他投资	亿美元	34.8	40.6	35.7	28.6	17.7	30.9
旅游							
入境过夜旅游者人数	万人次	4 681	4 991	5 472.0	5 304.9	5 087.5	5 566.5
国内旅游人数	亿人次	12.1	13.9	16.1	17.1	19.0	21.0
国际旅游外汇收入	亿美元	293.0	339.5	419.2	408.4	396.8	458.1
国内旅游总收入	亿元	5 285.9	6 229.7	7 770.6	8 749.3	10 183.7	12 579.8
教育、科技、文化、卫生							
普通高等学校学生数	万人	1 561.8	1 738.8	1 884.9	2 021.0	2 144.7	2 231.8
普通中等学校学生数	万人	8 581	8 451.9	8 243.3	8 050.4	7 876.9	7 703.2
小学学生数	万人	10 864	10 711.5	10 564.0	10 331.5	10 071.5	9 940.7
研究与试验发展经费支出	亿元	2 450	3 003.1	3 710.2	4 570.0	5 791.9	6 980
技术市场成交额	亿元	1 551	1 818.2	2 227	2 665	3 039	3 906
图书总印数	亿册（张）	64.7	64.0	62.9	68.7	70.4	74.0
杂志总印数	亿册	27.6	28.5	30.4	30.2	31.5	32.0
报纸总印数	亿份	412.6	424.5	438.0	445.3	439.1	448.0
医院、卫生院数	个	60 397	60 037	60 531	59 572	59 918	59 681
医生数	万人	193.8	199.5	201.3	208.2	232.9	241.3
医院、卫生院床位数	万张	313.5	327.1	343.8	374.8	408.1	440.1
城市公用事业							
城市房屋集中供热面积	亿米^2	25.2	26.6				
年末供水综合生产能力	万米^3/日	24 720	26 962				
年末污水处理能力	万米^3/日	7 990	9 734				
年末城市道路长度	千米	247 015	241 351				

注：① 由于计算误差的影响，按支出法国内生产总值不等于按生产法计算的国内生产总值。

② 本表价值量指标均按当年价格计算。

国民经济与社会发展结构指标摘要

单位：%

指　标	2005 年	2006 年	2007 年	2008 年	2009 年	2010 年
总人口	**100**	**100**	**100**	**100**	**100**	**100**
其中：城镇	43.0	43.9	44.9	45.7	46.6	49.7
乡村	57.0	56.1	55.1	54.3	53.4	50.3
国内生产总值	**100**	**100**	**100**	**100**	**100**	**100**
其中：第一产业	12.6	11.3	11.1	11.3	10.3	10.2
第二产业	47.5	48.7	48.5	48.6	46.3	46.9
第三产业	39.9	40.0	40.4	40.1	43.4	43.0
固定资产投资总额	**100**	**100**	**100**	**100**	**100**	**100**
城镇	84.6	84.9	85.6	86.0	86.3	86.8
农村	15.4	15.1	14.4	14.0	13.7	13.2
财政收入	**100**	**100**	**100**	**100**	**100**	**100**
其中：中央	52.3	52.8	54.1	53.3	52.4	51.1
地方	47.7	47.2	45.9	46.7	47.6	48.9
财政支出	**100**	**100**	**100**	**100**	**100**	**100**
其中：中央	25.9	24.7	23.1	21.4	20.0	17.8
地方	74.1	75.3	76.9	78.6	80.0	82.2
农、林、牧、渔业产值	**100**	**100**	**100**	**100**	**100**	**100**
其中：农业	49.7	51.1	50.4	55.7	50.7	53.3
林业	3.6	3.9	3.8	3.8	3.9	3.7
牧业	33.7	29.4	33.0	29.7	32.3	30.0
渔业	10.2	9.6	9.1	10.9	9.3	9.3
农、林、牧、渔、服务业	2.8	6.0	3.7		3.8	3.7
工业总产值	**100**	**100**	**100**	**100**	**100**	**100**
其中：轻工业	31.4	30.5				
重工业	68.6	69.5				

注：2002 年起轻重工业结构为国有及规模以上非国有工业企业口径。

自然资源状况

项 目	单位	2005 年	2006 年	2007 年	2008 年	2009 年	2010 年
国土面积	万千米^2	960	960	960	960	960	960
海域面积	万千米^2	473	473	473	473	473	473
大陆岸线长度	万千米	1.80	1.80	1.80	1.80	1.80	1.80
岛屿面积	万千米^2	3.87	3.87	3.87	3.87	3.87	3.87
降水量							
台湾中部山区	毫米	≥4 000	≥4 000	≥4 000	≥4 000	≥4 000	≥4 000
华南沿海	毫米	1 600～2 000	1 600～2 000	1 600～2 000	1 600～2 000	1 600～2 000	1 600～2 000
长江流域	毫米	1 000～1 500	1 000～1 500	1 000～1 500	1 000～1 500	1 000～1 500	1 000～1 500
华北、东北	毫米	400～800	400～800	400～800	400～800	400～800	400～800
西北内陆	毫米	100～200	100～200	100～200	100～200	100～200	100～200
塔里木盆地、吐鲁番盆地和柴达木盆地	毫米	≤25	≤25	≤25	≤25	≤25	≤25
耕地面积	万公顷	13 004	13 004	13 004	12 178	12 178	12 172
荒地面积	万公顷	10 800	10 800	10 800	10 800	10 800	
林业用地面积	万公顷	26 329	28 493	28 493	28 493	30 378	30 590
草地面积	万公顷	40 000	40 000	40 000	40 000	40 000	
活立木总蓄积量	亿米^3	136.2	136.2	136.2	136.2	136.2	149.1
森林面积	亿公顷	1.75	1.75	1.75	1.75	1.95	1.95
森林覆盖率	%	18.21	18.21	18.21	18.21	20.36	20.36
水资源总量	亿米^3	28 124	27 430	25 567	24 696	23 763	29 658
水力资源蕴藏量	亿千瓦	6.76	6.76	6.76	6.76	6.76	6.76
其中：可开发量	亿千瓦	3.79	3.79	3.79	3.79	3.79	3.79
内陆水域总面积	万公顷	1 747	1 747	1 747	1 747	1 747	1 747
其中：可养殖面积	万公顷	675	675	675	675	675	675
海水可养殖面积	万公顷	260	260	260	260	260	260
其中：已养殖面积	万公顷	109	109	109	109	109	109

注：① 森林资源为 1994—1998 年调查数；草地资源为 1991 年调查数；水面资源为 1988 年数；水力资源为 1999 年公报数；耕地面积为 1996 年农业普查数。

② 本表除国土面积、海域面积和降水量外，其余指标均未包括香港特别行政区、澳门特别行政区和台湾省数据。

9

主要统计指标解释

ZHUYAO TONGJI ZHIBIAO JIESHI

ANNUAL STATISTIC REPORT ON ENVIRONMENT IN CHINA

2010

【工业煤炭消费量】 指报告期内企业所用燃料煤和原料煤的总消费量。

【燃料煤消费量】 指报告期内企业厂区内用作燃料的煤炭消费量（实物量），包括企业厂区内生产、生活用燃料煤，也包括砖瓦、石灰等产品生产用的内燃煤，不包括炼焦等行业的原料用煤。

【原料煤消费量】 指报告期内企业在生产工艺中用作原料并能转换成新的产品实体的煤炭消费量。如转换为水泥、焦炭、煤气、碳素、活性炭、氮肥的煤炭称为原料煤。

【燃料油消费量（不含车船用）】 指报告期内企业用作燃料的各种油料总消费量（不包括车船交通用油量）。

【工业锅炉数】 指报告期内企业厂区内用于生产和生活的大于1蒸吨（含1蒸吨）的蒸汽锅炉、热水锅炉总台数和总蒸吨数，包括燃煤、燃油、燃气和燃电的锅炉，不包括茶炉。

【工业炉窑数】 指报告期内企业生产用的炉窑总数，如炼铁高炉、炼钢炉、冲天炉、烘干炉窑、锻造加热炉、水泥窑、石灰窑等。

【烟尘排放达标的工业炉窑数】 指烟尘浓度、烟气黑度均达到国家或地方排放标准的工业炉窑数[《工业炉窑大气污染物排放标准》（GB 9078）和《炼焦炉大气污染物排放标准》（GB 76171）或相应的地方排放标准值]。

【直排海的污水排放口数】 指企业污水排放口位于海边，使废水经过排放口直接排入海，而未经过城市下水道或其他中间体的污水排放口的数量。

【废水污染物在线监控仪器套数】 指报告期内企业对污染源主要污染物排放指标、流量以及污染物治理设施运行记录装置（包括黑匣子等）进行实时自动监测监控系统的数量。

【废气污染物在线监控仪器套数】 指报告期内企业对污染源废气排放参数、污染物排放量进行在线监测的仪器套数。监测同一废气排放口的为一套。

【废水治理设施数】 指报告期内企业用于防治水污染和经处理后综合利用水资源的实有设施（包括构筑物）数，以一个废水治理系统为单位统计。附属于设施内的水治理设备和配套设备不单独计算。已经报废的设施不统计在内。

【废水治理设施治理能力】 指报告期内企业的废水治理设施实际具有的废水处理能力。

【废水治理设施运行费用】 指报告期内企业维持废水治理设施运行所发生的费用。包括能源消耗、设备维修、人员工资、管理费、药剂费及与设施运行有关的其他费用等。

【工业废水处理量】 指经各种水治理设施实际处理的工业废水量，包括处理后外排的和处理后回用的工业废水量。虽经处理但未达到国家或地方排放标准的废水量也应计算在内。计算时，如遇有车间和厂排放口均有治理设施，并对同一废水分级处理时，不应重复计算工业废水处理量。

【工业废水排放量】 指报告期内经过企业厂区所有排放口排到企业外部的工业废水量。包括生产废水、外排的直接冷却水、超标排放的矿井地下水和与工业废水混排的厂区生活污水，不包括独立外排的间接冷却水（清浊不分流的间接冷却水应计算在内）。

【直接排入海的工业废水排放量】 指废水经过工厂的排污口直接排入海的量，而未经过城市下水道或其他中间体，也不受其他水体的影响。

【工业废水排放达标量】 指报告期内废水中各项污染物指标都达到国家或地方排放标准的外排工业废水量，包括未经处理外排达标的，经废水处理设施（包括企业内部的废水处理设施和工业区内设置的单纯处理工业废水的集中处理装置）处理后达标排放的，以及经污水处理厂（包括工业区废、污水集中处理装置）处理后达标排放的。国家排放标准见《污水综合排放标准》（GB 8978—1996）和相关行业废水排放标准。

企业的废水排入工业区内设置的单纯处理工业废水的集中处理装置进行处理的废水，应根据集中处理装置处理后实际达到的浓度来评价其是否达标。集中处理装置处理后排放的废水达到排放标准，则该企业的排放废水量计为达标废水量；如集中处理装置处理后排放的废水未达到排放标准，则该企业的排放废水不能计为达标废水量。

对于企业单位产生的废水排入城镇下水道并进入二级污水处理厂（包括工业区废、污水集中处理装置）进行处理的，其废水的达标问题，作出以下规定：

企业单位的废水排入城镇下水道并进入二级污水处理厂（包括工业区废水、污水集中处理装置）进行处理的废水，应根据污水处理厂处理后实际达到的浓度来评价其是否达标。污水处理厂处理后排放的废水达到排放标准，则该企业的排放废水量计为达标废水量；如污水处理厂处理后排放的废水未达到排放标准，则该企业的排放废水不能计为达标废水量。

【工业废水中污染物去除量】 指报告期内企业生产过程排出的废水，经过各种水治理设施处理后，除去废水中所含的挥发酚、氰化物、化学需氧量、石油类、氨氮和铅、汞、镉、六价铬等重金属污染物本身的纯重量。计算公式是：

污染物去除量（纯重量）= 处理的工业废水量 ×（处理前污染物的平均浓度 − 处理后污染物的平均浓度）

企业排出的工业废水经污水处理厂（包括工业区废水、污水集中处理装置）集中处理的，上述计算公式中的“处理后污染物的平均浓度”即为污水处理厂（包括工业区废水、污水集中处理装置）排放口的实际平均浓度。

【工业废水中污染物排放量】 指报告期内企业排放的工业废水中所含化学需氧量、氨氮、石油类、挥发酚、氰化物、砷等污染物和铅、汞、镉、六价铬等重金属本身的纯重量。它可以通过工业废水排放量和其中污染物的浓度相乘求得，计算公式是：

污染物排放量（纯重量）= 工业废水排放量 × 排放口污染物的平均浓度

企业排出的工业废水经污水处理厂（包括工业废水、污水集中处理装置）集中处理的，上述计算公式中“排放废水中污染物的平均浓度”即为污水处理厂（包括工业废水、污水集中处理装置）排放口的实际平均浓度。

凡安装连续自动在线监测设备（须由市级以上环境监测站标定的）并与当地环境监测站联网的单位，优先采用实时监测数据的汇总数作为排放量数据。未安装连续自动在线监测设备的单位，在采用实测法计算排污数据时，为保证监测数据能够准确地反映实际情况，须多次测定样品取值，并须经当地环境监测站认定。

【工业废气排放量】 指报告期内企业厂区内燃料燃烧和生产工艺过程中产生的各种排入空气中含有污染物的气体的总量，以标准状态（273K，101 325Pa）计。

工业废气排放总量 = 燃料燃烧过程中废气排放量+生产工艺过程中废气排放量

【燃料燃烧过程中废气排放量】 指燃煤、油、气锅炉、锻造加热炉、退火炉及其他工业炉窑在燃烧过程中所排废气的总量，以标准状态计。

【生产工艺过程中废气排放量】 指生产工艺过程中排放的废气总量。如化工、冶炼、建材（包括水泥行业）、化纤、造纸等行业生产工艺过程中排放的废气，以标准状态计。

【废气治理设施数】 指报告期末企业用于减少在燃料燃烧过程与生产工艺过程中排向大气的污染物或对污染物加以回收利用的废气治理设施总数，以一个废气治理系统为单位统计。附属于设施内的治理设备和配套设备不单独计算。已报废的设施不统计在内。锅炉中的除尘装置属于“三同时”设备，应统计在内。

【脱硫设施数】 指在治理设施中有专用（或兼用）的脱硫设备（或系统），其脱硫效率要达到40%及以上，脱硫后不再释放出二氧化硫，比如使系统中有足够的碱性物质与二氧化硫反应，生成稳定的盐类物质或采用活性炭吸附制酸等方法进行脱硫的设施数。

【废气治理设施处理能力】 指报告期末企业实有的废气治理设施的实际废气处理能力。

【废气脱硫设施脱硫能力】 指报告期末企业脱硫设施的实际去除二氧化硫的能力，计量单位为kg/h。

【废气治理设施运行费用】 指报告期内维持废气治理设施运行所发生的费用。包括能源消耗、设备折旧、设备维修、人员工资、管理费、药剂费及与设施运行有关的其他费用等。

【脱硫设施运行费用】 指报告期内维持脱硫设施运行所发生的费用。包括能源消耗、设备折旧、设备维修、人员工资、管理费、药剂费及与设施运行有关的其他费用等。

【二氧化硫去除量】 指报告期内燃料燃烧废气和生产工艺废气经过各种废气治理设施处理后去除的二氧化硫总量。

二氧化硫去除量＝燃料燃烧过程中二氧化硫去除量＋生产工艺过程中二氧化硫去除量

【燃料燃烧过程中二氧化硫去除量】 指燃料燃烧过程中产生的废气，经过各种废气治理设施后去除的二氧化硫数量。它可以根据实测、物料衡算或经验公式计算求得。

【生产工艺过程中二氧化硫去除量】 指生产工艺过程中产生的废气，经过各种废气治理设施后去除的二氧化硫量。

【二氧化硫排放量】 指报告期内企业在燃料燃烧和生产工艺过程中排入大气的二氧化硫总量。

二氧化硫排放量＝燃料燃烧过程中二氧化硫排放量＋生产工艺过程中二氧化硫排放量

【燃料燃烧过程中二氧化硫排放量】 指燃料燃烧过程中排入大气的二氧化硫量。它可以根据实测、物料衡算或排放系数计算求得。

凡安装连续自动在线监测设备并与当地环境监测站联网的单位，优先采用实时监测数据的汇兑数作为排放量数据。未安装连续自动在线监测设备的单位，在采用实测法计算排污数据时，为保证监测数据能够准确地反映实际情况，须多次测定样品取值，并须经当地环境监测站认定。

未安装连续自动在线监测设备的单位采用实测法计算的排放数据，须与使用燃料含硫率计算得出的排放数据进行对照验证。如计算结果偏差较大，应进行核实调整。

【生产工艺过程中二氧化硫排放量】 指企业在生产工艺过程中排入大气的二氧化硫量。它可以根据实测、物料衡算或经验公式计算求得。

【氮氧化物去除量】 氮氧化物去除量指报告期内企业利用各种废气治理设施去除的氮氧化物量。

【氮氧化物排放量】 指报告期内企业排入大气的氮氧化物量。

【烟尘去除量】 烟尘是指在燃烧过程中产生的烟气中夹带的颗粒物。烟尘去除量指报告期内企业利用各种废气治理设施去除的烟尘量。

烟尘去除量=（除尘设备进口烟尘平均浓度－除尘设备出口烟尘平均浓度）

×除尘系统排风量×烟尘设备运行时间

【烟尘排放量】 指报告期内企业排放烟尘量。烟尘排放量可以通过除尘系统的排风量和除尘设备出口烟尘浓度相乘求得。

【工业粉尘去除量】 工业粉尘指在生产工艺过程中排放的能在空气中悬浮一定时间的固体颗粒。如钢铁企业耐火材料粉尘、焦化企业的筛焦系统粉尘、烧结机的粉尘、石灰窑的粉尘、建材企业的水泥粉尘等。不包括电厂排入大气的烟尘。

工业粉尘去除量是指报告期内企业对生产工艺过程中产生的废气，经过各种废气治理设施处理后去除的粉尘总重量。它可以根据实测计算求得。

工业粉尘去除量=（除尘设备进口粉尘平均浓度－除尘设备出口粉尘平均浓度）

×除尘系统排风量×除尘设备运行时间

除尘系统进、出口的粉尘浓度，应以实测的数字为准。

【工业粉尘排放量】 指报告期内企业有组织排放的工业粉尘量。工业粉尘排放量可以通过除尘系统的排风量和除尘设备出口排尘浓度相乘求得。

【工业固体废物产生量】 指报告期内企业在生产过程中产生的固体状、半固体状和高浓度液体状废弃物的总量，包括危险废物、冶炼废渣、粉煤灰、炉渣、煤矸石、尾矿、放射性废物、脱硫石膏和其他废物等；不包括矿山开采的剥离废石和掘进废石（煤矸石和呈酸性或碱性的废石除外）。酸性或碱性废石是指采掘的废石其流经水、雨淋水的 pH 值小于 4 或 pH 值大于 10.5 者。

工业固体废物产生量＝危险废物产生量+冶炼废渣产生量+ 粉煤灰产生量
+炉渣产生量+煤矸石产生量+尾矿产生量+放射性废物产生量
+脱硫石膏产生量+其他废物产生量

工业固体废物产生量=（工业固体废物综合利用量-综合利用往年贮存量）
+工业固体废物贮存量+（工业固体废物处置量-处置往年贮存量）
+工业固体废物排放量

【危险废物】 指列入国家危险废物名录或者根据国家规定的危险废物鉴别标准和鉴别方法认定的，具有爆炸性、易燃性、易氧化性、毒性、腐蚀性、易传染性疾病等危险特性之一的废物。按原国家环保局、原国家经贸委、原外经贸部、公安部联合颁布的《国家危险废物名录》填报。其计量单位（t）保留 2 位小数。

【冶炼废渣】 指在冶炼生产中产生的高炉渣、钢渣、铁合金渣以及有色金属矿渣等。

【粉煤灰】 又称飞灰或烟道灰。指燃煤锅炉在燃烧过程中产生的烟气中的细灰，主要从烟道气体收集而得，应与其烟尘去除量基本相等。

完善指标解释：又称飞灰或烟道灰。指燃煤锅炉在燃烧过程中产生的烟气中的细灰，在引风机将烟气排入大气之前，这些细灰经过除尘器被分离、收集，因此粉煤灰应与其烟尘去除量基本相等。

【炉渣】 指企业燃烧设备从炉膛排出的灰渣，不包括燃料燃烧废气处理系统去除的烟尘。

【煤矸石】 指与煤层伴生的一种含碳量低、比煤坚硬的黑色岩石。通常由煤矿开采、洗煤及耗煤单位排出。是与煤层伴生的一种含碳量低、比煤坚硬的黑色岩石，是碳质、泥质和砂质页岩的混合物，具有低发热值，含碳 20%～30%。包括巷道掘进过程中的掘进矸石、采掘过程中从顶板、底板及夹层里采出的矸石以及洗煤过程中挑出的洗矸石。目前煤矸石主要被用于生产矸石水泥、混凝土的轻质骨料、耐火砖等建筑材料，此外还可用于回收煤炭，煤与矸石混烧发电，制取结晶氯化铝、水玻璃等化工产品以及提取贵重稀有金属，也可作肥料。

【尾矿】 指各种金属和非金属矿石经过选矿或水冶作业后，存于水溶液中的残留脉石、矿砂等，统称为尾矿。尾矿产生量是指选矿厂和水冶厂排出的尾矿量，包括赤泥。选矿厂包括各种金属和非金属矿石的选矿厂。赤泥指以铝土矿为原料的氧化铝厂的生产废料。

【放射性废物】 指含有天然放射性核素，并其比活度大于 2×10^4Bq/kg 的尾矿砂、废矿石及其他放射性固体废物（指放射性浓度或活度或污染水平超过规定下限的固体废物）。

【脱硫石膏】 指发电厂用于脱硫的湿式石灰石/石膏法工艺中，吸收剂与烟气中 SO_2 等反应后生成的的副产物。湿式石灰石/石膏法脱硫设备每处理 1 吨二氧化硫大约产生脱硫石膏 2.7 t.

【其他废物】 指工业垃圾、污泥及燃料燃烧过程中去除的烟尘等工业固体废物。工业垃圾指机械工业切削碎屑、研磨碎屑、废砂型等；食品工业的活性炭渣；硅酸盐工业和建材工业的砖、瓦、碎砾、混凝土碎块等。污泥指工业废水处理中排出的、以干泥量计的固体沉淀物。

【工业固体废物综合利用量】 指报告期内企业通过回收、加工、循环、交换等方式，从固体废物中提取或者使其转化为可以利用的资源、能源和其他原材料的固体废物量（包括当年利用的往年工业固体废物累计贮存量）。如用作农业肥料、生产建筑材料、筑路等。综合利用量由原产生固体废物的单位统计。

危险废物综合利用量是指按可能导致资源回收、再循环、直接再利用或其他用途的作业方式进行综合利用危险废物的实际量。

【工业固体废物贮存量】 指报告期内企业以综合利用或处置为目的，将固体废物暂时贮存或堆存在专设的贮存设施或专设的集中堆存场所内的量。专设的固体废物贮存场所或贮存设施必须有防扩散、防流失、防渗漏、防止污染大气、水体的措施。

粉煤灰、钢渣、煤矸石、尾矿等的贮存量是指排入灰场、渣场、矸石场、尾矿库等贮存的量。

【危险废物贮存量】 指将危险废物以一定包装方式暂时存放在专设的贮存设施内的量。专设的贮存设施指对危险废物的包装、选址、设计、安全防护、监测和关闭等符合《危险废物贮存污染控制标准》（GB 18597—2001）等相关环保法律法规要求，具有防扩散、防流失、防渗漏、防止污染大气和水体措施的设施。

【工业固体废物处置量】 指报告期内企业将固体废物焚烧或者最终置于符合环境保护规定要求的场所并不再回取的工业固体废物量（包括当年处置的往年工业固体废物贮存量）。

处置方式如：填埋（其中危险废物应安全填埋）、焚烧、专业贮存场（库）封场处理、深层灌注、回填矿井及海洋处置（经海洋管理部门同意投海处置）等。

【危险废物处置量】 指报告期内企业将危险废物焚烧或者最终置于符合环境保护规定要求的场所并不再回取的危险废物量（包括当年处置的往年危险废物贮存量）。

【工业固体废物排放量】 指报告期内企业将所产生的固体废物排到固体废物污染防治设施、场所以外的量。不包括矿山开采的剥离废石和掘进废石（煤矸石和呈酸性或碱性的废石除外）。排放方式如：

（1）向水体排放废油类、废酸碱及其他高浓度液态废物；

（2）在江河、湖泊、运河、渠道、海洋的滩场和岸坡倾倒、堆放和存贮废物；

（3）利用渗井、渗坑、渗裂隙和溶洞倾倒废物；

（4）向路边、荒地、荒滩倾倒废物；

（5）未经环保部门同意作填坑、填河和土地填埋固体废物；

（6）混入生活垃圾进行堆置的废物；

（7）未经海洋管理部门批准同意，向海洋倾倒废物；

（8）其他去向不明的废物。

工业固体废物排放量计算公式是：

工业固体废物排放量=工业固体废物产生量－贮存量－（综合利用量－综合利用往年贮存量）－（处置量－处置往年贮存量）

【危险废物排放量】 指报告期内企业将所产生的危险废物排到危险废物污染防治设施以外的量。

【污染治理项目本年完成投资及资金来源】 指在报告期内，企业实际用于环境治理工程的投资额。投资额中的资金来源，是指投资单位在本年内收到的用于污染治理项目投资的各种货币资金，包括排污费补助、政府其他补助、企业自筹。各种来源的资金均为报告期投入的资金，不包括以往历年的投资。

本年污染治理资金合计＝排污费补助＋政府其他补助＋企业自筹

【排污费补助】 指从征收的排污费中提取的用于补助重点排污单位治理污染源以及环境污染综合性治理措施的资金。

【政府其他补助】 指用于补助重点排污单位治理污染源以及环境污染综合性治理措施的除排污费补助以外的政府其他补助资金。

【企业自筹】 指除排污费补助、政府其他补助资金以外的其他用于污染治理的资金，包括国内贷款（不包括环保贷款）、利用外资、银行贷款等其他来源资金。

【银行贷款】 指企业向银行借入的用于污染治理项目建设投资的贷款，属于企业自筹资金。

【竣工项目设计或新增处理能力】 设计能力是指设计中规定的主体工程（或主体设备）及相应的配套的辅助工程（或配套设备）在正常情况下能够达到的处理能力。报告期内竣工的污染治理项目，属新建项目的填写设计文件规定的处理、利用“三废”能力；属改扩建、技术改造项目的填写经改造后新增加的处理利用能力，不包括改扩建之前原有的处理能力；只更新设备或重建构筑物，处理利用“三废”能力没有改变的则不填。

工业废水设计处理能力的计量单位为 t/d；工业废气设计处理能力的计量单位为（标态）m^3/h；工业固体废物设计处理能力的计量单位为 t/d；噪声治理（含振动）设计处理能力以降低分贝数表示；电磁辐射治理设计处理能力以降低电磁辐射强度表示（电磁辐射计量单位有电场强度单位：V/m、磁场强度单位：A/m、功率密度单位:W/m^2）。放射性治理设计处理能力以降低放射性浓度表示，废水计量单位为 Bq/L，固体废物计量单位为 Bq/kg。

【危险废物集中处置厂】 指将多个工业企业产生的危险废物或者将多个教学单位、第三产业或居民生活产生的危险废物集中起来进行综合利用或者焚烧、填埋等处置的场所。

目前统计范围仅限于集中式危险废物处置场所，没有扩充至所有取得综合危险废物经营许可证的企业。

【危险废物实际处置能力】 指危险废物集中处置厂每昼夜实际处理危险废物的能力。

【焚烧处置能力】 指处置厂每昼夜将危险废物焚烧或高温裂解处理的能力。

【填埋处置能力】 指处置厂每昼夜将危险废物以填埋方式进行处置的能力。

【危险废物处置量】 报告期内对危险废物采取焚烧、填埋的总量。通过危险废物焚烧、填埋和其他改变危险废物的物理、化学、生物特性的方法，达到减少已产生的危险废物数量、缩小危险废物体积、减少或者消除其危险成分的活动，或者将危险废物最终置于符合环境保护规定要求的填埋场的活动中，所消纳危险废物的量。

【危险废物焚烧量】 指报告期内对危险废物以焚烧或高温裂解方法进行处置的量。

【危险废物填埋量】 指报告期内对危险废物以填埋方法进行处置的量。

【处置工业危险废物量】 指报告期内采用各种方式处置的工业危险废物的总量。

【处置医疗废物量】 指处置厂在报告期内采用各种方式处置的医疗废物的总量。

【处置其他危险废物量】 指报告期内采用各种方式处置的除工业危险废物和医疗废物以外其他危险废物的总量，如教学科研单位实验室、机械电器维修、胶卷冲洗、居民生活、办公等产生的危险废物。

【危险废物综合利用量】 指报告期内通过回收、加工、循环、交换等方式，从危险废物中提取或者使其转化为可以利用的资源、能源和其他原材料的危险废物量。

【焚烧残渣量】 指报告期内危险废物经焚烧处置后生成的残渣，包括焚烧主体设备排出的炉渣以及烟气处理设备中收集的飞灰的重量。

【焚烧残渣利用量】 指对残渣（不包括飞灰）的再利用量，如用炉渣制水泥、混凝土砖及其他材料等的量。

【焚烧残渣填埋量】 指报告期内将炉渣、飞灰直接或经预处理后进行安全填埋的重量。

【本年运行费用】 指报告期内维持危险废物集中处置厂正常运行所发生的费用。包括能源消耗、设备折旧、维修、人员工资、管理费、药剂费及与危险集中处置厂运行有关的其他费用等。计量单位为万元，允许保留一位小数。

【污水处理厂】 指在城市或工业区，城市污水（生活污水、工业废水）和雨水通过排水管道集中于一个或几个处所，并利用由各种处理单元组成的污水处理系统进行净化处理，最终使处理后的污水和污泥达到规定的要求后排放水体或再利用。不包括氧化塘、渗水井和化粪池、改良化粪池。

【工业区废（污）水集中处理装置】 指提供社会化有偿服务、专门从事为工业园区、联片工业企业或周边企业处理工业废水（包括一并处理周边地区生活污水）的集中设施或独立运营的单位。不包括企业内部的污水处理设施。

【污水处理量】 指报告期内污水处理厂（或处理装置）实际处理的污水总量。一般用污水流量计测量。

【污水再生利用量】 指报告期内污水处理厂（或处理装置）处理后的污水中再利用的水量，包括直接用于工业冷却、洗涤、冲渣和生活冲厕、洗车等杂用，以及直接用于农田灌溉、绿地浇灌、养鱼、生态恢复等。

【污泥产生量】 是指报告期内污水处理厂（或处理装置）在整个污水处理过程中污泥的最终产生量。

【污泥排放量】 指报告期内不做处理利用而将污泥任意弃置到划定的污泥堆场以外的任何区域的量。

【本年运行费用】 指报告期内维持污水处理厂（或处理装置）正常运行所发生的费用。包括能源消耗、设备折旧、设备维修、人员工资、管理费、药剂费及与污水处理厂（或处理装置）运行有关的其他费用等。

【医院床位数】 指填报单位报告期内的固定实有床位（非编制床位）。包括正规床、简易床、监护床和正在消毒、修理的床位及因扩建或大修理而停用的床位（按扩建或大修理前的床位计算），但不包括产科的新生儿床、库存床、临时增设的床位、病人家属的陪床、接产室的待产床等。

【废水处理设施数】 指报告期内医院用于防治水污染和经处理后综合利用水资源的实有设施（包括构筑物）数，以一个废水治理系统为单位统计。附属于设施内的水治理设备和配套设备不单独计算。已经报废的设施不统计在内。

【废水处理设施能力】 指报告期内医院的废水治理设施实际具有的废水处理能力。

【废水处理设施运行费用】 指报告期内医院维持废水治理设施运行所发生的费用。包括能源消耗、设备折旧、设备维修、人员工资、管理费、药剂费及与设施运行有关的其他费用等。

【医疗废物处理设施数】 指报告期内医院用于处理医疗废物的实有设施数。

【医疗废物处理设施运行费用】 指报告期内医院维持医疗废物处理设施运行所发生的费用。包括能源消耗、设备折旧、设备维修、人员工资、管理费、药剂费及与设施运行有关的其他费用等。

【废水处理量】 指报告期内医院的各种废水处理设施实际处理的废水量。虽经处理但未达到国家或地方排放标准的废水量也应计算在内。计算时，如遇有病房和医院排放口均有治理设施，并对同一废水分级处理时，不应重复计算废水处理量。

【医疗废物产生量】 指报告期内医院在工作过程中产生的固体状、半固体状和高浓度液体状废弃物的总量。

【医疗废物处置量】 指报告期内医院将医疗废物焚烧或者最终置于符合环境保护规定要求的场所并不再回取的医疗废物量。

污染源普查解释：指对医疗废物采用符合环境保护和医疗废物管理规定的焚烧、高温蒸煮、微波消毒和化学消毒等方式处理医疗废物并最终实现医疗废物毁形、减容和消除其危害性，并对其残余物进行安全填埋的处置的过程。

【运往集中处置厂的量】 指将产生的医疗废物运往集中处置厂进行处置的数量。

【放射源数】 指用放射性物质制成的能产生辐射照射的物质或实体的数量，包括密封源和非密封源。其中，密封源是密封在包壳或紧密覆盖层里的放射性物质，如钴-60、铯-137、铱-192 等。非密封源是指没有包壳的放射性物质，如碘-131、碘-125、锝-99 等。

【集中管理的放射源数】 指医院里纳入集中控制管理的放射源的数量。

【退役放射源数】 指医院里已经不正常使用的放射源的数量。

【生活及其他污染】 指除工业生产活动以外的所有社会、经济活动及公共设施的经营活动产生的污染。

【生活及其他煤炭消费量】 指除工业用煤之外的生活和其他社会经济活动的煤炭消耗量。

计算公式是：

生活及其他煤炭消费量=煤炭消费总量－工业煤炭消费量

煤炭消费总量和工业煤炭消费量的数据都取自统计年鉴上的全社会数统计年鉴的工业煤炭消费量与环境统计汇总的工业煤炭消费量存在一个差值（理论上全社会工业燃煤量应大于环境统计的工业燃煤量），差值部分的燃煤排污量未在环境统计中计算。为将这部分差值统计进来，规定凡全社会工业燃煤量大于环境统计工业燃煤量的，差额部分归入生活燃煤统计，计算公式改为：

生活及其他煤炭消费量=煤炭消费总量（全社会数）－工业煤炭消费量（环境统计数）

如果有地区环境统计的工业燃煤汇总量比全社会工业燃煤量大。这种情况下，计算公式中的减数仍要用统计局的工业燃煤量。计算公式为：

生活及其他煤炭消费量=煤炭消费总量（全社会数）－工业煤炭消费量（全社会数）

指除工业用煤之外的生活和其他社会经济活动的煤炭消耗量。

【城镇生活污水排放量】 用人均系数法测算。测算公式为：

城镇生活污水排放量=城镇生活污水排放系数×城镇常住人口数×365

【城镇生活污水处理量】 指报告期内污水处理厂处理的城镇生活污水总量。由于污水处理厂不仅处理生活污水，也处理一部分工业废水，因此计算公式为：

城镇生活污水处理量=污水处理厂的污水处理总量—处理的工业废水量

【城镇生活污水处理率】 指报告期内城镇生活污水处理量占城镇生活污水产生量的百分率。计算公式是：

城镇生活污水处理率=城镇生活污水处理量÷城镇生活污水产生量×100%

【城镇生活污水中化学需氧量排放量】 指城镇生活污水中产生的化学需氧量（COD），经污水处理厂（包括工业区废、污水集中处理装置）处理去除部分量后，实际排放外环境的量。测算公式为：

城镇生活污水中 COD 排放量=城镇生活污水中 COD 产生量—污水处理厂去除生活污水中 COD 量

【城镇生活污水中氨氮排放量】 指城镇生活污水中产生的氨氮，经污水处理厂（包括工业区废、污水集中处理装置）处理去除部分量后，实际排放外环境的量。测算公式为：

生活污水中氨氮排放量=生活污水中氨氮产生量—污水处理厂去除生活污水中氨氮量

【生活及其他二氧化硫排放量】 指报告期内除工业生产活动以外的所有社会、经济活动及公共设施的经营活动中燃煤所排放的二氧化硫纯重量。

生活及其他二氧化硫排放量以生活及其他煤炭消费量及其含硫率为基础进行测算，其测算公式如下：

生活及其他二氧化硫排放量=生活及其他煤炭消费量×含硫率×0.8×2

【生活及其他烟尘排放量】 指报告期内除工业生产活动以外的所有社会、经济活动及公共设施的经营活动中燃煤所排放的烟尘纯重量。

生活及其他烟尘排放量以生活及其他煤炭消费量为基础进行测算。在测算中，要依照煤炭消费的不同用途和不同燃烧方式分别计算：

（1）集中供热锅炉房采暖用煤的烟尘排放量，按照工业锅炉燃煤排放烟尘的计算方法和排放系数计算；

（2）居民生活以及社会生活用煤的烟尘排放量，按照燃用的民用型煤和原煤，分别采用不同的计算系数：

①民用型煤的烟尘排放量，以每吨型煤排放 1～2kg 烟尘量计算，计算公式为：

$$\text{烟尘排放量}=\text{型煤消费量（t）}\times（1\sim2）‰$$

②原煤的烟尘排放量，以每吨原煤排放 8～10kg 烟尘量计算，计算公式为：

$$\text{烟尘排放量}=\text{原煤消费量（t）}\times（8\sim10）‰$$

【生活及其他氮氧化物排放量】 指报告期内除工业生产活动以外的所有社会、经济活动及公共设施的经营活动中燃料所排放的氮氧化物纯重量。

按照污染源分类，NO_x 排放量统计分为工业、生活和移动源三部分。生活 NO_x 排放量统计采取系数测算法。具体方法是以地市级为单位，根据该地区生活燃料消费量和生活 NO_x 排放系数，统计出该地区生活 NO_x 排放量。移动源 NO_x 排放量统计方法同生活源、其排放量包含在生活排放量中。

$$\text{生活 }NO_x\text{ 排放量}=\text{生活燃料消费量}\times\text{生活 }NO_x\text{ 排放系数}+\text{公路交通氮氧化物排放量}$$